# Beyond El Niño

## Decadal and Interdecadal Climate Variability

Springer

*Berlin*
*Heidelberg*
*New York*
*Barcelona*
*Hong Kong*
*London*
*Milan*
*Paris*
*Singapore*
*Tokyo*

Antonio Navarra (Ed.)

# Beyond El Niño

## Decadal and Interdecadal Climate Variability

With 168 Figures and 7 Tables

 Springer

**Editor**

Dr. Antonio Navarra
IMGA - CNR
Via Gobetti 101
I-40129 Bologna
Italy
*E-mail: navarra@rigoletto.imga.bo.cnr.it*

ISBN 3-540-63662-5 Springer-Verlag Berlin Heidelberg New York

Library of Congress Cataloging-in-Publication Data
Beyond El Niño : decadal and interdecadal climate variability / Antonio Navarra, (ed.)
p. cm. Includes bibliographical references and index.
    ISBN 3-540-63662-5
    1. Climatic changes. I. Navarra, A. (Antonio), 1956- .
    QC981.8.C5B46 1999    551.6 - - dc21    99-18126 CIP

© Springer-Verlag Berlin Heidelberg 1999
Printed in Germany

Typesetting: Camera-ready by editor
SPIN:10534661    32 / 3020 - 5 4 3 2 1 0 - Printed on acid -free paper

# Foreword

The interest and level of research into climate variability has risen dramatically in recent years, and major breakthroughs have been achieved in the understanding and modelling of seasonal to interannual climate variability and prediction. At the same time, the documentation of longer term variability and its underlying mechanisms have progressed considerably.

Within the European Commission's Environment and Climate research programs several important projects have been supported in these areas - including the "Decadal and Interdecadal Climate variability Experiment" (DICE) which forms the basis of this book. Within the EC supported climate research, we see an increasing importance of research into climate variability, as is evidenced in the upcoming Fifth Framework Programme's Key Action on Global Change, Climate and Biodiversity. This is because of the obvious potential socio-economic benefits from seasonal to decadal scale climate prediction and equally important for the fundamental understanding of the climate system to help improve the quality and reliability of future climate change and mankind's current interference with it.

The DICE group has performed important and pioneering work, and we hope this book will receive the wide distribution and recognition it deserves. We welcome the contributions from distinguished researchers from US, Japan and Canada to the EC's DICE group towards completing the scope of the book and as an example of international cooperation which is essential in such a high-level scientific endeavor.

Anver Ghazi and Ib Troen

Climate and Natural Hazards
European Commission
Brussels

# Preface

Major progress on observations and modelling of the El Niño/Southern Oscillation (ENSO) have been made in recent years. We can now monitor the evolution of the SST in real time and the modeling efforts have advanced to the point that several models used for climate studies now have intrinsic enso-like behavior on interannual scales and can simulate or predict ENSO impacts. The implications of such developments are enormous. Precipitation and surface temperature in the extratropics are influenced by the phase of ENSO and extratropical circulation anomalies are correlated with the tropical sea surface temperatures (SST).

However, little attention has been paid to scales longer than ENSO, but shorter than centuries. This segment of the climate variability is dominated by decadal fluctuations whose dynamical nature is more elusive than the ENSO cycle, but manifest themselves very clearly in the precipitation record, in the long-time statistics of the ENSO cycle and in the statistics of other atmospheric variables.

Decadal fluctuations show up clearly in the hydrological system, and regional rainfall regimes have been identified as undergoing wide variability with very large economic and societal impact. The decadal variations of precipitation in the tropics have been well documented, and recently some evidence of decadal variability of precipitation in Western Europe has also been proposed. The Indian monsoon undergoes large decadal oscillations that might affect Southern Europe via a possible connection with the Mediterranean surface pressure has been recently identified.

The picture emerging from these studies seems to indicate that decadal fluctuations in rainfall, though most pronounced in the tropics, exist also in mid-latitude areas and they are probably linked via some unknown teleconnection mechanism to the equatorial variability. The full impact of this variability for the mid-latitude climate remains to be assessed and the workings of the physical processes involved are still largely obscure.

The evolution of the GCMs and the availability of high quality SST data sets, have made possible the design of numerical experiments aimed at simulating drought events. These experiments, performed with coupled models and with atmospheric GCMs with prescribed SST, show that quantities as precipitation, soil moisture and soil temperature exhibit some degree of reproducibility in GCM simulations even outside the tropics. However, ensemble integrations, i.e. repeated GCM simulations with slightly perturbed initial conditions, are necessary to extract useful informations from the numerical experiments. The evidence is now sufficient to justify extensive investigations of the capability of GCMs to simulate decadal fluctuations of the climate variability.

Starting in 1994, the European Union supported a research project within the Environment Program of Fourth Framework Program of Research and Development called *DICE* (*Decadal and Interdecadal Climate variability: dynamics and predictability Experiments*) in which some of these puzzles have been investigated using global numerical models. The plan was to assess the reliability of present GCMs at simulating present climate variations on the decadal timescale, to evaluate the reproducibility of the simulations by performing ensemble integrations and to carry out model validation through comparison with the observed decadal fluctuations. The members of the DICE partnership included nine groups from seven european countries and the major effort was to construct and analyze ensemble simulations with the GCM of participating institutions, by perturbing initial conditions. The results of DICE are widely distributed in the literature in major scientific journals, but at the end of the project it was also felt that it would be useful to bring together in a single place the cultural advances that DICE had fostered. The international collaboration was an experience not only scientific, but also human and personal, forging lasting links of collaboration and friendship. This book represent the results of trying to put together the overall experience of DICE and to pass on "what we have learned" in a more coherent fashion than what is possible in specialized papers. The book is mostly based on contribution from DICE partners (IMGA-CNR, Italy; Max-Planck-Institute, Germany; Hadley Center, UK Meteorological Office; Laboratoire de Meteorologie Dynamique, France; KNMI, Holland; University of Alcala, Spain), but we have happily hosted contributions from Japanese and North American colleagues that have provided expertise not present in DICE, or different and complimentary views.

The first six chapter of the book contains observational studies on decadal variability. The variability of the arctic is discussed in first chapter by Larry Mysak, that illustrates how complex mechanisms can be at work in the arctic that could interact with the general circulation of the Atlantic, the following chapter, by Claude Frankignoul, extends the process in investigating the puzzling relationship between the atmosphere and the SST in the North Atlantic. The discussion of extratropical variability is completed by the Chapter 3, by Hisashi Nakamura and Toshio Yamagata, on the decadal variability in the North Pacific. Chris Folland and collaborators describes thoroughly the variability of the Sea Surface Temperatures, discussing the quality and availability of the data sets and the space-time behavior of the global patterns of variability of SST. The chapter is also giving a lucid discussion of the teleconnections between SST and rainfall at a global level and in selected areas at decadal time scales. An important component of the climate variability is the Indian Summer Monsoon, that is also a phenomenon for which we can have a very long time series of observations. Julia Slingo is discussing in Chapter 5 the decadal variability of the Indian Monsoon and some of its relation with the Pacific ENSO variability. One of the most famous problem involving decadal variability, the Sahel drought, is discussed in Chapter 6, by Neil Ward and collaborators. In this chapter the problem of the variability of rainfall over the Sahelian region is discussed in the wider context of the North African climate system.

The following chapter, by Antonio Navarra, introduces the chapters devoted to the description of numerical simulations of decadal and interannual variability. The chapter discusses the basic concepts that are important for these class of applications: ensembles, skill, reproducibility. The discussion is based on simulations performed with the T30 version of the ECHAM4 model for the period 1961-1994. Similar experiments are discussed in Chapter 8, by Z. Li, with the LMD finite difference model. Chapter 9, by Tett and collaborators, comparisons are made of observed behavior with simulations of interannual to decadal variability in coupled Hadley Center models. The same topic is the subject of Chapter 10, by Seiji Yukimoto, in which the performance of the Japan Meteorological Research Institute global coupled model is discussed. Chapter 11 contains an extensive discussion of the dynamics of the interdecadal variability in coupled ocean atmosphere models by Mojib Latif. The tropical variability is the subject of Chapter 12, by Ortiz-bevia and collaborators, that is discussed defining various indices of the ocean temperature in the Gulf of Guinea. Chapter 13, by Haarsma and collaborators, contains the discussion of an idealized coupled model developed at KNMI that is suitable for very long simulations. A more theoretical outlook of decadal variability is offered by Gu and Philander in Chapter 14. A discussion of the relevance of decadal variability in the thermohaline circulation by Stefan Rahmstorf in Chapter 15 concludes the book.

Working in DICE was an exciting experience and I would like to thank here the support of my partners in DICE (C. Frankignoul, M. Latif, T. Opsteegh, J.Slingo, C. Folland, S. Bassini, M. Sciortino, H. Le Treut and M. OrtizBevia) and have made the whole effort a pleasant enterprise. I would also like to thank the other contributors to the book that have joined enthusiastically the enlarged DICE community. It is also a pleasure to acknowledge the continous support and encouragement of the Commission Officers Ib Troen , R. Fantechi, and A. Ghazi.

A Navarra
Editor,
Coordinator of the European Union Project
"Decadal and Interdecadal Climate Variability: Dynamics and Predictability
Experiments" -- *DICE*

# Table of Contents

# List of Contributors

◆ **W. Cabos Narvaez**
University of Alcalá de Henares
Apdo 20, 28880 Alcalá de Henares, Spain

◆ **A. W. Colman**
Hadley Centre, Meteor. Office,
London Road,
RG12 2SY Bracknell, UK,
email: awcolman@govt.uk

◆ **Mike K. Davey**
Hadley Centre, Meteor. Office,
London Road,
RG12 2SY Bracknell, UK,
email: mkdavey@meto.govt.uk

◆ **Chris K. Folland**
Hadley Centre, Meteor. Office,
London Road,
RG12 2SY Bracknell, UK
email:ckfolland@meto.govt.uk

◆ **Claude Frankignoul**
LODYC,
Université Pierre & Marie Curie,
4, Place Jussieu, Tour 14-15, 2eme étage,
F - 75252 Paris cédex 05, France
e-mail: cf@lodyc.jussieu.fr

◆ **D. Gu**
Atmospheric and Oceanic Sciences Program,
Princeton Univesity,
Princeton, NJ, USA,
e-mail: gphlder@splash.Princeton.edu

◆ **Peter J. Lamb**
Cooperative Institute for Mesoscale
Meteorological Studies, School of Meteorology,
The University of Oklahoma,
Norman, Oklahoma 73019, USA.

♦ **Z. X. Li**
Laboratoire de Météorologie Dynamique du CNRS,
Ecole Normale Supérieure,
24, rue Lhomond,
F-75231 Paris cédex 05, France
e-mail: li@lmd.jussieu.fr

♦ **Reindert J. Haarsma**
KNMI,
De Bilt, Holland
e-mail: haarsma@knmi.nl

♦ **Mostafa El Hamly**
Moroccan Direction de la Météorologie Nationale
(DMN), Casablanca, Morocco

♦ **Sarah Ineson**
Hadley Centre, Meteor. Office,
London Road,
RG12 2SY Bracknell, UK
email: sineson@meto.gov.uk

♦ **Arie Kattenberg**
KNMI,
De Bilt, Holland,
e-mail: kattenbe@knmi.nl

♦ **Mojib Latif**
Editor, Monthly Weather Review
Max-Planck Institut fuer Meteorologie,
Bundestrasse 55, D-20146 Hamburg, Germany
e-mail: latif@dkrz.d400.de

♦ **Qing Liu**
KNMI,
De Bilt, Holland,
e-mail: liu@knmi.nl

♦ **Lawrence A. Mysak**
McGill University
805 Sherbooke Street West
Montreal, Quebec, Canada
email: mysak@zephyr.meteo.mcgill.ca

♦ **Hisashi Nakamura**
Department of Earth and Planetary Physics,
University of Tokyo,
Tokyo, 113-0033, Japan
e-mail: hisashi@geoph.s.u-tokyo.ac.jp

♦ **Antonio Navarra**
Istituto per lo Studio delle Metodologie
Geofisiche Ambientali (IMGA), Via Gobetti 101,,
I-40139, Bologna, Italy
e-mail: navarra@rigoletto.imga.bo.cnr.it

♦ **J. D. Opsteegh**
KNMI,
De Bilt, Holland
e-mail: opsteegh@knmi.nl

♦ **J. M. Oberhuber**
DKRZ
Bundestrasse 55,
D-20146 Hamburg, Germany

♦ **Maria J. Ortiz Bevia**
University of Alcalá de Henares
Apdo 20, 28880 Alcalá de Henares, Spain
email: fsortiz@alcala.es

♦ **David E. Parker**
Hadley Centre, Meteor. Office,
London Road,
RG12 2SY Bracknell, Berkshire, UK
email: deparker@meto.gov.uk

♦ **S. George H. Philander**
Atmospheric and Oceanic Sciences Program,
Princeton University,
Princeton, NJ, USA
e-mail: gphlder@splash.Princeton.edu

♦ **Diane H. Portis**
Cooperative Institute for Mesoscale
Meteorological Studies, School of Meteorology,
The University of Oklahoma, Norman, Oklahoma
73019, USA

♦ **Stefan Rahmstorf**
Potsdam Institute for Climate Impact Research,
14412 Potsdam, Germany
email: rahmstorf@pik-potsdam.de

♦ **Rachid Sebbari**
Moroccan Direction de la Météorologie Nationale
(DMN), Casablanca, Morocco

◆ **Frank M. Selten**
KNMI,
Den Bilt, Holland
e-mail: selten@knmi.nl

◆ **Julia Slingo**
University of Reading,
Reading, United Kingdom
e-mail: J.M.Slingo@reading.ac.uk

◆ **Simon F. B. Tett**
Hadley Centre, Met. Office,
London Road,
RG12 2SY Bracknell, UK
email: sfbtett@meto.gov.uk

◆ **Toshio Yamagata**
Department of Earth and Planetary Physics,
University of Tokyo,
Tokyo, 113-0033, Japan
e-mail: yamagata@geoph.s.u-tokyo.ac.jp

◆ **Seiji Yukimoto**
Climate Research Department,
Meteorological Research Institute,
1-1, Nagamine, Tsukuba, 305-0052, Japan
email: yukimoto@mri-jma.go.jp

◆ **Neil Ward**
IMGA - CNR,
Via Gobetti 101,
Bologna, Italy
e-mail: ward@imga.bo.cnr.it

◆ **Richard Washington**
School of Geography
University of Oxford
Mansfield Rd
OX1 3TB, Oxford, UK
e-mail: geogrw@unit.ox.ac.uk

# 1 Interdecadal Variability at Northern High Latitudes

LAWRENCE A. MYSAK
*Centre for Climate and Global Change Research and*
*Department of Atmospheric and Oceanic Sciences, McGill University,*
*805 Sherbrooke St. W, Montreal, QC H3A 2K6*

## 1.1 Introduction

Examination of the 130-year time series of observed surface air and sea surface temperature anomalies for the northern hemisphere reveals three fundamental timescales of variability: interannual, interdecadal and century. Many of the interannual signals can be associated with strong ENSO warming events originating in the tropical Pacific. The century-scale upward trend, on the other hand, is commonly believed to represent evidence of global warming due to the increase of greenhouse gases caused by the activities of mankind. The interdecadal signal, however, is less well understood; many investigators have suggested that it originates in middle-to-high latitudes and is related to decade-to-century scale ocean current fluctuations in the northern North Atlantic (Bjerknes, 1964; Bryan and Stouffer, 1991; Kushnir, 1994).

Interdecadal variability is the focus of a number of current large-scale climate research programs (e.g., CLIVAR), and a discussion of certain northern high-latitude (especially Arctic) aspects of this variability will be presented is this chapter. As a starting point, we refer the reader to the excellent Dickson et al. (1988) review paper which described the "Great Salinity Anomaly" (or GSA) and associated upper ocean cooling in the northern North Atlantic during the 1960s and 1970s. The GSA represented one of the most persistent and extreme decadal-scale variations of ocean climate in this century, and it was accompanied by anomalously cold surface air temperatures over the northern North Atlantic (Chapman and Walsh, 1993). The GSA consisted of a sequential freshening of the North Atlantic subpolar gyre, which was first observed north of Iceland in the late 1960s, where convection in the Iceland Sea was suppressed (Malmberg, 1969). The GSA was next detected in the Labrador Sea during the early 1970s (Lazier, 1980) and then found in the Norwegian Sea in the late 1970s (see Fig. 1.1). This ocean-climate event has captured the imagination of such a wide scientific audience that it now appears as a feature item in modern undergraduate textbooks in oceanography (Duxbury and Duxbury, 1994)!

Two fundamental questions concerning the GSA centre on its origin, and whether it was a singular event or part of an interdecadal high latitude climate fluctuation. If it is a recurrent event, then it is natural to ask whether there were similar

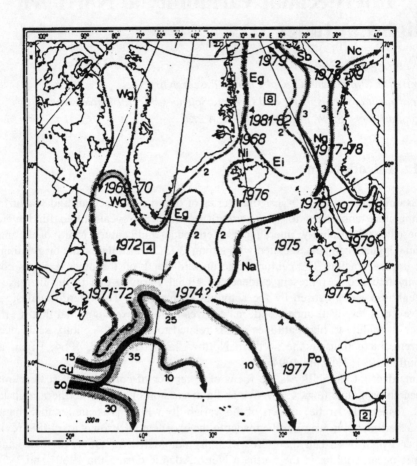

**Fig. 1.1** Transport scheme for 0-1000m layer of the northern North Atlantic (Dietrich et al., 1975) with date of the salinity minimum superimposed. (From Dickson et al., 1988.)

such freshening events in the past (Dickson et al., 1988). Similarly, will there be other GSA-type events in the future?

Earlier hypotheses concerning the generation of the GSA suggested that local effects were dominant: northerly wind anomalies over the Greenland Sea may have driven southward large amounts of fresh polar water there (Dickson et al., 1975), or the freshening may have been due to the excess of precipitation over evaporation (Pollard and Pu, 1985). Following on from the ideas of Aagaard and Carmack (1989), Walsh and Chapman (1990a) argued that fluctuations in the sea-level atmospheric pressure difference between Greenland and the northern Asian coast produced Arctic wind anomalies which resulted in enhanced export of sea ice (fresh water) from the central Arctic just prior to 1968, the year with lowest salinity.

In contrast to the above "regional" generation mechanisms, Mysak, Manak and Marsden (1990, hereafter referred to as MMM) argued that the GSA in part may have been remotely generated by prior anomalously large runoffs into the western Arctic from North America during the mid-1960s. MMM also proposed that the GSA may be part of a sequence of atmospheric, hydrological and oceanic events in the Arctic and northern North Atlantic that form an interdecadal (approximately 20-year) climate cycle that can be described in terms of a negative (or reversing) feedback loop (see section 2). Associated with the GSA were large sea-ice extents in the Greenland Sea (Mysak and Manak, 1989), which were found to be highly correlated with the Koch sea-ice severity index, i.e., the number of weeks per year when ice affected the coast of Iceland (Kelly et al., 1987). Fig. 1.2 shows that over

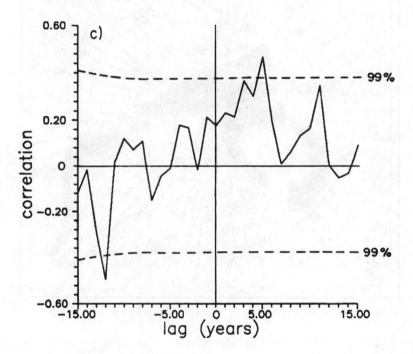

**Fig. 1.2** The lagged cross-correlation between North American runoff into the Arctic and the Koch ice index for the period 1915-75. A positive lag means runoff leads the sea-ice index. Thus runoff leads the sea-ice by 3 to 5 years, with the 5-year peak being above the 99% significance level. (From Mysak and Power, 1991.)

the period 1915-75, the high-latitude North American runoff fluctuations led the Koch ice index by about three-to-five years (maximum correlation at a 5-year lead).

There are two (sequential) mechanisms which are believed to account for the above lagged relationship between Greenland Sea ice cover and runoff. First, fluctuations in the Mackenzie River discharge tend to produce sea-ice cover anomalies of similar sign in the Beaufort and Chukchi Seas region about one year later (Manak and Mysak, 1989). Second, since the general pattern of ice drift in the Arctic consists of the Beaufort Gyre and the Transpolar Drift Stream (TDS) which supplies ice to the East Greenland Current  (Fig. 1.3), it is conceivable that any sea-ice

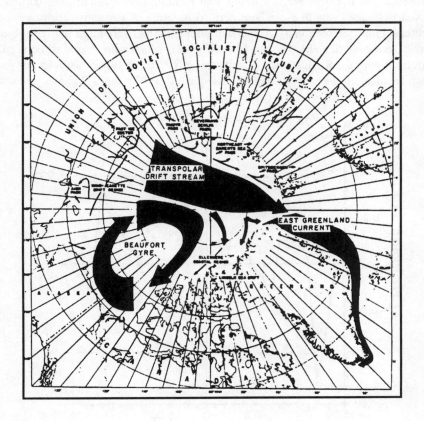

**Fig. 1.3**   Early schematic of the major drift patterns of sea-ice in the Arctic (Dunbar and Wittman, 1963). The general features are consistent with more recent data (see Colony and Thorndike (1984), for example).

anomaly that formed to the north of Yukon and Alaska would be exported out of the Arctic via the TDS into the Greenland Sea about three to four years later. Evidence for these mechanisms will be presented in section 1.3. The close relationship between positive sea-ice extents and freshwater anomalies in the Greenland Sea that characterizes the GSA have prompted Mysak and Power (1992) to identify this and other such ice/freshwater events as "GISAs", which stands for 'Great Ice and Salinity Anomalies'. Alternatively, since not all such anomalies are likely to be as large as the GSA, we might simply call such events "Ice and Salinity Anomalies", or "ISAs" for brevity.

The proposal by MMM and modified by Mysak and Power (1991, 1992) that GISAs may be part of an interdecadal climate cycle is indirectly supported by the fact that a spectrum of a 370-year Koch ice index time series shows a significant peak at 27 years (Fig. 1.4). Curiously enough, the spectra of ice-core data from

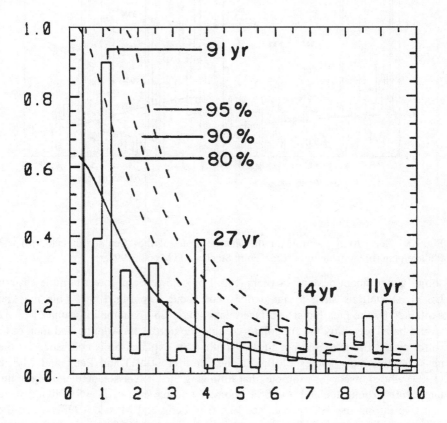

**Fig. 1.4** Normalized spectrum of the time series of the Koch ice index for Iceland (Koch, 1945; Lamb, 1977) from 1600-1970). The data were smoothed with a 5-year moving average. (From Stocker and Mysak, 1992)

Greenland also show an interdecadal peak of about 20 years (Hibler and Johnsen, 1979). However, such data may also be influenced by processes in the Labrador Sea, which are not explicitly included in the Arctic interdecadal cycle proposed by Mysak and his collaborators.

Mysak and Power (1992) also speculated that the interdecadal Arctic cycle might influence the climate at middle latitudes, especially that of Europe and eastern North America. The spectrum of the 318-year central England temperature record

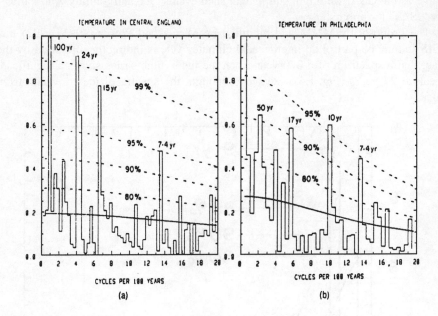

Fig. 1.5   Normalized spectrum of the annual means of the "Central England" (a) and Philadelphia (b) temperature record. (From Stocker and Mysak, 1992.)

shows a significant peak at 24 years (Fig. 1.5a): however, at lower latitude eastern US coast stations such a signal does not occur (see Fig. 1.5b). This may be explained by the fact that sea surface temperature (SST) anomalies around Iceland would be advected by the subpolar gyre into the northwest Atlantic and then eastward toward England, but not further south. This explanation is confirmed by the air-sea interaction studies of Bryan and Stouffer (1991) and Peng and Mysak (1993) who showed, for example, that following the GSA-associated cooling in the late 1960s around Iceland, the SST anomalies to the east of Newfoundland were negative during the 1970s (e.g., see Fig. 6 in Peng and Mysak, 1993). Since the SST anomalies in this region were positive during the 1950s, some investigators have suggested (e.g., Kushnir, 1994) that this may be observational evidence, albeit slim, of a natural 50-year period oscillation in the thermohaline circulation and climate of the North Atlantic that appears in a coupled atmosphere-ocean general circulation model (Delworth et al., 1993).

The approximately 15-year peak seen in both the central England and Philadelphia temperature spectra (Fig. 1.5), although less significant than the 25-year peak, may be a reflection of decadal-scale climate fluctuations at lower latitudes. For example, this signal may be due to long-term ENSO variability. Mann and Park (1994), in their singular value decomposition of 100 years of global temperature variations, found an interdecadal mode in the 15-to-18 years period range whose spatial structure can be related to the ENSO pattern of SST anomalies.

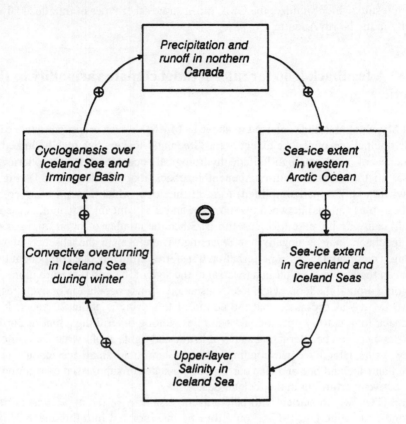

**Fig. 1.6** Negative (or reversing) feedback loop linking northern Canadian river runoff, Arctic sea-ice extent, Greenland-Iceland Sea ice extent, and salinity, convection and cyclogenesis around Iceland. This is a modified (and simplified) version of the ten-component loop originally proposed by Mysak et al. (1990) to account for interdecadal Arctic climate oscillations with a period of about 15-20 years. (From Mysak and Power, 1992)

The outline of the remainder of this chapter is as follows. Section 1.2 presents a simplified form of the feedback loop first proposed by MMM to account for inter-decadal climate cycles in the Arctic and northern North Atlantic. Section 1.3 describes the analysis of recent sea-ice and related climate data which suggests that another GISA (or ISA) event occurred in the Greenland Sea during the late 1980s, as predicted in MMM. In section 1.4 a brief report is given of the recent work by Bjornsson et al. (1995), who attempted to find the sources of the precipitation and runoff anomalies that help to drive part of the proposed feedback loop. Also, the related work of Walsh et al. (1994) and Slonosky et al. (1997) are presented. Finally, in section 1.5 a review is presented of selected ice/ocean/atmosphere modelling results which simulate the GSA, or illustrate other types of interdecadal variability in the North Atlantic.

## 1.2    A feedback loop for interdecadal climate variability in the Arctic

In Mysak et al. (1990, referred to above as MMM), it was proposed that decadal-scale fluctuations of sea-ice extent in the Greenland-Iceland Sea may be linked to a sequence of northern high-latitude hydrological, oceanic and atmospheric processes in the form a multi-component feedback loop (Kellogg, 1983). Fig. 1.6 shows a modified (and simplified) form of this loop, which has six basic components. A plus (minus) between two boxes A and B say, means that an increase in A would cause an increase (decrease) in B. Since the number of minus signs is odd, the feedback loop is negative or reversing. Therefore, in the absence of other strongly damping factors, a perturbation transferred from any one component to the next can theoretically result in a reversal of the sign of the initial perturbation. Thus by going around the loop twice, a self-sustained climate oscillation or cycle results. In MMM it was estimated that the period of this cycle is about 20 years. If this feedback loop is truly operational, it implies, among other things, that in northern Canada there can be alternating states of heavy and light runoff, which are followed a few years later by corresponding states of large and small ice extents in the Greenland-Iceland Sea and also corresponding states of suppressed convection and convective overturning in the Iceland Sea.

The GSA, which started with below average salinity in the upper waters of the Iceland Sea in the late 1960s, could thus be incorporated into this feedback loop and have originated, at least in part, from large runoffs into the western Arctic. Further, the feedback loop implies that the GSA could be a cyclic event. On the basis of the more complicated version of this loop, MMM predicted that another GSA-like event (a GISA) would occur in the Greenland-Iceland Sea in the late 1980s. This was confirmed in part by Mysak and Power (1991) in their analysis of monthly sea-ice cover maps for the Greenland Sea up to the end of 1988. In Mysak and Power (1992) it was found that the large Greenland Sea ice extents in the late 1980s shown in Mysak and Power (1991) could have been partly due to prior high Mackenzie river runoffs and the subsequent propagation of positive ice-cover anomalies from the Beaufort Sea into the Greenland-Iceland Sea. These results will be described in more detail in section 3.

The hypothesized links between convective overturning, cyclogenesis and precipitation (see the three boxes in the upper left part of Fig. 1.6) have not yet been clearly substantiated from data. During a GISA event, the occurrence of an anomalously fresh upper layer in the Iceland Sea has a stabilizing effect, and convective overturning during winter, which normally brings up warm Atlantic water to the surface, is suppressed and hence the SSTs are reduced. This, we hypothesize, would lead to reduced cyclogenesis in this general region, and through some as yet unidentified storm track or teleconnection pattern would lead to reduced precipitation and runoff over northern Canada. Walsh and Chapman (1990b) have shown that winter sea-level pressure anomalies at a point in northern Canada are strongly correlated with those in the Greenland-Iceland Sea (see their Fig. 12); however, exactly how this correlation relates to the hypothesized cyclogenesis-precipitation link remains an open question. In section 1.4we shall briefly describe recent attempts to establish the validity of these links.

## 1.3    Evidence for a Great Ice and Salinity Anomaly (GISA) event in the late 1980s

In Mysak and Power (1992) an analysis of 36 years of sea-ice concentration, sea-level pressure, runoff and salinity data was performed in order to search for the existence and origin of a late 1980s GISA event. In particular, a thorough study of the propagation of ice cover anomalies in the Arctic Ocean and Greenland Sea was made. Fig. 1.7 shows the subregions of the Arctic and marginal seas that were used in the lagged cross-correlation analysis of low-pass filtered areal sea-ice extent anomalies for the period 1953-88. Fig. 1.8 shows the filtered time series of the ice areal anomalies for the five contiguous subregions $B_1$ to $D_1$. Note that the sequence of subregions follows the ice drift pattern which leads to the export of sea ice out of the Arctic Ocean through Fram Strait (see Fig. 1.3).

The dashed lines with negative slope in Fig. 1.8 indicated that both positive and negative ice anomalies propagated from the Beaufort Sea region into the Greenland-Iceland Sea region. The line $L_2$ shows that the large positive GSA-related ice anomaly in the late 1960s seen in the bottom time series could have originated in part from the Beaufort Sea positive ice anomaly centered around 1965. The propagation of this anomaly has been successfully modelled by Tremblay and Mysak (1997)[1]. The line $L_4$ shows that the GISA event of the late 1980s in the Greenland Sea could have originated from the mid-1980s ice anomaly in the Beaufort Sea. Since negative salinity anomalies were observed in the Greenland Sea in the late 1980s (GSP Group, 1990), it is indeed appropriate to say that the peak just touching the lower end of $L_4$ is evidence of a GISA event. (Although all the time series shown in Fig. 1.8 end in 1988, Chapman and Walsh (1993) have analyzed an updated ice data set that extends through to December 1990 and have shown that the 1988 positive ice anomaly seen in the $D_1$ time series rapidly changes to a large negative anomaly by 1990 - see their Fig. 7a.) The Beaufort Sea mid-1980s positive ice anomaly in turn occurred because of the low salinity water on the Beaufort Sea continental shelf which was presumably caused by the above average runoff from the Mackenzie River in the mid-1980s (see Fig. 1.9, top. The fact that the late

---

[1] However, they argue that it was most likely generated by northerly wind anomalies in the western Beaufort Sea, rather than by runoff anomalies.

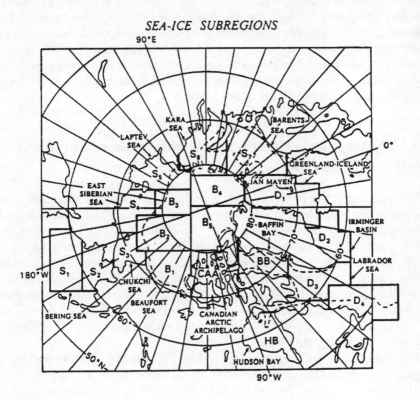

**Fig. 1.7** Subregions of the Arctic Ocean and marginal seas used in the lagged cross-correlation analysis of low-pass filtered areal sea-ice extent anomalies derived from monthly sea-ice concentration (SIC) data for the period 1953-88. The dashed curve denotes the 200 m isobath. (From Mysak and Power, 1992.)

1980s GISA event is characterized by a relatively small positive ice cover anomaly in the Greenland Sea is likely due to the fact that the wind anomalies at the time were in a direction that would have tended to drive the ice anomalies northward, in opposition to the mean ice outflow direction (Mysak and Power, 1992).)

It is interesting to note that the positive ice anomalies in the Beaufort Sea (Fig. 1.8, top) suggest the existence of an approximately 10-year cycle in this region, rather than a 20-year one. This 10-year cycle is roughly in phase with the decadal oscillation found in the Mackenzie runoff data by Bjornsson et al. (1995). However, it appears that the weaker or shorter lived Beaufort Sea ice anomaly signals (i.e., the ones that peaked in 1956 and 1975) were substantially reduced by the time they reached the Greenland Sea, perhaps because of the relatively thick ice in the central Arctic. Nonetheless, Slonosky et al. (1997) showed that this 10-year cycle is found in the first Empirical Orthogonal Functions (EOFs) of winter sea ice concentration (SIC), sea level pressure (SLP) and 850 hPa temperature (850 T)

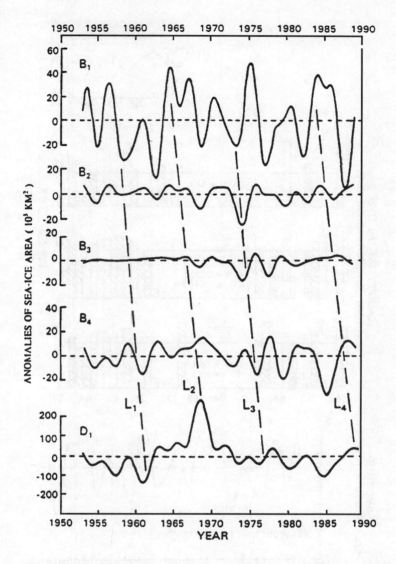

**Fig. 1.8** Low-pass filtered anomalies of areal sea-ice extent for subregions $B_1$, ... $D_1$ in the Arctic Ocean and Greenland-Iceland Sea (see Fig. 1.7); note that the vertical scale in the bottom time series (for subregion $D_1$) has been compressed by a factor of five. The distances between the zero-means of adjacent pairs of time series are proportional to the corresponding distances between the centers of the adjacent subregions shown in Fig. 1.7. The dashed lines with negative slope ($L_1$) indicate the propagation of ice anomalies from the Beaufort Sea (subregion $B_1$) through to the Greenland-Iceland Sea (subregion $D_1$). In particular, the line $L_4$ shows that the origin of the GISA in the Greenland-Iceland Sea in the late 1980s could have been partly due to the large anomaly in subregion $B_1$ in the mid-1980s. (From Mysak and Power, 1992.)

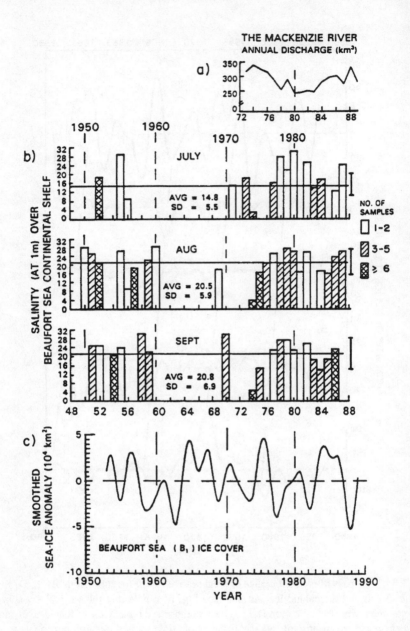

**Fig. 1.9**  a) Annual Mackenzie River runoff during 1973-89 at Arctic Red River (a city on the Mackenzie R.). (Data courtesy of R. Lawford.) b) Salinities at 1 m over the southeastern Beaufort Sea continental shelf (in subregion $B_1$) for July, August and September during 1950-87 (from Fissel and Melling, 1990). c) Low-pass filtered ice anomalies in the Beaufort Sea. (From Mysak and Power, 1992.)

fluctuations poleward of 45°N. The pattern of the first EOF of SIC (Fig. 1.10, top) describes well the main ice anomalies associated with the GSA in the late 1960s, and the time series of this EOF (Fig. 1.10, bottom) shows a fairly regular 10-year oscillation. The leading EOF of winter SLP (Fig. 1.11) also reveals, after the early 1960s, a decadal oscillation of the pressure over the polar and subarctic regions. In particular, this figure captures the high pressure anomaly cell over Greenland during the late 1960s that was observed by Dickson et al. (1975). Finally, the EOF 1 of winter 850 T (Fig. 1.12) most clearly shows a decadal oscillation superimposed on a long-term warming trend over the Greenland Sea and cooling trend over the Labrador Sea.

The Beaufort Sea decadal ice cover oscillations seen in Fig. 1.8 may also have implications for decadal-scale variability at lower latitudes. For example, they may force ice cover anomalies in the Canadian Arctic Archipelago (CAA) which in turn may excite ice anomalies in Baffin Bay and thus the Labrador Sea (see Fig. 1.8b in Mysak and Power, 1992). Hence through this route, the 10-year period oscillations in the Beaufort Sea may have contributed to the observed 10-year ice cover cycle observed in the Labrador Sea (Mysak and Manak, 1989; Deser and Blackmon, 1993; Wang et al., 1994; Mysak et al., 1996).

## 1.4    Possible sources for the precipitation and runoff anomalies in the Mackenzie basin: An attempt to verify the cyclogenesis-related links in the feedback loop

In Bjornsson et al. (1995), an analysis was presented of extratropical cyclone frequency and the 500 mb height standard deviation field in the northern hemisphere, together with precipitation data from 15 stations in the Mackenzie basin (Fig. 1.13) and the Mackenzie River runoff. Spatial and temporal variability in the data were examined for the period 1965-89, and a cross-correlation analysis was performed to determine the relationship between the runoff and the precipitation variations, and between the precipitation and the atmospheric anomalies.

It was found that the precipitation fluctuations in the Mackenzie basin are strongly linked to variations in the Mackenzie River discharge and to variations in the eastward moving Northeast Pacific storm tracks, with the timescales of variation ranging from interannual to decadal. In particular, a clear approximately decadal signal was found in the runoff time series, which may well explain the occurrence of the decadal signal seen in the Beaufort Sea ice cover time series (Fig. 1.8, top). If the Pacific is indeed the main source for the Mackenzie basin precipitation fluctuations, then this runs counter to the hypothesis implied by the upper left part of the feedback loop in Fig. 1.6, which implies that, at high latitudes, occasionally westward moving storms travelling from the northern North Atlantic to northern Canada bring moisture to the Mackenzie basin. Bjornsson et al. did find that there was some influence on the precipitation in the Mackenzie basin from cyclones that originated south of Iceland and migrated into the Labrador Sea, Baf-

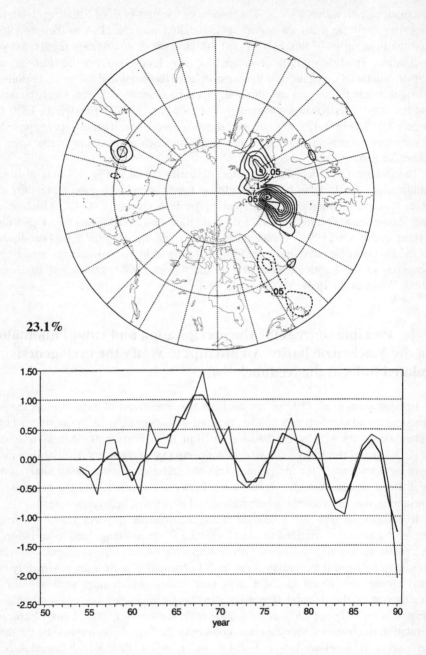

**Fig. 1.10** Spatial structure (top) and time series (bottom) of EOF 1 of winter sea ice concentration (SIC). Thick line in time series represents three-year running mean. (From Slonosky et al., 1997)

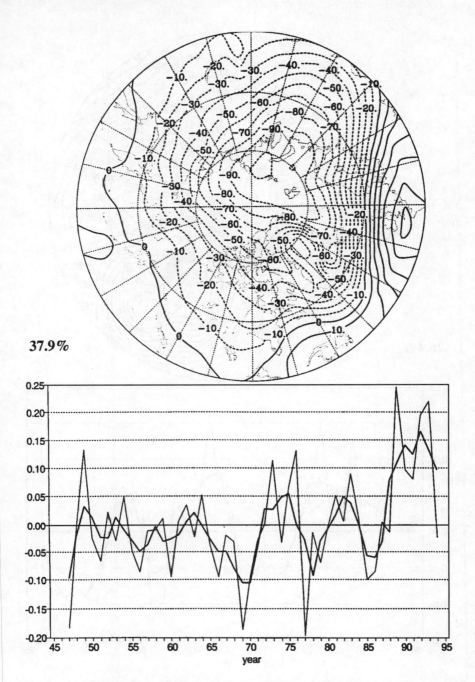

**Fig. 1.11** As in Fig. 1.10, but for EOF 1 of winter SLP. (From Slonosky et al., 1997.)

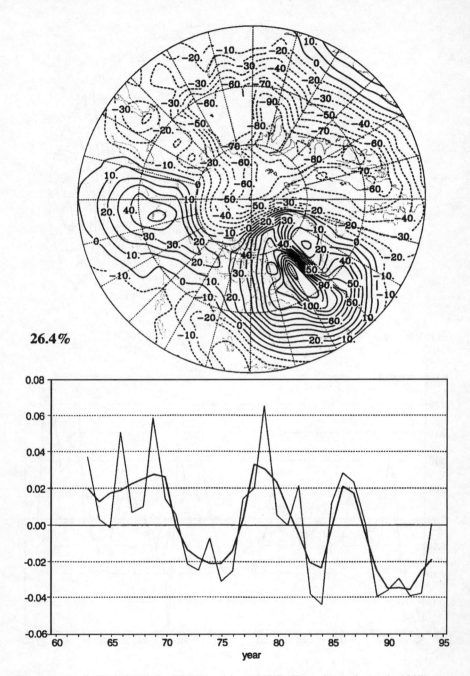

**Fig. 1.12** As in Fig. 10, but for EOF 1 on winter 850 T. (From Slonosky et al., 1997)

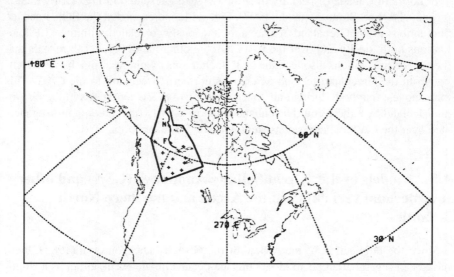

**Fig. 1.13** The Mackenzie River and its drainage basin (bold lines). The plus symbols indicate the positions of the precipitation stations. Runoff station Norman Wells is marked with "N" and Fort Simpson with "F". The cyclone frequencies were calculated for the whole area shown. (From Bjornsson et al., 1995)

fin Bay and northern Canada. However, the effect was small and unlikely to have contributed significantly to the runoff variations in the Mackenzie River.

In order to have the cyclogenesis part of the feedback loop fully operational, it appears that anomalous moisture fluxes from cyclonic activity around Iceland would have to enter the Arctic via the Greenland-Norwegian seas (as originally proposed in MMM), travel to northern Canada via the polar low vortex and then, through convergence of these fluxes, contribute to the precipitation anomalies there. This sequence of events is plausible in the light of the recent hydrological study by Walsh et al. (1994) and the cross-correlation analysis in Slonosky et al. (1997). Walsh et al. showed (see their Fig. 4) that during the period 1973-90, there were generally large northward fluxes of water vapor across 70°N centered at 10°E (the Norwegian Sea) and notable southward fluxes across 70°N centered around 260°E (central northern Canada-Mackenzie basin). Moreover, during the active precipitation season (May-July), the convergencies derived from the water vapor fluxes over the Mackenzie basin, which in theory are approximately proportional to P-E, were indeed highly correlated with the observed Mackenzie basin precipitation and the Mackenzie River runoff (see Fig. 14 in Walsh et al., 1994). In particular, prior to the late 1980s GISA event described in section 3, this figure shows that the flux convergence and runoff were both above average in the mid-1980s, which is consistent with the feedback loop. Further evidence of a link between atmosphere-ice interactions in the Greenland-Iceland Sea and atmospheric variability over northern Canada (especially over the CAA) is presented in Fig. 1.14 which is reproduced from Slonosky et al. (1997). Fig. 1.14a indicates that the EOF 1 sea ice fluctuations in the Greenland Sea are simultaneously correlated with the SLP fluctuations both locally and over the CAA. On the other hand, Fig. 1.14b reveals that the sea ice fluctuations also lead the SLP fluctuations by one year in a band stretching from the Greenland-Iceland Sea across northern Greenland to the CAA. This may suggest that both during and after anomalous sea ice conditions in the Greenland-Iceland Sea, there could be enhanced storm activity (and precipitation) generated over the CAA, which is consistent with the feedback loop.

## 1.5 Models of the "Great Salinity Anomaly" (GSA) and other interdecadal variability in the Arctic and northern North Atlantic

Since the publication of the review of the GSA by Dickson et al.(1988), there have been several attempts to model this ice-ocean climate event, either as a singularly occurring feature or as part of an interdecadal climate cycle. At about the same time, there have been many modelling studies of decadal-scale internal variability of the North Atlantic thermohaline circulation (THC), which were prompted by the accidental discovery by Weaver and Sarachik (1991) of an 8.6-year oscillation in the strength of the THC in a 3-D general circulation model under steady forcing. In this section a brief discussion of a number of GSA modelling papers will first be given. This will be followed by a survey of simple ice-ocean models, ocean GCMs and a coupled atmosphere-ocean GCM which contain interdecadal signals in the North Atlantic that are intimately connected with variations in the THC.

### 1.5.1 Simulations of the GSA

The first successful simulation of the increased sea-ice cover associated with the GSA was performed by Darby and Willmott (1993), who used an extended version of the simple Willmott and Mysak (1989) 2-D ice-ocean model for the Greenland-Iceland-Norwegian (GIN) Sea. Since the model consists of a limited domain (a box lying between Greenland and Norway, with open boundaries at 60 and 80°N) and contains only a single active layer in the ocean part, the postulated Arctic forcing of the GSA and the observed suppression of convection in the Iceland Sea could not be simulated. These shortcomings were partly removed in the paper by Häkkinen (1993) who used a fully prognostic Arctic-GIN Sea 3-D ice-ocean model to study the interannual variability of the sea-ice cover during the period 1955-75, which spans the GSA years. She found that prior to large sea-ice extents in the Greenland Sea associated with the GSA, there were large pulses of ice and fresh water exports through Fram Strait due to wind field changes in the Arctic, a result which supports the GSA generation conjecture of Aagaard and Carmack (1989).

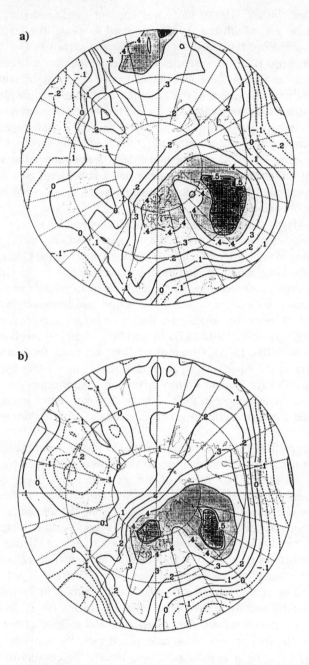

**Fig. 1.14** Map of temporal correlation coefficients between the time series of EOF 1 of winter SIC (see Fig. 1.10) and winter SLP anomalies for the period 1954-1990 for a) zero lag, and b) lag of +1 (ice leading atmosphere by one year). Lightly shaded areas represent correlations that are significant at the 95% level, dark shading represents areas of 99% significance. (From Slonosky et al., 1997)

In Darby and Mysak (1993), the novel concept of Boolean delay equations (BDEs; see Ghil and Mullhaupt, 1985) was used to model the feedback loop in MMM as modified in Mysak and Power (1992) and shown in Fig. 1.6 above. In their simplest form, BDEs represent evolution equations for a vector of discrete variables which can take on the values of 0 or 1 (a "low" or "high" state), but which also depend on previous values of the variables. BDEs are thus ideally suitable for modelling systems with threshold behavior and feedbacks. The delays represent the interaction times between pairs of variables and thus are a crude representation for the time constants that would arise in a differential equation approach to modelling the climate system. The BDE approach, of course, is not intended as a replacement for traditional differential equation models; but it is a useful tool for answering conceptual questions concerning interactions in the climate system and for developing more sophisticated models.

Darby and Mysak (1993) introduced six Boolean variables to represent the state of: (1) precipitation over northern Canada, (2) surface salinity in the western Arctic, (3) ice cover in the western Arctic, (4) surface salinity in the Greenland Sea, (5) ice cover in the Greenland Sea, and finally, (6) convection in the Greenland Sea. In the model, a high state (the variable equals 1) refers to positive anomalies of precipitation, salinity or ice cover, or active convection, and conversely a low state (the variable equals 0) refers to negative anomalies or suppressed convection. From a set of six BDE equations ((1)-(6) in Darby and Mysak (1993)) involving these variables, it was shown that 15- to 20-year oscillations can occur for a variety of realistic time delays in the equations. In particular, it was found that by allowing for different timescales for ice and salinity advection from the western Arctic to the Greenland Sea, the ice extent in the Greenland Sea can persist longer than an ice anomaly in the western Arctic, in general agreement with the observations (e.g., see Fig. 1.8).

In Robitaille et al. (1995, hereafter referred to as RMD), a simple dynamical system model of the Arctic Ocean and marginal seas was developed by applying the Martinson et al. (1981) box model of a high latitude two-layer ocean to four regions connected together: the Greenland Sea, the Norwegian Sea, the Arctic Ocean and the Greenland Gyre (see Fig. 1.15) . The model for each region consists of a thermodynamic ice layer that covers two layers of saline water which can, under specific conditions, become gravitationally unstable and hence create a state of active overturning (Fig. 1.16) . Also, under sufficiently large ocean to atmosphere heat fluxes, the ice can disappear (see states 1 and 2 in Fig. 1.16). The system was forced by monthly mean atmospheric temperatures in the four regions, by continental runoffs and by inflows from adjacent oceans (Fig. 1.15). The model predicted the ice thickness, and the temperature and salinity of the water in the upper layer of the four regions. Also determined were the water temperature and salinity of the lower layer in the Arctic box. Finally the convective state of each region was computed as a continuous function of time.

In the control run (climatological forcing with a seasonal cycle), the Arctic Ocean was found to always have continuous ice cover, whereas the Greenland Sea

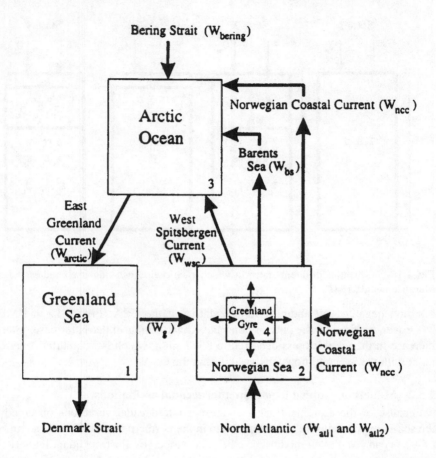

**Fig. 1.15**  The links between the four regions in the model and the rest of the world oceans. The numerical values of the water transports W are given in Table 1 in Robitaille et al. (1995). All the *arrows* indicate water or ice transports, except between the Norwegian Sea and Greenland Gyre where the *double headed* arrows indicate diffusion between the two regions. (From Robitaille et al., 1995.)

and Gyre had ice cover only during winter, and the Norwegian Sea was always ice free. Another important feature of the control run was the wintertime occurrence of convective overturning in the upper 200 m of the Gyre region. The model was also used for a number of anomaly experiments, including the insertion of various negative salt anomalies into the Norwegian Sea to simulate the GSA. In the latter set of experiments, convective overturning in the Greenland Gyre in winter was greatly affected. Instead of a well-defined period when the system went from state 4 to state 3 (see Fig. 1.16) and back to state 4 during winter as in the control run, the system oscillated rapidly back and forth between states 3 and 4 for the case of a weak anomaly (Fig. 9 in RMD) and had no convection during winter for the case of

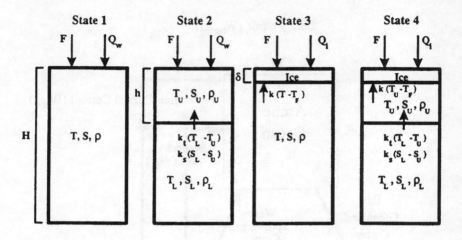

**Fig. 1.16**    The four different states in which each region can find itself (adapted from Martinson et al., 1981).

a strong negative salt anomaly representative of the GSA (Figure 11 in RMD). Moreover, because of the lack of convective overturning in the latter case, a large increase in the ice thickness occurred in this region, which is in qualitative agreement with the ice conditions observed during the GSA.

### 1.5.2   Models of coupled ice-ocean interdecadal oscillations

Because of the aforementioned concurrent interdecadal variations observed in the sea-ice cover, salinity and temperature in the northern North Atlantic, a number of ice-ocean circulation models have been developed to illustrate quantitatively the mechanisms which might explain such variability. The models all bring into play feedbacks between the sea ice and the thermohaline circulation (THC). Yang and Neelin (1993) studied the interaction between a thermodynamic sea-ice model due to Welander (1977) and a zonally-averaged THC model similar to Marotzke et al. (1988). They found that a self-sustaining interdecadal oscillation in ice thickness, SST and salinity arises through feedbacks between salinity anomalies induced by the sea-ice melting/freezing process and anomalous meridional heat transport associated with the THC (see the feedback loop in Fig. 6 in Yang and Neelin). For the parameters chosen, the period of the oscillation was found to be about 13.5 years, which was set not by an oceanic timescale but by the nature of the ice-THC coupling.

In Zhang et al. (1995) on the other hand, a THC-ice model was developed which contains an interdecadal ice-ocean oscillation that is due to a feedback between ice cover and ocean temperature, which was originally described by Saltzman (1978). The ocean model is based on the 3-D planetary-geostrophic ocean circulation model of Zhang et al. (1992), and the ice model is the classical thermodynamic one of Semtner (1976). Ice forms in the model in high latitude regions where the ocean rapidly loses heat to the atmosphere; this results in an increased sea-ice extent which then acts as insulator over the ocean and therefore makes the ocean warmer. The THC, which transports heat poleward, tends to enhance this ocean warming which then leads to ice melt. The resulting open water then loses heat which in turn leads to ice formation and the process repeats itself. Salinity rejection/dilution associated with ice formation/melting was found to be of secondary importance in this oscillation, which had a period of about 17 years (as seen in the strength of the THC and the sea-ice extent).

### 1.5.3 Interdecadal variability in ocean GCMs

In their long-time integrations of a 3-D ocean GCM of box geometry the size of the North Atlantic, Weaver and Sarachik (1991) found that under steady mixed surface boundary conditions (a restoring condition for temperature and a flux condition for salinity), self-sustained 8.6-year period oscillations spontaneously occurred in the strength of the THC and other variables. This serendipitous discovery inspired many other important modelling studies of internal variability of the THC, some of which will be described below. The variability was linked to the turning on and shutting off of high latitude convection and the subsequent generation and removal of east-west steric height gradients which caused the THC to intensify and weaken over a decadal timescale. The timescale was set by the time taken for warm saline anomalies to travel from the mid-ocean, between the subpolar and subtropical gyres where there was a region of net evaporation, to the eastern boundary and then, as subsurface flow, towards the polar boundary. In a more a complicated geometry with topography and realistic coastlines, the timescale of the oscillation is longer, about 20 years (Weaver et al., 1994).

In contrast to the above studies which involved steady forcing, Weisse et al. (1994) examined the decadal-scale response of a North Atlantic 3-D ocean circulation model to a time varying white noise freshwater forcing flux superimposed on steady mixed boundary conditions. They detected a 10 to 40-year broad band signal in the output of the model which involved the generation of quasi-periodic salinity anomalies in the Labrador Sea that subsequently propagated into the North Atlantic. According to their analysis, this signal represented the local (ocean) integration of the high-frequency (weather) freshwater flux forcing.

It should be noted that in the modelling studies described in the above two subsections, there were no attempts to relate the results to observed interdecadal variability, or to events like the GSA. However, recently a multidecadal oscillation in the THC was detected in a coupled atmosphere-ocean GCM which could be related to the observed climate variability of this century. We shall close section 5 with a discussion of this remarkable oscillation.

### 1.5.4 The 50-year oscillation in the North Atlantic THC

In a long-time integration of a fully coupled atmosphere-ocean GCM under present day greenhouse gas radiative forcing, Delworth et al. (1993) noted the existence of irregular oscillations in the strength of the THC and the pattern of SST in the North Atlantic. The period of the oscillations was about 50 years. They appeared to be driven by density anomalies in the northern sinking region of the THC combined with much smaller density anomalies of opposite sign in the broad rising region to the south. The spatial pattern of the SST anomalies associated with the oscillation closely resembled the SST changes observed in the North Atlantic during this century (Kushnir, 1994). The SST anomalies also induced model surface air temperature anomalies over the northern North Atlantic, the Arctic and northwestern Europe.

Greatbatch and Zhang (1995) have recently performed an elegant ocean-only modelling study which helps to give a deeper understanding of the multidecadal oscillation discovered by Delworth et al. (1993). Using the 3-D ocean circulation model of Zhang et al. (1992), Greatbatch and Zhang showed that under a constant zonally-uniform surface heat flux and a zero salt flux (in the first numerical experiment the salinity in the model was maintained at a constant value of 33 psu), the strength of the THC had about a 50-year oscillation which was very similar to that found by Delworth et al. The model SST anomaly pattern was also like the observed pattern seen for the period 1950-1984 (Kushnir, 1994). Since the surface heat flux was constant, the oscillation is an internal one due to a balance between convergence in the oscillatory part of the poleward heat transport and changes in local heat storage. A similar balance applies to the coupled model used by Delworth et al. where changes in surface heat flux weakly oppose the oscillation. The inclusion of salt flux forcing in the Greatbatch-Zhang model tended to weaken the oscillation, but did not change its form. Greatbatch and Zhang suggested that an oscillation of this kind may have played a role in the observed warming of the North Atlantic surface waters during the 1920s and 1930s and the subsequent cooling in the 1960s.

**Acknowledgments.** The author is grateful to a number of agencies who have supported his research program and graduate students during the past several years: the Canadian Natural Sciences and Engineering Research Council, the Canadian Atmospheric Environment Service, the Quebec Fonds FCAR, and the USA Office of Naval Research.

# 2 Sea Surface Temperature Variability in the North Atlantic: Monthly to Decadal Time Scales

CLAUDE FRANKIGNOUL
*LODYC,*
*Université Pierre and Marie Curie, Paris, France.*

## 2.1 Introduction

In his seminal study of sea surface temperature and sea level pressure variations in the Atlantic ocean, Bjerknes (1964) showed that the year to year sea surface temperature (SST)changes in the North Atlantic could be explained by the variations in the air-sea fluxes which are associated with changes in the winds. The interannual SST variability was mostly attributed to the variability of the local latent heat transfers, and the changes in evaporation linked to the strength of the surface westerlies and easterlies. Bjerknes characterized the strength of the westerlies by a zonal index given by the Azores-Iceland pressure difference (an index of the North Atlantic Oscillation, hereafter referred to as NAO), and he pointed out that its variations involved changes in the vorticity of the surface wind stress, resulting after some time lag in changes of the wind driven circulation. As the heat advection by the ocean currents significantly influences the upper ocean heat budget in regions of large currents, changes in the advective warming or cooling were presumed to modulate the SST near the Gulf Stream System and the Labrador current on time scales of a few years, becoming a significant cause for decadal changes. The decadal fluctuations in SST during the cold season are dominated by a dipole pattern and were investigated further by Deser and Blackmon (1993), who noted the similarity in the wind-SST relationship at the yearly and decadal scales and emphasized the importance of the local atmospheric forcing, yet suggesting that a positive feedback between the atmosphere and the ocean may also be present. Kushnir (1994) suggested that the yearly to decadal changes were primarily maintained through local heat exchanges.

Pluridecadal SST changes were also investigated by Bjerknes (1964). He argued that they were again primarily driven by the atmosphere and related to changes in the surface heat fluxes and the wind-driven circulation, but speculated furthermore that ocean circulation changes would alter the contribution of the North Atlantic ocean to the meridional heat flux so that, if the net meridional heat flux was to remain constant, large climate changes could be associated with the compensating changes in the atmospheric heat transport. Kushnir (1994) and Deser and Blackmon (1993) showed that the dominant SST anomaly pattern was basin scale and mainly of one polarity, and that the association with the atmosphere was different from that on shorter time scales, which led Kushnir (1994) to suggest that the SST changes were largely governed by ocean circulation changes and the atmosphere possibly responded to the SST changes. Levitus (1989) showed that the pluridec-adal SST variations coincided with temperature and salinity changes in roughly the first 1000m of the ocean.

Starting in the mid-sixties, much effort was also devoted to studying the association between large-scale SST anomalies and the atmospheric circulation on the monthly time scale with a view on using the SST anomaly persistence to improve the long-range forecasts. Ratcliffe and Murray (1970) suggested that SST anomalies south of Newfoundland could be used to predict the atmospheric patterns the following month over western and northern Europe but, as reviewed by Frankignoul (1985), subsequent studies mostly stressed the active role of the atmosphere in generating the midlatitudes SST anomalies while failing to show significant back interaction on the atmosphere. Nonetheless, the response of some atmospheric general circulation models (GCMs) to prescribed SST anomalies indicated that North Atlantic SST anomalies could have a weak influence on the atmospheric circulation (e.g. Palmer and Sun, 1985). The stochastic climate model provides a useful framework for understanding the origin of the SST variability (Hasselmann, 1976; Frankignoul and Hasselmann, 1977). Atmospheric variables have a time scale of a few days and are mostly unpredictable on the longer SST anomaly time scale. Hence the air-sea fluxes act on the ocean as a white noise forcing, creating growing "random walk" SST anomalies whose amplitude is only limited by dissipation and feedback processes.

In this chapter, we review recent observational and modeling studies of the large-scale extratropical SST anomalies in the North Atlantic and discuss their dominant causes, as well as their possible influence on the atmosphere. Our main analysis tool is the stochastic climate model, which is well-adapted to interpreting the statistical properties of the SST anomalies, either to test the null hypothesis that the ocean only reacts passively to the weather changes or to seek its more active role. We focuss on two main regimes of SST variability: the monthly to yearly variability, which primarily reflects the response of the oceanic surface mixed layer to the natural variability of the local air-sea fluxes, and the decadal variability, which is also linked to the response of the wind-driven circulation to changes in the atmospheric forcing. The pluridecadal variability is not discussed, as the observations are limited and its understanding requires a more extensive discussion of ocean circulation theory than can be given here (see chapter 15). The ocean seems mostly, but not entirely, passive on the monthly to decadal scales, but it may be more active on longer time scales, even though the link with the atmosphere remains to be clarified. As the time scale of SST variability increases, larger portions of the ocean come into play. At the monthly to yearly time scale, the air-sea interactions are primarily local and only affect the oceanic surface layer. At the decadal scale, the adjustment of the oceanic gyres to changes in the air-sea fluxes plays a role, but the air-sea interactions can still be considered to be local on the gyre scale. At the pluridecadal scale, the changes may be found in the whole water column and involve the thermohaline circulation in the entire basin. Longer time scales thus involve more complex oceanic dynamics. In view of the short adjustment time of the atmosphere, they are unlikely to involve different atmospheric dynamics. It is the nature of the air-sea interactions and the strength of the air-sea coupling that may differ.

## 2.2   Short-term variability

In a simple slab mixed-layer model where the temperature $T$ and the horizontal current $u$ are constant within a mixed layer of depth $h$, the vertically-integrated temperature equation can be written

$$h\left(\frac{\partial T}{\partial t} + u \cdot \nabla T\right) + w_e(T - T^-) = \frac{Q - Q^-}{\rho C_p} + \kappa h \nabla^2 T \qquad (2.1)$$

where $k$ is the horizontal mixing coefficient, $Q$ the surface heat flux into the ocean (positive downward) given by the sum of latent heat flux $Q_L$, sensible heat flux $Q_S$, short wave radiation $Q_{SW}$ and longwave radiation $Q_{LW}$; $Q^-$ is the heat flux at the mixed layer base, and the minus-index indicates values just below the mixed layer. The vertical entrainment velocity $w_e$ is defined by

$$w_e = \Gamma\left(\frac{\partial h}{\partial t} + \nabla(hu)\right) \qquad (2.2)$$

where $\Gamma = 1$ for $\partial h / \partial t + \nabla(hv) > 0$ and $\Gamma = 0$ otherwise. If the mixed layer is deepening, there is entrainment of colder water, while if it is shallowing, there is detrainment and fluid is left behind without changing the SST. On short time scales, $h$ is primarily determined by the turbulent kinetic energy budget, which is dominated by the energy transfer from the wind, surface heat exchanges, and turbulent dissipation. On long timescales, Ekman pumping, lateral fluxes and geostrophic flow convergence also play a role. Except at high latitude (section 2.3), salinity generally has little influence on the mixed layer dynamics, and we can study the interannual SST variations independently from the surface salinity variations.

Each field in (1) can be decomposed into a seasonally varying ensemble mean (denoted by an overbar) and an anomaly (denoted by a prime). To a good approximation, the equation for large scale SST anomalies can be written (Frankignoul, 1985)

$$\frac{lT'}{dt} \approx -\frac{(hu) \cdot \nabla \bar{T}'}{\bar{h}} - \frac{h'}{\bar{h}}\frac{\partial}{\partial t}\bar{T} - \frac{[\Gamma(w_e)w_e(T - T^-)]'}{\bar{h}} + \frac{Q' - Q^-'}{\rho C_p \bar{h}} + \kappa \nabla^2 T' \qquad (2.3)$$

$$\phantom{xxxxxx} A \phantom{xxxxxxxx} B \phantom{xxxxxxxxxx} C \phantom{xxxxxxxxxx} D \phantom{xxxxx} E$$

where

$$\frac{d}{dt} = \frac{\partial}{\partial t} + \bar{u} \cdot \nabla \qquad (2.4)$$

is the time derivative following the mean motion. Term A represents the temperature advection by anomalous currents, term B the main influence of mixed layer depth anomalies, term C the entrainment anomalies, term D the heat flux anomalies and term E horizontal mixing. Note that the anomalies in terms A, B, and C can be produced by variations in both the atmospheric forcing and the geostrophic motions. The stirring by the mesoscale eddies primarily contributes a small scale noise in regions of large eddy activity (Frankignoul, 1981; Halliwell et al., 1991), but variations in the gyre circulation may contribute to the low frequency SST changes.

In extratropical latitudes, the atmospheric forcing is dominated by the day-to-day changes in the weather and has a primarily white frequency spectrum at low frequencies. The main SST anomaly forcing is by surface heat exchanges and vertical entrainment. Within the Gulf Stream and to the north of it, advection by Ekman and geostrophic currents plays an important role (Kelly and Qiu, 1995). The atmospheric variability is most pronounced in winter. The turbulent heat fluxes dominate the heating anomalies in fall and winter, while $Q_{SW}'$ is of comparable magnitude in spring and summer. Heat flux forcing dominates wind stirring from late fall to early summer, whereas the reverse is true from mid-summer to mid-fall (e.g., Cayan, 1992; Alexander and Deser, 1995; Battisti et al., 1995). Because of its small mechanical inertia, the oceanic mixed layer responds rapidly to this forcing and the characteristic time scale of $h'$ is also of a few days. At low frequencies, the atmospheric forcing terms in (3) can thus be represented by a white noise. The latter creates growing SST anomalies whose amplitude is limited by dissipation and feedback processes (Frankignoul and Hasselmann, 1977).

As discussed by Frankignoul (1985), the main oceanic feedback involves entrainment and mixing, and perhaps the effective diffusion by surface current fluctuations (Molchanov et al., 1987). The atmosphere primarily feeds back onto the SST anomalies, once they have been generated, via the latent and sensible heatfluxes (the turbulent fluxes), although in the AGCM experiments of Palmer and Sun (1985) the atmospheric adjustment was hypothesized to induce a positive feedback via Ekman advection. A small positive feedback may also be associated with SST anomaly-induced changes in the radiation fluxes. Using the bulk formulae and assuming that the wind speed and the relative humidity $R_h$ are not affected by $'T'$ leads to the sensible and latent heating feedbacks, denoted by $\lambda_{Q_s}$ and $\lambda_{Q_L}$, which can be written in SI units

$$\rho C_p \bar{h} \lambda_{Q_s} \approx \rho^a C_p^a C_s u^a \frac{\partial}{\partial T}(T' - \langle T^{a'} \rangle) \qquad (2.5)$$

$$\rho C_p \bar{h} \lambda_{Q_L} \approx \rho^a L C_L u^a \frac{1.210^{10}}{\bar{T}^2} e^{-5388/\bar{T}} \frac{\partial}{\partial T'} [T' - R_k \langle T^{a'} \rangle] \qquad (2.6)$$

Here the superscript $a$ indicates atmospheric variables at 10 m, $L$ is the latent heat of evaporation, $q$ the specific humidity, $q_s$ the saturation specific humidity at the sea surface, $C_S$ and $C_L$ are bulk exchange coefficients. The mean surface air temperature adjustment $<T^{a'}>$ to $T'$ (an average over many realizations of the atmospheric fields for given $T'$) strongly conditions the strength of the feedback.

For small SST anomalies, dissipation and feedback can be represented to a good approximation by linearizing the (in part hidden) $T'$-dependence of the right hand side of (3) so that, in regions of small mean current, the SST anomaly equation takes the simple form

$$\frac{\partial}{\partial t} T' = F' - \lambda T' \qquad (2.7)$$

where $F'$ denotes the stochastic part of the atmospheric forcing (that is the part which is not significantly affected by the SST anomalies but controlled by the dynamics of the free atmosphere), and $\lambda$ the net feedback (which may be scale-dependent and is positive when the feedback is negative). If the seasonal modulation is neglected and the forcing $F'$ white, (2.7) represents a first-order autoregressive or Markov process. The frequency spectrum of the SST anomalies is red and given by

$$F_{TT}(\omega) = \frac{F_{FF}(0)}{\omega^2 + \lambda^2} \qquad (2.8)$$

for $\omega \ll \tau_F^{-1}$, where $\tau_F$ is the short correlation time of the stochastic forcing, and their autocovariance function is

$$R_{TT}(\tau) \approx \frac{\pi}{\lambda} F_{FF}(0) e^{-\lambda|\tau|} \qquad (2.9)$$

for $\tau \gg \tau_F$. Here $F_{FF}(0)$ is the white noise level of the forcing, $R_{XY}(\tau) = \langle X'(t+\tau)Y'(t) \rangle$ denotes the covariance between $X'$ and $Y'$ at lag $\tau$, and angle braces ensemble mean. The stochastic forcing model provides a valid first order representation of the SST anomalies over the North Atlantic on weekly to yearly time scales (Reynolds, 1979; Halliwell and Mayer,1996; Deser and Timlin, 1997; Frankignoul et al., 1998). The SST anomaly decay time $\lambda^{-1}$ is of the order of 3 months (see Fig. 2.5 below). However, the wintertime SST anomalies are capped in the spring by the seasonal thermocline and tend to recur during the following fall when the mixed layer deepens again (e.g., Alexander and Deser, 1995), so that they also have a longer time scale which is not represented in (2.8) and (2.9).

The main SST anomaly patterns are illustrated in Fig. 2.1 by an empirical orthogonal function (EOF) analysis. The SST anomaly patterns primarily reflect

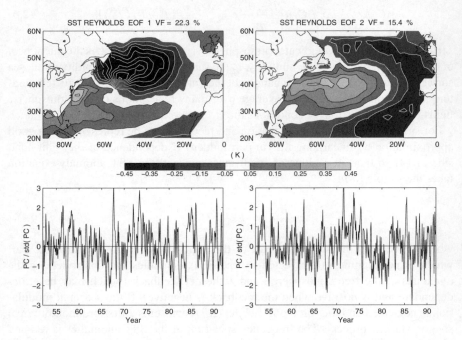

**Fig. 2.1** First two EOFs (top) and normalized principal components (bottom) of the monthly SST anomalies in the North Atlantic during 1952-1992, representing 22% and 15% of the variance, respectively. Based on the SST data by Reynolds and Smith (1994).

those of the atmospheric forcing, and indeed EOF 1 is associated with the NAO and EOF 2 with the Western Atlantic pattern (see discussion in Wallace et al., 1990; Cayan, 1992; Zorita et al., 1992; and Fig. 2.6 below). Note that the frequency spectrum of the dominant SST modes remains slightly red at low frequencies (Fig. 2.2), rather than white as predicted by (2.8), so that there is a substantial SST variability at the decadal scale (section 2.3).

Since the heat flux contributes both to the forcing and the feedback, its contribution to (6) can be singled out by writing

$$\frac{\partial}{\partial t}T' = H' + m' - \lambda_0 T' \qquad (2.10)$$

with

$$H' = \frac{Q'}{\rho C_p \bar{h}} = q' - \lambda_a T' \qquad (2.11)$$

where $q'$ represents the stochastic forcing part of the heat flux anomalies, $\lambda_a$ the heat flux feedback, $m'$ all the other stochastic forcing terms (which are primarily associated with wind stress changes), and $\lambda_0$ the remaining feedback contributions.

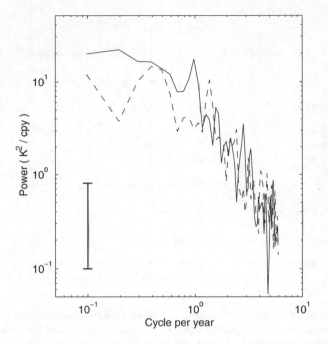

**Fig. 2.2** Power spectrum of first (dotted line) and second (continuous line) principal components in Fig. 2.1, estimated from four independent pieces. The 95% confidence interval is indicated.

One has $\lambda = \lambda_a + \lambda_0$ and $F' = q' + m'$. Eq. (2.10) and (2.11) provide clear statistical signatures of the air-sea interactions. For instance, the covariance between $T'$ and $q'$ obeys

$$\frac{\partial R_{Tq}}{\partial \tau} = R_{qq} + R_{mq} - \lambda R_{Tq}. \qquad (2.12)$$

When SST leads, the correlation between $T'$ and $q'$ is negligible (causality), except for lags of the order of the atmospheric persistence time; when it lags, it has a positive maximum at small lag and then decreases slowly on the SST anomaly time scale (Fig. 2.3, continuous line). In addition, the temporal smoothing associated with the use of monthly anomalies shifts the maximum to lag zero or one (Fig. 2.3, dashed line). Hence, the correlation at zero lag between smoothed SST anomalies and atmospheric variables that are not (or only very little) affected by the SST anomalies results from the atmospheric persistence and reflects the atmospheric forcing of the ocean, as first shown by Frankignoul and Hasselmann (1977) (see also Zorita et al., 1992 and Deser and Timlin, 1997). Empirical studies that interpret synchronous correlations or composites as indicative of the atmospheric

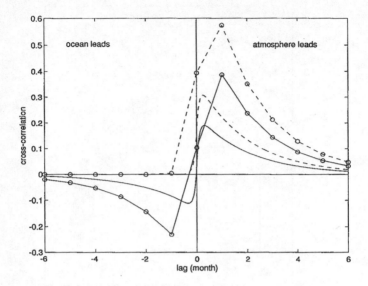

**Fig. 2.3**  Predicted correlation between $T'$ and $H'$ for no atmospheric feedback (dashed line) and for negative atmospheric feedback ($\lambda_a = (4 \text{ month})^{-1}$, continuous line) when estimated from un averaged (smooth curves) and monthly averaged (circle) data, for $\lambda = (2 \text{ month})^{-1}$. See Frankignoul et al. (1998) for details.

response to SST anomalies (e.g., Palmer and Sun, 1985; Peng et al., 1995) are thus misleading.

The covariance between the heat flux term $H'$ and $T'$ is given by

$$R_{TH}(\tau) = R_{Tq}(\tau) - \lambda_\alpha R_{TT}(\tau) \qquad (2.13)$$

critically depending on the sign of the atmospheric feedback. For $\lambda_a > 0$ (negative feedback), $R_{TH}(\tau)$ takes an antisymmetric appearance, with negative values when $T'$ leads, positive ones when it lags, and zero crossing near zero lag. As before, smoothing increases the correlations and shifts the maxima toward lags of plus and minus one month (Fig. 2.3, continuous line), in agreement with the observations (Fig. 2.4, continuous line). This behavior is also seen in coupled model results (Delworth,1996; Zorita and Frankignoul, 1997; Bladé, 1997). For $\lambda_a < 0$ (positive feedback), the covariance would be always positive and peak when T' lags by one month.

The local turbulent heat flux feedback has been estimated from (12) in the central and eastern North Atlantic by Frankignoul et al. (1998), who used monthly heat flux and SST anomalies and found a negative feedback averaging to about 20 $W m^{-2} K^{-1}$ (Fig. 2.5). Thus, the surface air temperature adjustment in (2.5) and (2.6) is about half the SST anomaly and the atmospheric feedback accounts for about half the net feedback in (2.7). In the western North Atlantic, advection by the

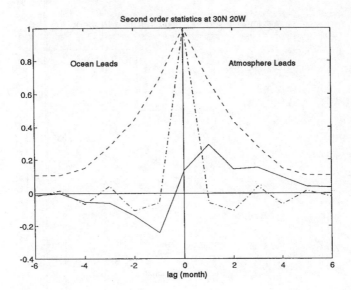

**Fig. 2.4** Autocorrelation of observed SST (dashed line) and turbulent heat flux forcing (dashed-dotted line) anomalies at30°N, 20°W, and cross-correlation between the two variables (continuous line). (From Frankignoul et al., 1998).

mean current should be taken into account, but the main patterns of air-sea interaction suggested that the feedback remains negative (Fig. 2.6). The feedback varies seasonally, being largest in fall or early winter, depending on location, and smaller and more uniform in summer, with no indication that it can become significantly positive and reinforce the SST anomalies. These estimates are consistent with the weak negative heat flux feedback seen in the GCM experiments of Palmer and Sun (1985) and Kushnir and Held (1996), but they do not support the GCM results of Peng et al. (1995) who found that the heat flux feedback was strongly positive in November conditions, and even stronger, but negative in January conditions. However, even if the local heat flux feedback is negative, that acting on a particular SST anomaly pattern may not be so, as shown in coupled model data by Zorita and Frankignoul (1997), who found no heat flux feedback at the monthly timescale for the SST anomaly patterns that were associated with the dominant modes of decadal variability in the North Atlantic.

The anomalous heating caused by the midlatitude SST anomalies affects the atmospheric boundary layer and should thus generate a small but persistent perturbation in the tropospheric circulation. However, because of the difficulties in separating causes and effects in the observations, there has been no reliable estimate of the atmospheric response to North Atlantic SST anomalies. GCM experiments with prescribed North Atlantic SST anomalies suggest that there is some response in fall and winter, but they show much disparity and the response seems model-depen-

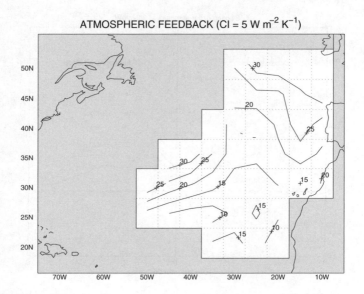

**Fig. 2.5** Estimated atmospheric heat flux feedback in $Wm^{-2}K^{-1}$. Positive values indicate negative feedback. (From Frankignoul et al., 1998).

dent. In Kushnir and Held (1996), the response is weak and baroclinic, while in Palmer and Sun (1985) and Peng et al. (1995) the SST anomaly primarily displaces the storm track and alters the upper tropospheric eddy vorticity flux and the equivalent barotropic structure of the atmosphere. However, in Peng et al. (1995) the atmospheric response is only significant for a positive SST anomaly. Ensembles of GCM experiments using the observed, time-varying SST have also been considered, but the influence of North Atlantic SST anomalies is more difficult to identify. Note also that the atmospheric circulation in the Atlantic/European region might be slightly influenced by tropical Pacific SST changes (Lau and Nath, 1994; May and Bengtsson, 1996).

## 2.3   Decadal variability

### 2.3.1   Observations

The main mode of decadal SST variability in the North Atlantic resembles the first EOF in Fig. 2.1 and is dominated by a dipole pattern with anomalies of one sign east of Newfoundland and anomalies of the other sign off the southeast coast of the United States, accompanied by weaker sign reversals to the north and south. As discussed by Deser and Blackmon (1993), the SST fluctuations are irregular and have a weak spectral peak at about 10-15 year period and their relationship to the atmospheric anomalies in the wintertime is broadly similar to that found on shorter

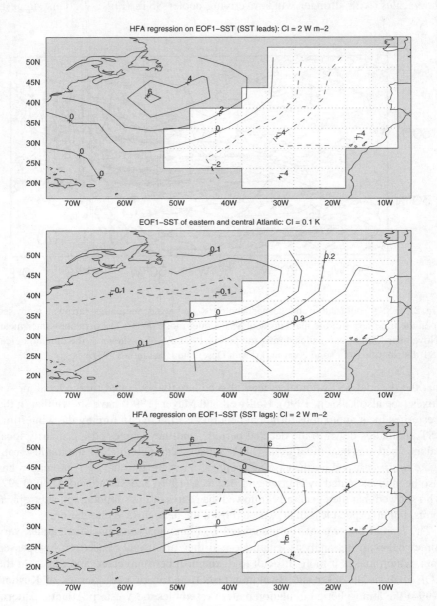

**Fig. 2.6** Middle: first EOF of the SST anomalies in the central and eastern North Atlantic. The SST pattern has been extended to the western part of the basin (shaded) by linear regression. The units are in K, and the principal component (not shown) normalized. Top: Associated turbulent heat flux pattern one month later. Bottom: Associated turbulent heat flux pattern one month earlier, both in $Wm^{-2}K^{-1}$. (From Frankignoul et al., 1998)

time scales, with stronger winds overlying cooler SSTs (Fig. 2.7). This suggests

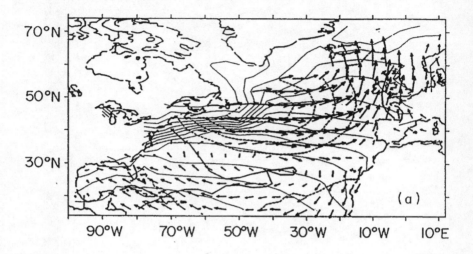

**Fig. 2.7** SST (bold solid and dashed contours) and wind anomalies (arrows) regressed upon the time series of the main decadal SST winter mode (EOF).The data are winter-mean (November-March) anomalies during the period 1900-89. Also shown is the climatological SST distribution (thin solid contours) (From Deser and Blackmon,1993).

that the decadal SST anomalies may also be largely generated by the local air-sea fluxes (see also Luksch, 1996). Halliwell and Mayer (1996) have shown that in the westerlies local anomalous turbulent heat flux is effective in forcing the wintertime SST anomalies down to the decadal periods, and they suggested a primarily local balance between heat flux forcing and feedback, although the SST anomalies prop-agate to the east and northeast. The propagation of the decadal SST anomalies has also been investigated by Hansen and Bezdeck (1996) and Sutton and Allen (1997) who showed that they generally followed the routes of the subpolar and subtropical gyres, but at a speed less than expected from the near-surface circulation.

The spatial patterns of the associated SST and sea level pressure modes vary somewhat with the analysis technique and they are usually based on low-passed data, which makes it more difficult to distinguish between cause and effect. In the EOF analysis of Deser and Blackmon (1993) and in two composites of Kushnir (1994) the atmospheric circulation pattern resembles the Western Atlantic pattern, as defined by Wallace and Gutzler (1981), while in a third composite and in the canonical correlation analysis of Grötzner et al. (1998) it resembles the NAO. This sensitivity may reflect the existence of different modes of variability and the limita-tions of representing propagating patterns with techniques that emphasize standing oscillations.

Although the observations suggest that the decadal SST anomalies are largely forced by the local air-sea fluxes and advected by the mean current, they do not preclude their reinforcement by some positive ocean-atmosphere feedback nor their dependence on the ocean circulation and water mass changes. To explore the latter, it is useful to look at the vertical structure and the decorrelation scale of the temperature anomalies, although the amount of subsurface data is limited, especially at depth. Levitus et al. (1994) showed that at 125-m depth the spatial patterns of the decadal temperature fluctuations were similar to those at the surface, but it is expected since the winter mixed-layer often reaches deeper levels than 125 m. At ocean station S (32°N, 64°W) near Bermuda, Joyce and Robbins (1996) found that the interannual SST variations were significantly correlated with the temperature fluctuations in the 18°C mode water found below the winter mixed layer, remained somewhat correlated with the temperature variations below, down to about 1500 m, but were unrelated with the long-term, secular changes below (Fig. 2.8). At the surface, the changes in temperature and salinity were decorrelated, but in the thermocline the decadal temperature and salinity fluctuations were in-phase, which could be explained by large vertical or meridional displacements of potential density surfaces (see also Reverdin et al., 1997). The thermocline variability near Bermuda was investigated by Sturges and Hong (1995), who showed that the observed decadal sea level changes could be well-explained by the baroclinic oceanic response to the North Atlantic wind stress fluctuations along the same latitude (Fig. 2.9). As discussed below, this can be viewed as the dynamical response of the thermocline to stochastic wind stress forcing (Frankignoul et al., 1997). Note that Houghton (1996) argued that the quasi-decadal wintertime SST changes at station S were primarily caused by turbulent heat exchanges and were independent of the variations of the gyre baroclinic transport. However, he observed in XBT data that in the subtropical gyre west of 40°W the temperature variations near the surface seemed to be leading those below the winter mixed layer by 3 to 4 years. One interpretation is that this delay is due to the lag between surface layer and thermocline response to large scale changes in the atmospheric forcing.

In some areas of the northern North Atlantic, like north of Iceland and along the slope off Labrador and the Grand Banks, salinity has the dominant influence on density variations on yearly and longer timescales (Fig. 2.10), hence it may control the SST changes by convective mixing. Along the Labrador continental margin, Houghton (1996) found that the quasi-decadal wintertime SST variations were again primarily surface-forced and could be traced to approximately the same depth as the penetration of the seasonal cycle. The surface salinity variations were in-phase with the SST anomalies because of the entrainment flux and convective mixing with the warmer, saltier water found at depth, and the mixing strength seemed to be modulated by fresh water anomalies due to precipitation, river runoff and sea ice. At sites in the slope current around the Labrador Sea, Reverdin et al. (1997) found that the coherence with the surface changes extended well below the base of the winter mixed layer, presumably because the changes are associated with the outflow of fresh water from the Arctic. A thorough discussion of the decadal variability of temperature and salinity at various sites of the northern North Atlantic is given in Reverdin et al. (1997), who give ample evidence of their eastward

38                                                                    Claude Frankignoul

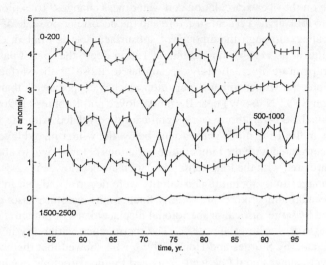

**Fig. 2.8** Yearly temperature anomalies for five layers at station S as a function of time. The deepest layer is plotted as indicated by the axes and each successive shallower layer has been offset by 1°C. An estimate of the standard error is shown for each yearly mean. (From Joyce and Robbins, 1996)

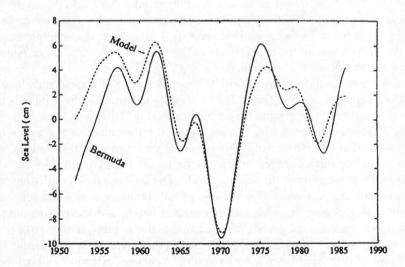

**Fig. 2.9** Comparison between observed and computed low-passed sea level variations at Bermuda. The model data is predicted by a linear baroclinic longwave model forced by the wind stress curl along 32°N, as estimated from the COADS winds (From Sturges and Hong, 1995).

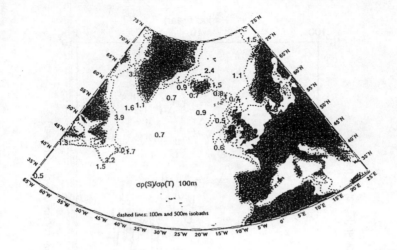

**Fig. 2.10** Ratio of the temperature and salinity root-mean-square contributions to the yearly density variability in the upper ocean (at 100 m). (From Reverdin et al., 1997).

advection/diffusion by the North Atlantic currents (see also Taylor and Stephens, 1980). Dickson et al. (1988) followed the propagation of the fresh "great salinity anomaly" and its associated cold SST anomaly along the subpolar gyre from 1968 to 1982, and they suggested that a large low-salinity anomaly could shut off convection in the regions where salinity controls the surface density, thereby strongly affecting the winter SST. This would account for the observed persistence and propagation of the SST anomalies in the subpolar regions although, as discussed by Reverdin et al. (1997), changes in the oceanic circulation, atmospheric forcing and sea ice also play a role in the hydrographic variability.

### 2.3.2  Coupled ocean-atmosphere models

Coupled GCMs do not suffer from data limitations and provide a more complete view of the air-sea interactions at the decadal scale. However, the results are model-dependent and caution is required since model resolution has been limited, the physics over simplified, and flux corrections often applied to avoid climate drift. In the coupled model developed at the Geophysical Fluid Laboratory, the dominant mode of variability in the North Atlantic is a multidecadal fluctuation of the thermohaline circulation, but Delworth (1996) has shown that the dominant mode of interannual SST variability remains slightly red down to the decadal scale. The mode is primarily forced by white noise surface heat flux anomalies, with some influence of oceanic advection by Ekman currents and, to the south of Greenland, by changes in the oceanic convection, so that its dynamics is broadly consistent with the stochastic model of section 2 and Halliwell and Mayer's (1996) analysis.

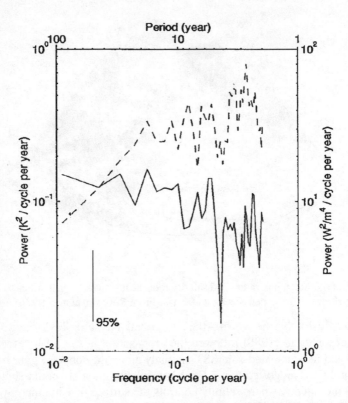

**Fig. 2.11** Frequency spectrum of yearly SST (solid line) and turbulent heat flux (dashed line) at 30°N, 39°W as estimated from the last 810 years of the ECHAM1/LSG run. The spectra were estimated from 5 overlapping pieces, using a Barlett window.

In a long simulation with another low-resolution model, the ECHAM1/LSG model developed at the Max-Planck Institute für Meteorologie (MPI), the frequency spectrum of the interannual SST variability in the North Atlantic is also found to be very slightly red down to the decadal scale (Fig. 2.11, solid line). On the other hand, the temperature fluctuations in the upper kilometer or so of the water column have a red spectrum (as $\omega^{-2}$) at periods less than decadal and a nearly white one at longer periods (as in Fig. 2.14 below). Although these fluctuations form a continuum, Zorita and Frankignoul (1997) identified two distinct modes of ocean-atmosphere variability. The most energetic mode had a dominant period of about 20 years. The simultaneous evolution of sea level pressure, SST and temperature in the thermocline is shown over a half cycle by the extended EOF analysis in Fig. 2.12. The starting year was chosen arbitrarily, and the other half cycle is similar, but with the reversed sign. The sea level pressure pattern starts as a well developed positive anomaly in the mid-Atlantic centered at about 50° N, flanked by negative anomalies over Iceland-Greenland and in the subtropical Atlantic. The anomalies migrate northwards so that the positive anomaly is replaced by a negative one coming from the south. The SST evolution is characterized by a positive anomaly off the American coast centered near 40° N, which slowly wanders northwards up to the coast of Newfoundland while decaying and being replaced by a negative anomaly; there are also weaker anomalies of the opposite sign in the eastern side of the basin. The sea level pressure and SST anomalies evolve in phase and reach their largest amplitude (albeit small) in the mid-Atlantic at year 1 (positive) and 10 (negative), and their weakest one at year 5. Their space-time structure is consistent with a response of the upper ocean to the atmospheric forcing. For instance, SST is cold in areas where the anomalous wind comes from colder regions during winter, the surface anomaly currents are southward, and mechanical mixing is enhanced. Initially, the temperature anomalies in the thermocline are dominated by a dipole with large positive values north of 30° N. The latter strengthens slightly through years 1 to 4, then weakens while rotating clockwise, with minimum amplitudes around year 7. By year 10, the dipole has the reversed sign than at year 1. The amplitude modulation thus follows that of sea level pressure and SST by a few years.

The extended EOF analysis is powerful to detect propagation and patterns of covariation, but it does not easily distinguish between cause and effect because of its inherent narrow band filtering. However, lag correlation analysis clearly shows that the mode primarily reflects the passive response of the ocean to the atmospheric forcing, with no significant correlations when the ocean leads. This is illustrated in Fig. 2.13 where several fields are correlated at different lags to the principal component (PC) associated with the dominant EOF of the yearly anomalies in the Ekman pumping (top left). Note that the PC has a white frequency spectrum but large spatial scales that reflect the dominant atmospheric fluctuations, which are also seen in the associated pattern (no lag) of the surface heat flux (top center) and the sea level pressure (top right). On the yearly time scale, the Ekman pumping acts as a white noise forcing and alters the intensity of the interior currents, described here by the baroclinic pressure at 250 meter depth[1] (bottom panels, left). The baroclinic response is strong at lag 0, peaks at lag 1, and slowly decays there after, with a hint of westward propagation in the subtropical gyre, presumably linked to Rossby wave dynamics. As mechanical and thermal forcings are correlated (top), surface layer and ocean interior respond in a coherent manner. However, the SST anomaly (bottom panels, center) is mostly due to surface heat exchanges and Ekman advection (clearly seen in the lag 0 correlation of the surface current at left), and its persistence is smaller than that of the interior response, as expected from the whiter SST anomaly spectrum. Thus, the decadal time scale in Fig. 2.12 arises from the interior response, although horizontal advection by geostrophic current plays a significant role in the SST changes, as discussed by Zorita and Frankignoul

---

[1]  The baroclinic pressure was estimated from the baroclinic currents and is thus defined on a smaller domain

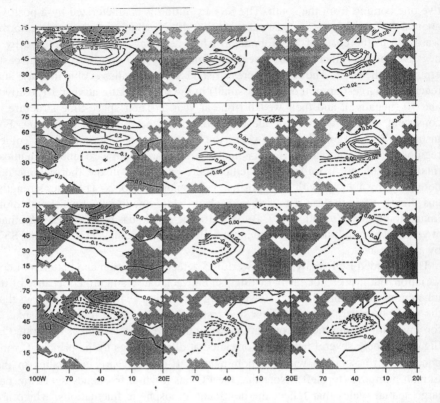

**Fig. 2.12** Anomaly patterns associated with the combined extended EOF pair describing the ECHAM1/LSG 20-year mode at (from top to bottom) year 1, 4, 7 and 10. Left: sea level pressure (mb); middle, SST (K), and right, temperature at 450 m (K) (the principal components are normalized so the amplitudes are indicative of the magnitude of the yearly fluctuations). (From Zorita and Frankignoul, 1997)

(1997). The latter is seen at positive lags in Fig. 2.13 (right), after the influence of the white noise Ekman current has vanished.

In the higher resolution ECHO model also developed at MPI, Grötzner et al. (1998) found a similar, but more energetic, North Atlantic mode that resembled the observed one and had a dominant period of about 17 years. The SST anomalies intensified as they traveled along the Gulf Stream extension and weakened thereafter, and they were associated with a NAO-like pattern in the extreme phase of the oscillation. Again, surface heat exchanges, advection, wind forcing and Rossby wave propagation were found to play a role in the SST changes. Furthermore, Grötzner et al. (1998) argued that the decadal oscillation was sustained by a positive ocean-atmosphere feedback, as suggested for the North Pacific by Latif and Barnett (1994) (see below).

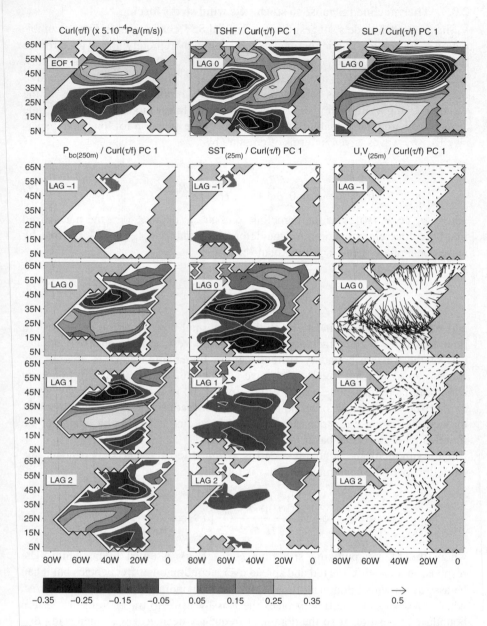

**Fig. 2.13** Top row: First EOF of curl($\tau$/f) over the North Atlantic in the last 810 years of the ECHAM1/LSG run (left), and correlation of the corresponding (normalized) principal component (PC1) with the surface heat flux (center) and sea level pressure (right). Bottom rows: lagged correlation of the 250 m baroclinic pressure (left), SST (center), and surface current (right) with PC1. Based on yearly data; curl($\tau$/f) leads at positive lags. The 5% level for no correlation is 0.06 (one-sided test).

### 2.3.3  Thermocline response to stochastic wind stress forcing

Since coupled models suggest that the decadal SST changes result from a combination of local atmospheric forcing and modulation by remotely forced geostrophic fluctuations, it is of interest to single out some of the mechanisms at play. Frankignoul et al. (1977) investigated the response of the oceanic pressure field to the stochastic wind stress forcing associated with the day-to-day changes in the weather. Using a simple non dissipative flat-bottom linear model with a basic state at rest, the baroclinic variability was modeled by the equation for long non dispersive Rossby waves, given in terms of pressure by

$$\frac{\partial p_{bc}}{\partial t} + c_{bc}\frac{1}{acos\theta}\frac{\partial p_{bc}}{\partial \varphi} = -\frac{\rho f^2}{H}\phi(0)R_{bc}^2\, w_e \qquad (2.14)$$

where $p_{bc}(\phi, \theta, t)\phi(z)$ is the baroclinic pressure at depth $z$, $\phi(z)$ the normalized first baroclinic eigenfunction, $R_{bc}$ the deformation radius, $c_{bc} = -\beta R_{bc}^2$ the Rossby wave zonal phase speed, $w_e$ the Ekman velocity,

$$w_e = \frac{1}{\rho}\left(\nabla \times \frac{\tau}{f}\right)_z , \qquad (2.15)$$

and $\tau$ the atmospheric windstress. With a no normal flow condition at the eastern boundary at $x = 0$ and a radiation condition in the west, so the model only applies east of the western boundary current regions, the solution consists of a directy forced part to which a free Rossby wave is added to satisfy the eastern boundary condition. Note that model (2.14) differs from (2.7) in that the system has eigenmodes (the Rossby waves) but no dissipation nor feedback. Here the amplitude of the oceanic response to stochastic atmospheric forcing is limited by fetch, i.e. the finite width of the ocean basin.

If for simplicity the wind stress curl is assumed to be zonally-independent, the baroclinic pressure spectrum at depth $z$ is found to be

$$F_{p_{bc}p_{bc}}(x, z, \omega) = 4\phi^2(z)\frac{\phi^2(0)\rho^2 f^4 R_{bc}^4}{H^2\omega^2}\frac{1}{2}\left(1 - cos\frac{\omega x}{\beta R_{bc}^2}\right)F_{w_e w_e}(0) \qquad (2.16)$$

where the Ekman pumping spectrum $F_{w_e w_e}(0)$ is assumed to be white. The response at a location $x$ is red and spread over a continuum of frequencies but it can be associated with a dominant time scale[2] corresponding to frequency $\omega^* = \pi/\Delta t$, where $\Delta t = x/c_{bc}$ is the time it takes the Rossby wave to travel from the eastern boundary to position $x$, so the dominant frequency decreases with increasing distance from the eastern boundary. At high frequency, (2.16) decays as $\omega^{-2}$ with a level that is $x$-independent and at low frequency it flattens toward a level that

---

[2]  Just as $\lambda^{-1}$ is the dominant time scale of the spectrum (7)

increases quadratically with $x$. The spectrum is zero at frequencies $\omega_n = 2n\pi c_{bc}/x$, but it should not be seen in observed or GCM data since including $x$-dependent forcing, adding vertical modes or adding friction would smooth the spectrum. Since the power at $\omega^*$ increases quadratically with $x$, it is the effective width of the basin which determines the period that dominates overall, namely $2\pi/\omega^*$ in the western part of the domain. This matches the decadal scale found in observations and coupled models.

The spectrum (2.16) compares well with the variability found in the interior of the North Atlantic subtropical gyre in ECHAM1/LSG (Fig. 2.14). It is also broadly

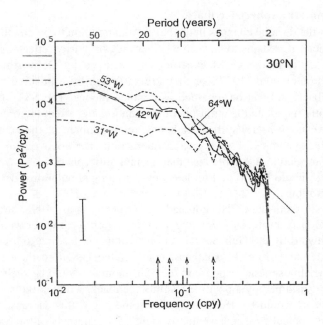

**Fig. 2.14** Frequency spectrum of the ECHAM1/LSG baroclinic pressure at 250 m and 30°N for various longitudes. The model predictions (2.16) are given in thin line for the high frequency slope, the zero frequency power, and the value of $\omega^*$, which is indicated by an arrow. The 95% confidence interval is indicated.(From Frankignoul et al., 1997).

consistent in shape and level with the frequency spectrum of observed sea level and thermocline temperature near Bermuda (Frankignoul et al., 1997). Hence, stochastic wind stress curl forcing could explain a substantial part of the decadal variability in the midlatitude thermocline. The model (2.14) is very coarse, however, and it does not apply in the North Atlantic subpolar gyre because the mean current exceeds the Rossby wave phase speed and, as shown by Qui et al (1997), Rossby waves become strongly dissipated at high latitudes. A basic state with a non-uniform thermocline structure should be considered, as in Liu (1993, 1996) who suggested that thermocline variability could be strongly affected by advection in ventilated regions or in pools of homogenized potential vorticity.

To link this work with the SST variability, the modulation of the surface mixed layer by the atmospherically forced geostrophic variability needs to be considered. Although it has been assumed in some simplified coupled models that the SST anomalies were simply proportional to the thermocline depth (see Chapter 15), the coupled GCM data suggest that horizontal advection is important for the decadal SST changes, as discussed above (see also Robertson, 1996). Hence, a more refined representation of the link between the surface and subsurface response is needed.

### 2.3.4  Ocean-atmosphere feedback

If most of the decadal changes in the NorthAtlantic seem to be directly forced by the atmosphere, it remains to be seen whether they primarily reflect a passive oceanic response, as in the GCM experiments discussed by Delworth (1996) and Zorita and Frankignoul (1977), or they are sustained by a positive ocean-atmosphere feedback and can be regarded as coupled modes. The SST changes may indeed slightly modulate the atmospheric circulation and affect the transient atmospheric storm track in a way that leads to a reinforcement of the oceanic signal. However, the dynamics of such air-sea interactions are not well understood, as pointed out in section 2.2. Note also that a small atmospheric response to decadal SST changes should be difficult to detect even in long simulations, hence more so in the observations.

In their analysis of the ECHO coupled run, Grötzner et al. (1998) suggested that in an anomalously strong subtropical gyre, more subtropical water would be transported northward by the Gulf Stream and its extension, leading to a positive SST anomaly extending from the southeast coast of the U.S. into the central North Atlantic. The atmospheric response to the SST anomaly would be such that the latter would be sustained by a positive heatflux feedback, while at the same time the wind stress curl would be decreased. After some delay, this decrease would spin down the subtropical gyre and eventually yield a SST anomaly of the reversed sign. This scenario was first proposed by Latif and Barnett (1994) for the North Pacific and supported by response studies with an atmospheric GCM. However, this was not done for the North Atlantic, and cause and effect could not be unambiguously separated in Grötzner et al.'s (1998) analysis. Note that the North Atlantic fluctuations in ECHO were coherent with the North Pacific ones, hence they may well have been forced (or enhanced) from the Pacific by an atmospheric teleconnection. This also occurs in the coupled model discussed in Chapter 11. In contrast, the decadal oscillations in the North Atlantic and the North Pacific seem independent in the observations (Deser and Blackmon, 1993) and in ECHAM1/LSG (Zorita and Frankignoul, 1997).

It is also possible that the observed decadal fluctuations reflect a more global mode of low frequency atmospheric variability due to nonlinear interactions which would even be present without SST variability, as in the simplified GCM of James and James (1989). The NAO has a marked low frequency variability and it has been observed to remain for decades in one extreme phase during the winters (Hurrell, 1995). Since the NAO forces a dominant mode of SST variability in the North Atlantic (EOF 1 in Fig. 2.2), its enhanced persistence contributes to the mode redness and thus explains part of the decadal SST changes. However, the enhanced persistence of the NAO may itself be due to ocean-atmosphere coupling, and there is indeed a general increase in low frequency atmospheric variance when an atmospheric GCM is coupled to an ocean model (e.g., Bladé, 1997).

Such enhancement does not necessarily imply a positive air-sea feedback but it could be simply associated with the negative heat flux feedback discussed in section 2. Using a simple one-dimensional coupled ocean-atmosphere model that generalizes the stochastic model (4), Barsugli and Battisti (1998) have suggested that at low frequencies the oceanic mixed layer would be adjusted to the surface air temperature, thereby attenuating the damping of the atmospheric fluctuations by surface heat exchanges. This would increase the variance and persistence of the anomalies in the two media, while decreasing the heat flux variance at very low frequencies. This was verified by Bladé (1997) in a low resolution atmospheric GCM coupled to a motionless oceanic mixed layer but, as the SST could only vary because of the heat exchanges, the strong decrease in the heat flux spectrum that was found at periods larger than a few years is unlikely to be observed in realistic conditions. In ECHAM1/LSG, some decrease in the spectral density of the turbulent heat flux is seen at very low frequencies (Fig. 2.11, dotted line), but the sea level pressure spectrum remains white (not shown), so that the surface adjustment may contribute to enhancing the low frequency variability of the SST, but have a negligible impact on that of the atmosphere.

## 2.4   Conclusions

On time scales ranging between a week and several years, the SST variability in the North Atlantic has a red frequency spectrum and primarily reflects the oceanic response to the natural variability of the atmosphere, which acts as a white noise forcing. It can be simulated by the response of the surface mixed layer to stochastic forcing by the local air-sea fluxes, although in regions of large currents the advection by the gyre circulation must be represented. The surface heat exchanges contribute both to generating the SST anomalies and to damping them once they are generated, and the ocean behavior is primarily passive.

On the decadal time scale, the SST anomalies are also directly forced by the air-sea fluxes, but they seem to be affected by the variability of the gyre circulation, which largely reflects the response of the ocean interior to the wind stress fluctuations. Even when the wind stress forcing is white, the geostrophic response is red and the dominant time scale decadal, which corresponds to the time it takes a first mode baroclinic Rossby wave to propagate across the basin. The mixed layer modulation by the geostrophic fluctuations might thus explain in part the enhanced variance found at near-decadal periods in the North Atlantic SST fluctuations.

However, the role of the surface heat exchanges at very low frequencies needs to be clarified and it remains to be established whether the ocean only responds passively to the atmospheric forcing or some positive air-sea feedback enhances the variability in the two media, as suggested for the North Pacific by Latif and Barnett (1994).

This review has primarily focussed on the midlatitudes. At high latitudes, the SST anomalies are closely linked to the changes in the surface salinity and the characteristic time scales are larger, since the winter mixed layer is deeper and there is no atmospheric damping of the salinity changes. Also, advection effects become more important (Hall and Manabe, 1997), hence the model (2.7) needs to be refined. The interactions with sea ice may also need to be considered, although the observed correlation between SST and sea ice variations appears to arise from their respective response to the same large scale atmospheric forcing (Houghton, 1996; Reverdin et al., 1997). Thus, significant departures from the "null hypothesis" of a primarily passive North Atlantic ocean may well require considering the climate changes on the multidecadal and longer time scales.

*Acknowledgments*. The author would like to thank A. Czaja, E. Kestenare, N. Sennéchael, and F. Besset for fruitful discussions and/or preparing some of the figures. This research was supported by the EC Environment Research Programme under contract EV5V-CT94-0538

# 3 Recent Decadal SST Variability in the Northwestern Pacific and Associated Atmospheric Anomalies

HISASHI NAKAMURA AND TOSHIO YAMAGATA
*Institute for Global Change Research,*
*Frontier Research System for Global Change, Tokyo, Japan,*
*and*
*Department of Earth and Planetary Physics,*
*University of Tokyo, Japan.*

## 3.1 Introduction

It has been well known that the North Pacific atmosphere/ocean system fluctuates with periods of 2~6 years associated with the El Niño/Southern Oscillation (ENSO) events (e.g., Horel and Wallace 1981; Rasmusson and Wallace 1983). Recently evidence has been presented that the system also fluctuates with decadal and even interdecadal timescales (e.g., Nitta and Yamada 1989; Trenberth 1990; Tanimoto et al. 1993; Nakamura et al. 1997 and references therein). A marked decadal and interdecadal climatic event (DICE) occurred around 1976. At that time, sea surface temperature (SST) in the central North Pacific dropped significantly, and SST tended to be lower than the long-term average until the late 1980's. During that period, the Aleutian low tended to be more intense than before, the Pacific stormtrack was likely to be shifted southward (Trenberth and Hurrel 1994), and the atmospheric intraseasonal fluctuations over the North Pacific including blocking activities tended to be weaker than before (Nakamura 1996). In addition to its own scientific significance, importance of studying North Pacific DICEs lies in their influence upon epipelagic ecosystems (Mantua et al. 1997) and its potential to mask the anthropogenic climatic trend.

Still, we do not fully understand the cause and mechanisms of the North Pacific DICEs. Many of the previous studies suggest that it is caused by tropical forcing. Over the tropical Pacific, a period of relatively cool SST ceased in 1976 and a persistent warm period followed immediately that activated tropical convection (Graham 1994; Nitta and Kachi 1994). They hypothesized that the enhanced convective activities triggered a particular anomaly pattern in the midlatitude atmospheric circulation (i.e., "teleconnection pattern"), which could lower SST over the central North Pacific (Graham et al. 1994; Kachi and Nitta 1997). This hypothesis is analogous to the "atmospheric bridge" (Lau and Nath 1994), a scenario proposed to explain the observed response of the extratropical SST to ENSO events (Horel and Wallace 1981).

Yet, recent global ocean/atmosphere model simulations suggest that the coupled system in the extratropical North Pacific may excite interdecadal variability by itself (Latif and Barnett 1994; Robertson 1996). An active role the North Pacific climate system possibly plays in DICEs has now started to be recognized through observational data analyses. Tourre et al. (1998) identified a slowly-evolving mode in the North Pacific climate system with a period of ~20 years. They showed that, in association with the mode, anomalies in the atmosphere and underlying SST over the extratropical Pacific are out of phase with tropical SST fluctuations and hence the former is unlikely to be forced by the latter via a fast atmospheric tele-connection. Nakamura et al. (1997; hereafter referred to as NLY97) found that the observed decadal variability in the wintertime SST over the North Pacific basin is concentrated in the subarctic and subtropical frontal zones (hereafter referred to as SAFZ and STFZ, respectively), the two major frontal zones in the northern extrat-ropical Pacific. This distribution is contrasting to the anomaly pattern for shorter time scales that is determined primarily by the atmospheric anomalies in response to ENSO. The decadal SST anomalies in SAFZ are associated with changes in the strength of the Aleutian Low, whereas those in STFZ are associated with the fluctu-ations of the subtropical high. They also found that the decadal SST fluctuations in STFZ exhibit strong negative simultaneous correlation with the tropical fluctua-tions but those in SAFZ do not. They argued that cooling within SAFZ in the mid-1970's which occurred in advance of the tropical warming cannot be explained by the tropical influence via the "atmospheric bridge". Their findings suggest that the observed fluctuations in the subpolar gyre and the atmospheric circulation aloft could be associated with internally-generated variability. Recent observational studies identified a mode of the interannual SST variability over the North Pacific that is statistically independent of ENSO signal and resembles their first SST mode (Deser and Blackmon 1995; Zhang et al. 1996). Seemingly, the northern North Pacific is not just a slave of the tropical Pacific with respect to its decadal variabil-ity.

In this chapter, the low-frequency SST variability observed over the Northwest-ern (NW) Pacific during the recent decades and associated tropospheric anomalies are documented. The analysis domain includes SAFZ (~42°N), where the decadal SST fluctuations are strongest, and the frontal zone in the Sea of Japan. It also includes the Kuroshio and its extension, where marked ocean/atmosphere thermal interactions occur with enormous heat loss from the ocean due to winter Asian monsoon (Tanimoto et al. 1997). Also, the monsoonal cooling lowers SST along the Chinese coast, yielding intense SST gradients over the East China Sea (Fig. 3.1). High-resolution SST data are necessary to resolve these SST fronts and associated SST anomalies. They are, to some extent, capable of resolving complex structure of SAFZ as depicted in Yasuda et al. (1996) and Yuan and Talley (1996).

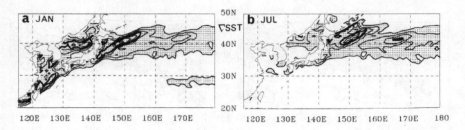

**Fig. 3.1** Magnitude of the climatological-mean SST gradient (°C/100 km) over the Northwestern Pacific for boreal (a) winter and (b) summer, based on the JMA data. Shaded for 0.8 or greater and contoured for every 0.4 (heavy lines for 2.0). The tropics are excluded where the gradient is weaker than 0.8 in both seasons.

## 3.2 Data

We use high-resolution, monthly SST data for 1950-1991 that almost covers the entire NW Pacific [0°-53°N, 110°E-180°]. The data based on ship observations have been archived on a 1°× 1° latitude-longitude grid for every 10-day period by the JMA (Japan Meteorological Agency). For a basin-wide view of the variability, we also use ship-measurement data of monthly SST in the COADS (Comprehensive Ocean-Atmosphere Data Set) archive, arranged onto a 2°× 2° latitude-longitude grid at the NOAA (U.S. National Oceanic and Atmospheric Administration) Geophysical Fluid Dynamics Laboratory (GFDL) by Pan and Oort (1990). The data for 1951-1988 were prepared at GFDL, and we updated them up to 1992 with data provided by the JMA. The data between 80°N and 40°S have been retained for our analysis.

We also use monthly fields of sea-level pressure (SLP) and 500-hPa geopotential height (Z500) for 1950-1992 archived in the NCAR (U.S. National Center for Atmospheric Research) Data Library, which are based on the NMC (U.S. National Meteorological Center; currently National Centers for Environmental Prediction) operational analyses. The atmospheric data cover the region poleward of 17°N. As a proxy for the surface wind, we use 1000-hPa geostrophic wind (U1000) obtained from SLP with the latitude-dependent Coriolis parameter (Hanawa et al. 1996). We also use 500-1000 hPa thickness as a measure of the lower tropospheric temperature.

Separately for boreal winter (December-January-February) and summer (June-July-August), we prepared 3-month mean maps of the aforementioned variables for individual years (December 1950 was included in 1951 winter etc.). Anomalies are defined as deviations from their 30-year means for 1961-1990. For each of the variables, an empirical orthogonal function (EOF) analysis was performed over the entire data domain, and we retained the fewest number of the leading modes that account for at least 90% of the total variance within the domain.

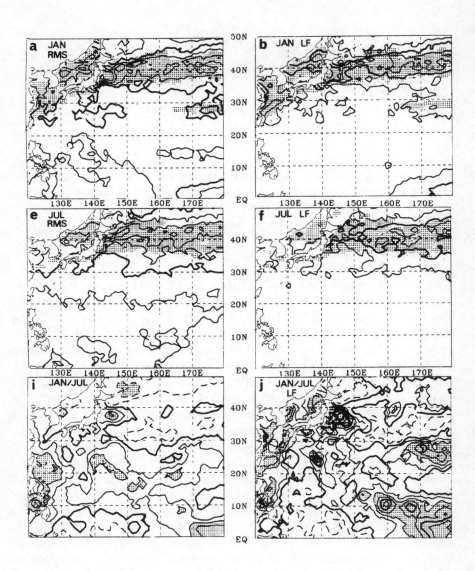

**Fig. 3.2** SST variability as measured by its standard deviation (σ; contoured for σ= 0.3, 0.5, 0.7, 1.1 and 1.3°C; heavy lines for σ= 0.5°C) with (a) total (unsmoothed) interannual, (b) decadal and (c) ENSO time scales for boreal winter. Shaded are regions where the climatological-mean SST gradient > 0.8 (°C/100 km). (d): Local ratio (r) of (b) to (c); contoured for every 0.25 (dashed for r< 1, heavy lines for r= 1.0, 1.5 and 2.0) and shaded for r< 0.5 or r> 1.5. (e)~(f): As in (a)~(d), respectively, but for boreal summer. (h) As in (d) but for summer. Winter-to-summer ratio of s with (i) total interannual [i.e., (a)/(e)], (j) decadal [(b)/(f)] and (k) ENSO [(c)/(g)] time scales. For (h)~(k), contour and shading conventions are the same as for (d). All statistics are based on the JMA data.

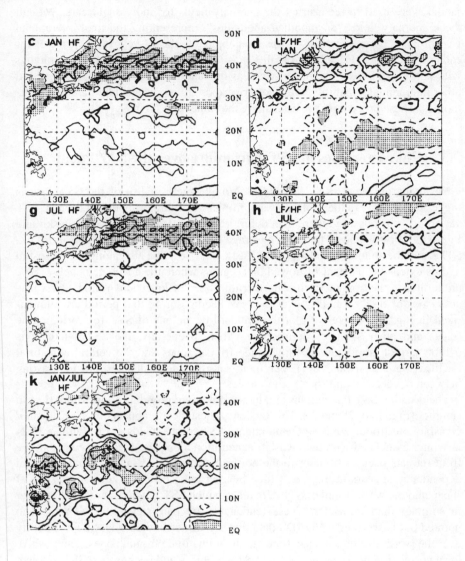

**Figure 3.2 (continued)** .

The EOFs and 42-year time series of the principal components (PCs) of these modes were used to reconstruct the anomaly fields for individual years. We call these anomaly fields "unsmoothed anomalies", representing the total interannual variability. A harmonic analysis was then applied to the 42-year PC time series for each of the modes. Decadal variability was represented by the 6 lowest harmonics with periods of 7 years or longer, whereas fluctuations associated with ENSO and other interannual variability with shorter periods are represented by the higher harmonics. These procedures are essentially the same as in NLY97.

## 3.3   Characteristics of SST Variability over the NW Pacific

We focus on the period 1967-1991, during which quality of the SST data is higher particularly in the subtropics and tropics and more pronounced decadal SST variability was observed there than before (NLY97; also Fig. 3.5f). Ostrovskii and Piterbarg (1995) showed that data coverage for this period exceeds 80% in SAFZ, STFZ and the southeastern half of the East China Sea, and also along the Kuroshio and its extension. They also showed that regions where the coverage is below 50% are limited to the tropical Pacific, northernmost Sea of Japan and Yellow Sea. Characteristics of SST variability over the NW Pacific during that period are summarized in Fig. 3.2, where we plot the standard deviations of SST for various time scales and other statistics derived from them. Generally, the SST fluctuations exhibit largest amplitudes in midlatitude frontal zones and the central equatorial Pacific (Fig. 3.2a~c and e~f), but their seasonal dependencies are contrasting between the decadal and shorter time scales.

The decadal SST fluctuations are, in general, more pronounced in winter than in summer (Fig. 3.2b, 2f and 2j). This tendency most clearly appears in the Kuroshio/ Oyashio interfrontal zone, where an intense SST front extends as far south as 35°N in winter associated with the Oyashio penetration along the Japanese coast but the front retreats back in summer to the north of 40°N (Fig. 3.1). The same seasonal dependency appears in the East China Sea, an intense frontal zone in the Sea of Japan and the western end of STFZ, in all of which SST gradients are much weaker in summer than in winter. These tendencies are consistent with NLY97, who pointed out based on the COADS data that the wintertime decadal SST variability over the North Pacific is concentrated in the midlatitude frontal zones. The positive local correlation between the decadal SST variability and background SST gradient does not apply to the equatorial Pacific, where the fluctuations are substantially stronger in boreal winter than in summer despite weak background SST gradients throughout the year.

Seasonality of SST fluctuations with the ENSO time scales exhibits an apparent latitudinal dependency (Fig. 3.2c, 2g and 2k). To the south of 30°N, they are stronger in winter than in summer due probably to the stronger atmospheric response to ENSO in winter. To the north, the fluctuations are strong in winter but they tend to be even stronger in summer, although the atmospheric forcing is generally stronger in winter. This counterintuitive-looking seasonality may reflect the shallow oceanic mixed layer in summer, by which the SST is probably sensitive to year-to-year changes in summertime insulation and the storminess, even though summertime storms are weak on average. It may also reflect such changes in the current system as meandering in the Kuroshio extension whose seasonal dependency must be weaker than the atmospheric forcing because of the longevity of the former (Mizuno and White 1983). The summertime enhancement of the SST variability with the ENSO time scales is so dominant in SAFZ that this seasonal tendency appears even in the unsmoothed, total variability Fig. 3.2a, 2e and 2i), as pointed out by Ostrovskii and Piterbarg (1995).

The amplitude ratios of the SST fluctuations between the decadal and shorter time scales for boreal winter and summer are shown in Fig. 3.2d and 2h, respectively. In winter, the decadal SST variability is dominant in the midlatitude frontal zones, whereas the variability with shorter periods is dominant in much of the lower latitudes. As the background SST gradients are relaxed and the mixed layer shallows in summer, the year-to-year variability becomes dominant over the decadal variability in midlatitudes.

## 3.4   Dominant Decadal Variability in Winter

### 3.4.1   Spatial Structure

It  was shown in the preceding section that DICE-related SST fluctuations over the NW Pacific are pronounced particularly in the midlatitude frontal zones and near the equator. To examine whether the variability in those regions is coherent, a conventional  EOF analysis was applied to the whole domain of the JMA data for 1967-1991. We performed the analysis separately for winter and summer, recognizing the apparent seasonal dependencies of the variability. The three leading EOFs of the wintertime decadal SST variability are shown in Fig. 3.3, in which the linear regression coefficient is plotted between the decadal SST anomaly timeseries at every grid point and the PC timeseries for each of the corresponding EOFs (Fig. 3.5). The SST anomaly pattern over the entire Pacific for each of these modes (Fig. 3.4d, 4h and 4l) is identified by computing the linear regression coefficients of the COADS SST anomalies with the corresponding PC. Similarly, the anomaly pattern of a given atmospheric variable associated with each of those SST modes is identified in the linear regression map of the variable with the corresponding PC (Fig. 3.4).

The first mode for winter represents the fluctuations in SAFZ (Fig. 3.3a). This mode, which explains 41% of the decadal SST variance over the NW Pacific, indeed dominates over any other modes. From a wider view (Fig. 3.4d), this mode is characterized by a prominent seesaw in the decadal SST fluctuations between SAFZ and the Gulf of Alaska, but it exhibits no significant simultaneous correlation with the tropical and subtropical variability. Our first EOF is virtually identical to the counterpart of NLY97 based on the COADS data. Consistent with NLY97

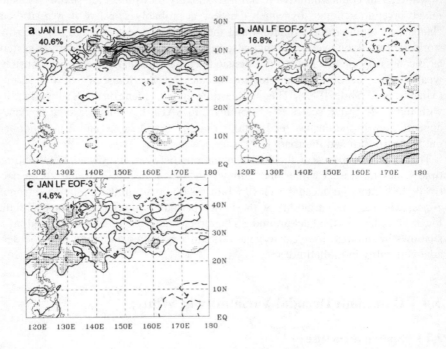

**Fig. 3.3** The (a) first, (b) second and (c) third EOFs of the decadal variability in the Northwestern Pacific wintertime SST for 1967-1991, based on the JMA data. Linear regression coefficients are plotted between decadal SST anomalies and PCs of these modes. The coefficient corresponds to a local change in SST (°C) when a given PC increases by its unit standard deviation. Contoured for every 0.1, thickened for every 0.5, dashed for negative, zero lines omitted. Shaded for regions where correlation between SST and a given PC exceeds 90% significance with 3 degrees of freedom assumed (see Fig. 3.5).

the SST anomalies in SAFZ tend to occur in conjunction with the anomalous Aleutian Low at the surface (Fig. 3.4a) and the Pacific/North American (PNA) teleconnection pattern aloft (Fig. 3.4c) as defined by Wallace and Gutzler (1981). Polarities of these anomalies are such that, during warm periods in SAFZ, the surface westerlies (U1000; not shown) tend to be weakened over the extratropical North Pacific in accordance with the weak Aleutian Low, and so does the upper-level westerly jet with the anticyclonic anomalies over the northern North Pacific as the most dominant component of the PNA pattern. During warm SAFZ periods, air temperature above SAFZ tends to be higher than average (Fig. 3.4b). The opposite is the case during cool periods. As discussed later, these atmospheric anomalies act to reinforce the SST anomalies in SAFZ, and hence this mode may be regarded as a manifestation of the variability generated autonomously in the coupled system of the North Pacific subpolar gyre and the atmosphere above.

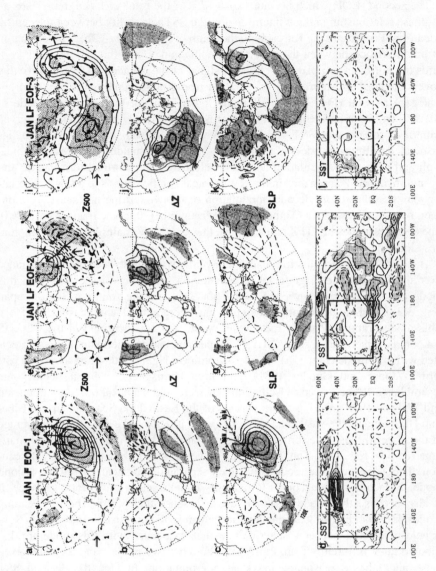

**Fig. 3.4** Atmospheric and SST anomalies associated with the 3 leading modes of the decadal SST variability over the Northwestern Pacific for winter. Linear regression coefficients (dashed for negative; zero lines omitted) of (a) Z500 (every 5m; thickened for every 20m), (b) 500-1000 hPa thickness (every 5m, thickened for every 20m), (c) SLP (every 0.5 hPa, thickened for every 2 hPa) and (d) the COADS SST (every 0.1°C, thickened for every 0.5°C) are plotted with PC for the first mode (see Fig. 3.5). Shaded for regions where correlation with PC exceeds 90% significance (as in Fig. 3.3). (e)~(h): As in (a)~(d), respectively, but for the second mode. (i)~(l): As in (a)~(d), respectively, but for the third mode. Anomalies of Z500 and SLP are multiplied by [sin 45°N/sin (lat.)] to mimic streamfunction-like anomalies. The arrows with Z500 are $W$ scaled as inidicated ($m^2s^{-2}$).

The second  EOF, which accounts for ~17% of the total variance, represents an in-phase relationship in the wintertime decadal SST variability between the central equatorial Pacific and the Kuroshio to the south of Japan (Fig. 3.3b). There is a hint of an alternation between the equatorial Pacific and STFZ. Indeed, this EOF captures the northwestern part of a coherent signal of the decadal SST variability that covers the northern subtropical gyre and the tropics and even the southern subtropical gyre as well (Fig. 3.4h). This mode looks identical to the second mode of NLY97, and it perhaps contributes to a gross equatorial symmetry in the observed pattern of the decadal SST variability (White and Cayan 1997). SST anomalies associated with this mode are accompanied by the anomalous surface subtropical high and associated circulation anomalies aloft over the Northeastern Pacific and by significant SLP anomalies in the NW tropical Pacific (Fig. 3.4e and 4f). The latter may be related to a decadal modulation of the wintertime northeasterly monsoon over Southeast Asia (Yamagata and Masumoto 1992; Kachi and Nitta 1997). The SST anomalies in STFZ tends to have the same sign as air temperature anomalies aloft (Fig. 3.4f).

The third mode accounts for nearly 15% of the wintertime decadal SST variance over the NW Pacific. Significant, coherent SST anomalies associated with this mode can be seen only over the East China Sea and the southern Sea of Japan (Fig. 3.3c), with no other coherent signals anywhere in the Pacific basin (Fig. 3.4l). This mode, trapped near the East Asian coast, was not identified in the COADS data on a coarser grid used by NLY97. Upper-level atmospheric anomalies associated with this mode (Fig. 3.4i) include a prominent signal of the Eurasian (EU) pattern (Wallace and Gutzler 1981), and in a hemispheric view it bears some resemblance to the "Arctic Oscillation" (Kodera and Yamazaki 1990; Thompson and Wallace 1998). During warm periods of the East China Sea, the surface Siberian High tends to be weakened (Fig. 3.4k) and it tends to be warmer than average in the lower troposphere throughout the midlatitude East Asia and adjacent maritime regions. Fig. 3.4j). The third mode, however, should be interpreted with caution. This is because the mode is not well separated statistically from the second mode, as each of them accounts for a nearly equal fraction of the total variance. It is also because this mode is not robust in a sense that, unlike the two leading modes, it cannot be extracted cleanly through an EOF analysis for 1950-1991. Nevertheless, the SST and atmospheric anomalies associated with this mode are distributed quite differently from those with the other two modes, and we have given those anomalies a reasonable physical interpretation. In fact, the decadal SST anomaly averaged over the East China Sea is not correlated significantly with the decadal SST fluctuations anywhere within the Northwestern Pacific basin except just to the east of Taiwan and Okinawa Islands. Moreover, as shown below, the PC of this mode well represents the low-frequency SST variability over the East China Sea. We hence consider that this mode, which represents the influence of a decadal modulation of the winter East Asian monsoon upon the marginal seas, is perhaps physically meaningful.

To shed light on where the DICE-related atmospheric circulation anomalies are forced, we estimated wave activity flux of stationary Rossby waves on the zonally-varying climatological mean flow $U=(U,V)$, applying a formula developed by Takaya and Nakamura (1997) to Z500 anomalies in the linear regression maps for the individual SST modes. Only the zonal and meridional components of the flux ($W$) were evaluated, as given in a vectorial form by

$$W = |U|^{-1} \cdot [U(v'^2 - \psi'v'_x) - V(u'v' - \psi'v'_y),$$
$$-U(u'v' + \psi'v'_y) + V(u'^2 + \psi'u'_y)] \qquad (3.1)$$

where suffixes $x$ and $y$ indicate zonal and meridional derivatives, respectively, $\psi'$ denotes anomalous geostrophic streamfunction and $u'$ and $v'$ are the corresponding wind anomalies. It has been shown in the limit of small-amplitude, plane waves on a slowly-varying, unforced non-zonal flow that $W$ is independent of wave phase, even if unaveraged, and parallel to the local group velocity. $W$ is a flux of pseudo-momentum that is conserved while a stationary Rossby wave packet propagates under the non-acceleration conditions. Hence, $W$ should be divergent where atmospheric wave-like anomalies are forced. In Fig. 3.4 $W$ is plotted with arrows, superimposed on the DICE-related $\psi'$ anomalies in the mid-troposphere. Note that reversing the signs of the SST anomalies and associated atmospheric anomalies yields no change in $W$. It is evident for the first SST mode that $W$ associated with the PNA pattern is strongly divergent in midlatitudes where a vigorous stormtrack resides and strong SST anomalies are observed underneath in SAFZ. Equatorward $W$ in the subtropical Pacific implies that the PNA pattern must be forced in midlatitudes. For the second SST mode, $W$ is again divergent above STFZ and predominantly poleward farther to the north, suggestive of SST-related wave forcing in the subtropics. In the subtropics, the direction of wave propagation is hard to infer unambiguously from $W$, but there appears to be a hint of reinforcement of wave activity coming in from the lower latitudes. For the third SST mode, southeastward $W$ is converging over Mongolia and northern China. In a hemispheric view (not shown), $W$ appears to be originated in the north Atlantic.

### 3.4.2 Time Evolution

Time evolution of the three leading modes of the decadal SST variability over the NW Pacific is presented in Fig. 3.5 with the time series of their PCs, which have been extended into the period before 1967 by projecting the smoothed SST anomalies for individual years onto each of the EOFs 1[1]. In Fig. 3.5 we also plotted the

---

[1] The "extended PCs" are no longer mutually uncorrelated for 1950-1966. An overall correspondence between the unsmoothed anomaly time series for the selected key regions and the corresponding extended PCs suggests that the leading modes represents the decadal SST variability observed within the key regions over the last four decades reasonably well.

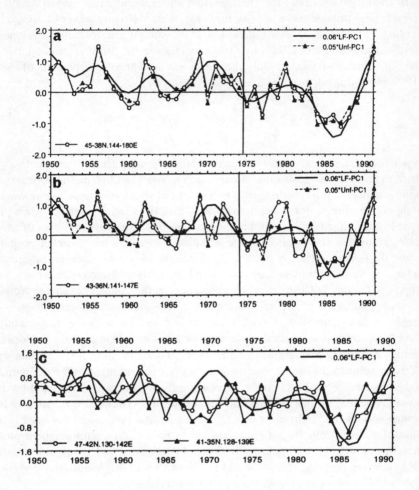

**Fig. 3.5** Time series of first PCs for decadal (× 0.06; heavy solid line) and unsmoothed (× 0.05; dashed line with triangles) SST variability over the Northwestern Pacific, with unsmoothed SST anomalies (°C; ΔT) averaged over SAFZ [45°-38°N, 144°E-180°; solid lines with circles]. (b): As in (a) but with ΔT averaged over KOIZ [43°-36°N, 141°-147°E; circles]. (c): First decadal PC (× 0.06; heavy solid), with ΔT averaged over the northern and southern Sea of Japan [(47°- 42°N, 130°-142°E) and (41°-35°N, 128°-139°E), respectively; circles and triangles, respectively].

unsmoothed SST anomalies averaged over several key regions in which these modes exhibit coherent, significant signals.

Our first PC is similar to the counterpart of NLY97 based on the COADS data, but ours traces the decadal SST variability in SAFZ even better than theirs (Fig. 3.5a). Our first PC also traces the variability in the Kuroshio/Oyashio inter-

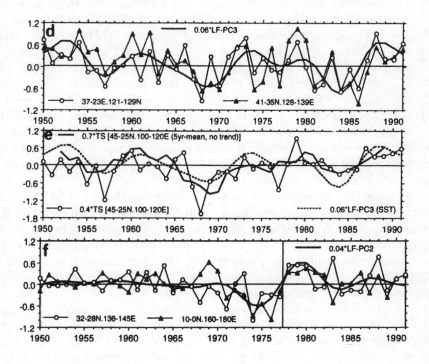

**Figure 3.5 (continued)** (d): Third decadal PC (× 0.06; heavy solid), with ΔT averaged over the southern Sea of Japan (circles) and over the East China Sea [37°-23°N, 121°-129°E; triangles]. (e): Third decadal PC of SST (× 0.06; heavy dashed), with surface air temperature anomalies over China unsmoothed (× 0.4; circles) and smoothed by 5-year running mean (× 0.7; heavy solid; linear trend removed). (f): Second decadal PC (× 0.04; heavy solid), with ΔT averaged along the Kuroshio and central equatorial Pacific [(32°-28°N, 136°-145°E) and (10°N-0°, 160°E-180°), respectively; circles and triangles, respectively]. All time series are for boreal winter.

frontal zone (KOIZ) reasonably well (Fig. 3.5b), although the warming and cooling appear to occur slightly earlier in KOIZ than in SAFZ. As pointed out by NLY97, cooling within SAFZ in the mid-1970's preceded tropical warming by about 2 years. For KOIZ the lag is nearly 3 years. After a warm period around 1970, SAFZ underwent a cool period that began in 1975 and continued until the late 1980's with a brief intermission around 1980. Yet, it is not until 1977 that a cool period ended and SST rose to above-normal in the tropics (Fig. 3.5f). These results are consistent with Watanabe and Mizuno (1994). Unlike the "1988/89 shift" (Koide and Kodera 1997), the "1976/77 climate shift" did not manifest itself dramatically in SAFZ (Miller et al. 1994; NLY97).

On decadal time scales, SST in the northern half of the Sea of Japan varied more or less coherently with that in SAFZ and hence with the first PC (Fig. 3.5a~c). These regions both underwent warm and cool periods almost simultaneously over the last 40 years, except a warm period in SAFZ around 1970 that has no counterpart for the Sea of Japan. Coherency of the decadal SST fluctuations is rather low between the north and south of an intense SST front in the Sea of Japan (Fig. 3.5c). Rather, the variability to the south of the front looks more coherent with that in the East China Sea, both of which are represented reasonably well by the third PC (Fig. 3.5d). The decadal SST fluctuations over the East China Sea exhibit marked positive correlation with those in surface air temperature over China, once a secular trend is removed from the latter[2] (Fig. 3.5e). This result is consistent with the negative correlation between SST over the East China Sea and the lower-tropospheric temperature (Fig. 3.4j).

Our second PC represents the in-phase relationship in the decadal SST anomalies between the central equatorial Pacific[3] and the Kuroshio region to the south of Japan (Fig. 3.5f). A similar in-phase relationship with interannual time scales has been found by Yamagata et al. (1985). In those two regions dramatic warming occurred almost simultaneously around 1977, which is concomitant with cooling in STFZ (NLY97). The warming is consistent with the increase in the Kuroshio transport observed in the same period (Kawabe 1995).

At least two different time scales appear to contribute significantly to the low-frequency interannual fluctuations in the wintertime SST observed over the NW Pacific during the last 4 decades. In SAFZ (Fig. 3.5a and 5b), strong decadal fluctuations with periods of ~10 years are superimposed on the interdecadal variability that was manifested as the relatively warm 1950's and 1960's and a cool period after the mid-1970's. This interdecadal component seems to be the same as the 50~70-year variability identified by Minobe (1997). This signal, however, is not apparent in the midlatitude marginal seas (Fig. 3.5c and 5d). The interdecadal warming trend in surface air temperature over China observed since the late 1960's is in contrast with the cooling trend in SST along SAFZ during the same  period.

---

[2] We use monthly station air-temperature ($T_s$) data arranged onto $5° \times 5°$ latitude/longitude grid by Jones et al. (1986). For individual winter seasons, the December-January-February mean $T_s$ anomalies were averaged over China [45°-25°N, 100°-120°E] to yield an time index for 1950-1991. No enough data are available to the north. The index was then smoothed by 5-year running mean to represent its decadal signal, which exhibits a linear warming trend that is much stronger than that in the third PC and SST over the East China Sea. The marked linear warming trend in $T_s$ over China has been pointed out by Yatagai and Yasunari (1994).

[3] The time index for the central equatorial Pacific is based on the unsmoothed SST anomalies averaged over [20°N-0°, 160°E-180°]. Since ENSO-related SST fluctuations tend to exhibit a seesaw within this domain (Fig. 6a), they are somewhat suppressed in this index.

### 3.4.3   Comparison with ENSO-Scale Variability

Most of the previous studies emphasized the resemblance in SST and atmospheric anomaly patterns over the North Pacific between the ENSO and decadal time scales. Yet, Tanimoto et al. (1993, 1997) and NLY97 argued that there are some subtle but significant differences between them. In this subsection, we refine their findings by using SST data with a higher spatial resolution.

Fig. 3.6a shows the first EOF of wintertime SST variability with the ENSO time scales over the NW Pacific. The EOF, which accounts for 30% of the variance, is characterized by a seesaw between the region including the Kuroshio, KOIZ and the East China Sea and the rest of the subtropical gyre over the NW Pacific. The pattern is very similar to that obtained by Hanawa et al. (1989a, 1989b). Unlike the first decadal EOF, there is no indication of the anomalies in the first ENSO-scale EOF being confined in the oceanic fronts. One can identify this mode representing a remote SST response to ENSO, in recognition of pronounced positive and negative peaks of the PC at major events of El Niño and La Niña, respectively (Fig. 3.6d), and also of the strongest SST signal confined in the equatorial Pacific (Fig. 3.7c). The leading mode of the ENSO-related SST variability over the NW Pacific is accompanied by a meridional dipole in Z500 anomalies over the Far East (Fig. 3.7a) that resembles the Western Pacific (WP) pattern (Wallace and Gutzler 1981). The meridional axis of the near-surface counterpart of that dipole is shifted eastward, particularly in the subtropics (Fig. 3.7b). This indicates a tendency that warm SST anomalies around Japan and along the Kuroshio are associated with the weaker surface northwesterlies than normal and warm temperature anomalies aloft. It also indicates that cool SST anomalies in the subtropical NW Pacific tend to be under stronger trades than normal and cool temperature anomalies aloft. The near-surface atmospheric anomalies presented here are consistent with those in Hanawa et al. and Kutsuwada (1991). These tendencies are strongly indicative of forcing by the atmospheric anomalies over the NW Pacific in remote response to ENSO, which probably leads to the formation of the underlying SST anomalies as depicted in the EOF.

On the ENSO time scales, the seesaw-like SST anomalies over the NW Pacific that characterize our first EOF are also evident in the first EOF of Tanimoto et al. (1993, 1997). However, their EOF also includes zonally-elongated SST anomalies over the central and eastern North Pacific with a sign opposite to that of the anomalies around Japan. In addition to the WP pattern, their atmospheric anomalies also include a strong seesaw between Gulf of Alaska and western Canada. These differences in the anomaly patterns can be attributed probably to the influence of the eastern North Pacific that is included in their analysis but not in ours. Very recently, Kodera (1998) showed that the extratropical atmospheric response to ENSO differs substantially from one event to another. During some events the upper-level anomaly pattern looks like the PNA pattern, but it looks more like the WP pattern during other events. The latter type is likely to be emphasized in our analysis based only on the NW Pacific anomalies, while those two types are perhaps mixed up in Tan-

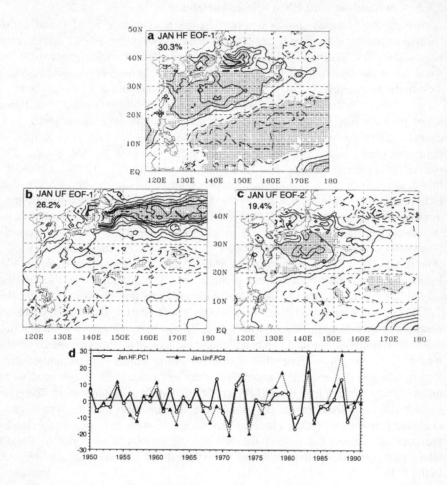

**Fig. 3.6** (a) The first EOF of the ENSO timescale SST variability. (b) First and (c) second EOFs of the total interannual variability. Linear regression coefficients are plotted between the SST anomalies and corresponding PCs. The coefficient corresponds to a local change in SST (°C) when a given PC increases by its unit standard deviation. Contoured for every 0.1, thickened for every 0.5, dashed for negative, zero lines omitted. Shaded for SST-PC correlation exceeding 95% significance. (d) Time series of the first PC of ENSO-timescale SST [thick solid line with circles; EOF in (a)] and the second PC of total interannual SST variability [dotted line with triangles; EOF in (c)]. All plots are for 1967-1991, based on the Northwestern Pacific wintertime SST.

imoto et al. whose analysis domain extends over the entire extratropical North Pacific.

We have shown that the first mode for each of the decadal and ENSO time scales indeed dominates over higher modes, and that these first EOFs represent distinct SST anomaly patterns from one another. The latter fact and distinction in the corre-

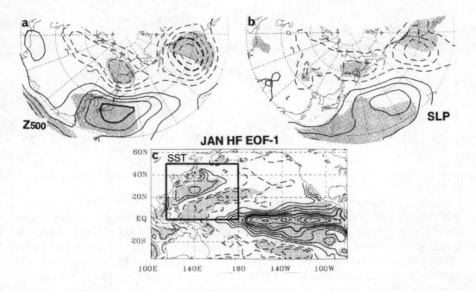

**Fig. 3.7** Atmospheric and SST anomalies associated with the first mode of the ENSO-timescale SST variability over the Northwestern Pacific for winter. Linear regression coefficients with PC (dashed for negative; zero lines omitted) of (a) Z500 (every 5m; thickened for every 20m), (b) SLP (every 0.5 hPa, thickened for every 2 hPa) and (c) the COADS SST (every 0.1°C, thickened for every 0.5°C) are plotted with the PC. Shaded for regions where correlation with PC exceeds 95% significance.

sponding atmospheric anomalies both *a posteriori* justify the timescale separation we applied in prior to our analysis. Also, they suggest that physical processes involved in the formation of those anomaly patterns are different, at least to some extent, between the two time scales. Consistent with the aforementioned findings, the total interannual SST variability to which the two time scales both contribute is characterized by the two dominant modes that resemble the respective first modes. The decadal SST fluctuations in SAFZ are dominant in the first mode of the total variability (Fig. 3.6b). Indeed, the PC is almost perfectly correlated with the unsmoothed SST fluctuations in SAFZ, on which marked decadal fluctuations are embedded (Fig. 3.5a). The second EOF of the total variability is similar to the first ENSO-scale EOF (Fig. 3.6c), and their corresponding PCs vary coherently peaking at major ENSO events in the 1970's and 1980's (Fig. 3.6d).

## 3.5 Dominant Decadal Variability in Summer

In this section we examine the dominant SST anomalies with decadal time scales over the NW Pacific and associated atmospheric anomalies for boreal summer. The first mode (Fig. 3.8a), which explains 39% of the total variance, represents the dec-

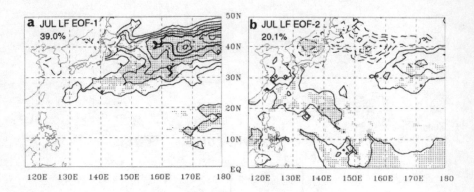

**Fig. 3.8** The (a) first and (b) second EOFs of the decadal variability in the Northwestern Pacific summertime SST for 1966-1990, based on the JMA data. Linear regression coefficients are plotted between decadal SST anomalies and PCs of these modes. The coefficient corresponds to a local change in SST (°C) when a given PC increases by its unit standard deviation. Contoured for every 0.1, thickened for every 0.5, dashed for negative, zero lines omitted. Shaded for regions where correlation between SST and a given PC exceeds 90% significance with 3 degrees of freedom assumed (see Fig. 3.10).

adal SST anomalies in SAFZ. With respect to this aspect, the first EOF for summer is similar to the leading wintertime EOF. However, a detailed comparison reveals some differences between them. First, compared with the winter EOF (Fig. 3.3a), confinement of the SST anomalies in SAFZ is rather weak in summer with substantial variability in the Kuroshio extension to the south of the front. Second, unlike in winter, summertime SST does not vary coherently along SAFZ on decadal time scales. Most of the decadal SST variability to the east of 160°E is captured in the leading EOF, but to the west the variability is scattered into higher modes. Presumably, these differences may reflect the facts that in SAFZ to the east of 160°E, SST gradients are stronger and the frontal zone is wider in summer than in winter and that the Oyashio penetration front off Japan is weaker in summer than in winter (Fig. 3.1). In a wider view, the first mode represents the decadal SST variability in SAFZ in the North Pacific (Fig. 3.9c), associated with no significant coherent SST anomalies in the tropics except weak ones in the central tropical Southern Pacific. In the time series plot (Fig. 3.10a), this mode represents the summertime decadal variability in SAFZ very well, particularly to the east of 160°E. Overall tendencies in the decadal variability are nearly the same between the two seasons, with the relatively warm 1950's and 1960's followed by a cool period since the mid-1970's with a brief warm period near the end of the 1970's. Yet, the decadal signal in summer is somewhat distorted by the pronounced fluctuations with shorter time scales. Comparing Fig. 3.10b with Fig. 3.5a, we recognize that in the mid-1970's the cool period in SAFZ began in 1974 summer, well in advance of the tropical warming in 1976/77.

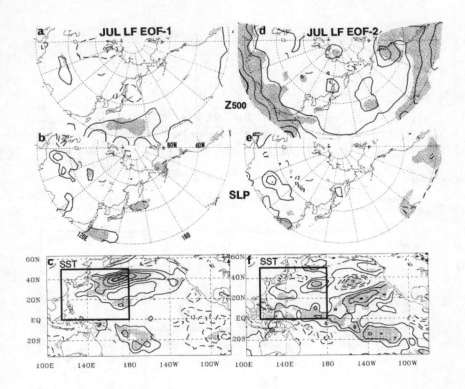

**Fig. 3.9** Atmospheric and SST anomalies associated with the 2 leading modes of the decadal SST variability over the Northwestern Pacific for summer. Linear regression coefficients (dashed for negative; zero lines omitted) of (a) Z500 (every 5m; thickened for every 20m), (b) SLP (every 0.5 hPa, thickened for every 2 hPa) and (c) the COADS SST (every 0.1°C, thickened for every 0.5°C) are plotted with PC for the first mode (see Fig. 3.10). Shaded for regions where correlation with PC exceeds 90% significance. (d)~(f): As in (a)~(c), respectively, but for the second mode. Anomalies of Z500 and SLP are multiplied by [sin 45°N/sin (lat.)] to mimic streamfunction-like anomalies.

The second mode (Fig. 3.8b), which accounts for 20% of the total variance, appears to represent a remote signal that co-varies with the decadal summertime variability over the tropical Pacific (Fig. 3.9f). Compared with the counterpart for winter (Fig. 3.3b and Fig. 3.4f), the amplitudes are generally smaller except in the western tropical Pacific and seesaws in SST anomalies between the tropics and STFZ are less pronounced, especially in the Southern Hemisphere. In contrast to winter, the decadal SST fluctuations around Japan associated with this mode are negatively correlated with those in the tropics.

Atmospheric anomalies associated with the decadal SST variability are generally much weaker in summer than in winter. Virtually no significant coherent SLP anomalies in the extratropics are associated with the two leading SST modes, indicative of virtually no surface wind anomalies. Virtually no significant Z500 anoma-

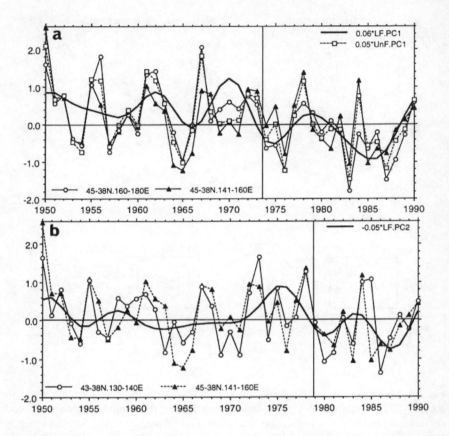

**Fig. 3.10** (a): Time series of first PCs for decadal (× 0.06; heavy solid line) and unsmoothed (× 0.05; dashed line with squares) SST variability over the Northwestern Pacific, with unsmoothed SST anomalies (°C; ΔT) averaged over the eastern and western SAFZ [(45°-38°N, 160°E-180°) and (45°-38°N, 141°-160°E), respectively; solid lines with circles and triangles, respectively]. (b): Time series of second decadal PC (× 0.05; sign reversed, heavy solid line), with unsmoothed SST anomalies (°C; ΔT) averaged over the western SAFZ (triangles) and over the Sea of Japan [43°- 38°N, 130°-140°E; circles]. All time series are for boreal summer.

lies are associated with the first SST mode, either. The second SST mode, in contrast, is accompanied by coherent significant anomalies in Z500, but they are mostly in lower latitudes. The Z500 anomalies seem to be a manifestation of a hemispheric mode in the lower-tropospheric temperature variability found by Wallace et al. (1993) that is dominated by decadal and interdecadal signal. These results suggest that the decadal SST anomalies over the extratropical Pacific in summer are unlikely to be forced either by the turbulent momentum and heat fluxes above the ocean surface, or through the anomalous temperature advection by the Ekman transport. Rather, they are likely to be generated by vigorous interactions with the atmosphere during winter. Yet, if so, it remains to be solved how the summertime SST anomalies observed in SAFZ tend to be more or less in phase with the winter anomalies on decadal time scales, in the presence of surface warming in summer that acts to obscure the latter (Namias and Born 1970; Alexander and Deser 1995). It may also be likely that changes in the gyre circulation generated by wintertime wind stress anomalies persist throughout the year and cause persistent thermal anomalies in the upper-ocean.

The pattern of the most dominant SST variability with ENSO time scales, as

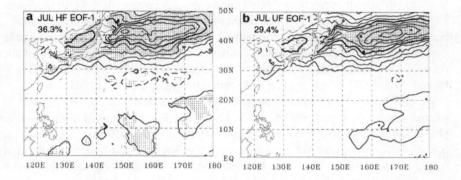

**Fig. 3.11** The first EOFs of the (a) decadal and (b) total interannual variability in the Northwestern Pacific summertime SST for 1966-1990, based on the JMA data. Linear regression coefficients are plotted between decadal SST anomalies and PCs of these modes. The coefficient corresponds to a local change in SST (°C) when a given PC increases by its unit standard deviation. Contoured for every 0.1, thickened for every 0.5, dashed for negative, zero lines omitted. Shaded for regions where correlation between SST and a given PC exceeds 95% significance.

depicted in the first EOF (Fig. 3.11a), is characterized by the anomalies confined strongly in SAFZ, which is in a sharp contrast to the first winter EOF (Fig. 3.6a). The leading mode of the unsmoothed summertime SST variability also represents the fluctuations in SAFZ, which reflects the fact that the variability with either time scale tends to be trapped in SAFZ. The confinement is even more pronounced on the shorter time scales (Fig. 3.8a). As the associated atmospheric anomalies are quite weak (not shown), the SST anomalies in SAFZ are unlikely to be a remote response to ENSO via an atmospheric teleconnection in summer. The marked confinement of the summertime SST variability in SAFZ may be attributed to the shallow mixed layer that is substantially warmer than underneath. The confinement may also be attributed to changes in the current system in the frontal zone (Mizuno and White 1983).

## 3.6   Discussion

Results presented in this chapter for winter are all consistent with NLY97 as mentioned in the first section. We have added some new findings to theirs. First, we found the decadal SST variability inherent to the midlatitude marginal seas, which appears to be independent of the variability over the Pacific basin. The variability is manifested as the relaxing or enhancement of intense winter-mean SST gradients over the East China Sea and the southern part of the Sea of Japan, in response to the decadal modulation of the East Asian northwesterly winter monsoon. Decadal anomalies in SST over these marginal seas are, in fact, correlated positively with those in surface air temperature over China and negatively with those in the intensity of the Siberian High. The associated upper-level atmospheric anomaly pattern is the EU pattern, a teleconnection pattern in the form of a wavetrain propagating across over the Eurasian continent. Unlike in SAFZ, this particular mode of the decadal SST variability almost disappeared in summer, during which the surface is heated up and SST becomes nearly uniform throughout the East China Sea.

We also found strong wintertime SST variability with decadal time scales in an intense frontal zone centered at ~40°N across the Sea of Japan. It tends to be in phase with that in SAFZ but not around 1970, one of the warmest periods for SAFZ. During that time the frontal zone in the Sea of Japan underwent the first cool period in the postwar era. Interestingly, it just coincides with the coolest period over China during the last 40 years. It is conjectured that during that period vigorous cold air outbreaks out of the continent associated with the abnormally strong winter monsoon maybe cooled the entire Sea of Japan.

Our result is consistent with one of the important findings of NLY97 that the tropical influence through the atmospheric teleconnection is limited to the subtropical gyre and hence it cannot explain the decadal SST fluctuations in SAFZ. In fact, we have shown that cooling in SAFZ occurred 2~3 years in advance of the tropical warming in 1976/77, in agreement with Watanabe and Mizuno (1994). Indeed, the mean SST difference map of Koide and Kodera (1997) based on the two adjacent 5-year periods before and after 1977 resembles our second EOF (Fig. 3.3b and Fig. 3.4h) but not our first EOF (Fig. 3.3a and Fig. 3.4d). Another mean SST difference map they presented for the two adjacent 5-year periods before and after 1989 resembles our first EOF. These facts are consistent with data analyses by Miller et al. (1998) and Tourre et al. (1998), who suggested an active role of the midlatitude atmosphere-ocean coupled system in the decadal variability over the North Pacific.

Our analysis using the high-resolution SST data reveals that local characteristics of the decadal SST variability are sensitive to the rather fine structure of the oceanic fronts. In specific, a pronounced seasonal contrast in magnitudes of the decadal SST fluctuations in the Kuroshio/Oyashio interfrontal zone appears to be linked to the seasonal migration of the Oyashio penetration front. Also, summertime weakening of the decadal SST variability in the central Sea of Japan is well correlated with the weakening of the front there.

The findings mentioned above are suggestive of an important role that oceanic frontal zones play in DICEs over the North Pacific, as suggested by NLY97, Miller et al. (1998), White and Cyan (1998) and Tourre et al. (1998). Indeed, SST anomalies associated with our first decadal mode for winter are concentrated in the narrow SAFZ, despite the fact that the associated change in the intensity of the Aleutian Low tends to accompany the surface wind anomalies that are distributed more coherently over the North Pacific basin. Suppose the Aleutian Low somehow becomes significantly deeper than in the climatological mean. Then, the associated anomalous westerly wind stress along SAFZ augments the Ekman mass transport significantly and hence the temperature advection across the front. The consequent anomalous cooling should be confined within SAFZ (Frankignoul and Reynolds 1983). Although the maximum Ekman transport occurs at the surface, the influence of the cooling must be spread into the entire depth of the mixed layer by mechanical mixing in the presence of the substantial surface westerlies along SAFZ. NLY97 presented a rough estimate that an increase in the surface westerly speed as observed in the mid-1980's leads to a cooling that could replenish at least 1/3 of the observed SST anomalies in SAFZ within the winter half of the year. Moreover, the anomalously deep Aleutian low enhances the southward Oyashio current by driving the subpolar gyre more strongly (Sekine 1988), which contributes to the anomalous cooling in SAFZ and also alters the position and intensity of minor frontal systems. The rest of that cooling is attributed to such processes as mechanical mixing within the oceanic mixed layer and turbulent heat fluxes above the ocean surface (Miller et al. 1994), both of which tend to be enhanced in the presence of the stronger surface westerlies than the long-term average. Lower tropospheric temperature anomalies as observed above SAFZ (and STFZ) also act to reinforce the SST anomalies below by altering downward long wave radiation.

As discussed above, the decadal SST anomalies in SAFZ tend to be reinforced by the atmospheric anomalies observed concurrently with them, which is suggestive of a feedback loop in the coupled system over the North Pacific subpolar gyre that vitalizes the variability of the system with decadal time scales. The feedback loop we hypothesize here may be closed, if we assume the observed atmospheric anomalies to be maintained in response to the midlatitude SST anomalies, as simulated in some atmospheric general circulation models (GCMs; Palmer and Sun 1985; Latif and Barnett 1994; Peng et al. 1997). In particular, the wave activity flux diagnosis in Section 4 suggests that the PNA pattern associated with the decadal SST anomalies in SAFZ must be forced in the midlatitude Pacific but not in the tropics. Yet, the atmospheric response to the midlatitude SST anomalies is generally too weak to be simulated unambiguously in many GCM experiments (e.g., Kushnir and Lau 1992). Nevertheless, to examine the midlatitude SST forcing of the atmosphere above, we have to rely on GCM experiments. This is due to extreme difficulties in identifying the forcing in the observed atmospheric data, which arise from the short response time of the extratropical atmosphere to the oceanic forcing, its vigorous internal variability and strong coupling with the underlying SST.

   An important issue still left to be solved is what mechanism switches the polarities of anomalies in the feedback loop of the atmosphere-ocean coupled system over the northern North Pacific by terminating their amplification so that the system acts as a delayed-action oscillator. We are not sure at this stage whether the polarities of the anomalies are switched in association with the gyre adjustment to the anomalous wind stress curl in the form of westward propagation of oceanic Rossby waves (e.g., Latif and Barnett 1994), or with the advection of oceanic thermal anomalies with the opposite sign by the mean gyre circulation (e.g., Tourre et al 1998). As atmospheric anomalies in response to the oceanic forcing can be formed within a single winter season, the delayed mechanism that forces the coupled system to oscillate with decadal time scales should be oceanic no matter which of the two mechanisms mentioned above is really operative. To our deeper understanding of North Pacific DICEs, it is hence important to investigate how the thermal anomalies generated in the oceanic mixed layer along SAFZ through the vigorous interactions with the atmosphere are subducted into the pycnocline level and how they interact with the mean gyre circulation.

   **Acknowledgments**. We thank Dr. Y. Masumoto for helping us with processing the JMA SST data set and Mr. M. Watanabe for preparing Jones' surface temperature data. Dr. Y. Tanimoto reviewed the first draft thoroughly and gave us valuable comments. Discussion with Drs. N. Iwasaka and N. Maximenko was helpful in interpreting some of our findings. One of the authors (HN) is supported in part by the Grant-in-Aid for Scientific Research on Priority Areas (08241104) of the Japanese Ministry of Education, Science, Sports and Culture.

# 4  Large Scale Modes of Ocean Surface Temperature Since the Late Nineteenth Century

C. K. FOLLAND, D. E. PARKER, A. W. COLMAN
*Hadley Centre, Meteorological Office, Bracknell, U.K.*

R. WASHINGTON
*School of Geography, University of Oxford, U.K.*

## 4.1  Introduction

There is increasing evidence of coherent patterns of variability on near quasi-bidecadal time scales in a range of climatic data from many parts of the world. Folland et al. (1984) found peaks at periods of 16 and 21 years respectively in spectra of globally- averaged sea surface temperature (SST) and night marine air temperature (NMAT) for 1856-1981. Newell et al. (1989) found variations near a period of 21 years in global and Southern Hemisphere NMAT for 1856-1986, and to a lesser extent in Northern Hemisphere NMAT and global and hemispheric SST. Ghil and Vautard (1991) drew attention to variations on approximately 20-year time-scales in globally-averaged anomalies of combined land surface air temperature and SST for 1854-1988, though Allen and Smith (1996) question the statistical significance of their results. Mann and Park (1994) found a 15-18 year mode in fields of mainly land surface air temperature anomalies for 1891-1990. They suggested that this mode, which had a pattern similar to that of the thermal signature of the interannual El Niño-Southern Oscillation (ENSO), may be a manifestation of long timescale modulation of ENSO as well as being the reason for Ghil and Vautard's (1991) global-average result. Latif and Barnett (1996) discussed near bidecadal variations in SST and atmospheric circulation over the North Pacific in both observations and a coupled model, and the consequential variations of temperature and precipitation over North America. In the Southern Hemisphere, Venegas et al. (1996) found a coupled mode in South Atlantic SST and mean sea level pressure (MSLP) data for 1953-1992, with significant variations on near 15-year timescales and provided evidence that the atmosphere was forcing the ocean. The fluctuations of this mode were significantly correlated with those of the interdecadal joint mode of Northern Hemisphere surface temperature and MSLP found by Mann and Park (1996). Chang et al. (1997) found a 13-year spectral peak in their analysis of a dipole of SST anomalies straddling the equator in the Atlantic, a result close to that of previous authors. The dipole was strengthened by a positive feedback with wind-induced heat fluxes, and damped by ocean currents. However, they found no close relationship with the leading global scale mode of SST. In addition, New Zealand

land air temperatures were compared with nearby SST and NMAT by Folland and Salinger (1995). New Zealand Temperature (NZT) for 1871-1993 displayed prominent one to two decade variations peaking near a period of 16 years, which were well replicated in neighboring (30°-50°S, 165°-180°E) SST and NMAT data. However these variations were only marginally coherent with SST variations on quasi-bidecadal time-scales in the tropical Eastern Pacific or variations averaged over the Southern Hemisphere as a whole. Note that 14-18 years is a shorter period than the 21-year peak noted by Folland et (1984) or Newell et al. (1989). In tropical East Pacific SST there is a near 14-15 year peak (Figs 2n,o), though a peak near 20-30 years is more prominent in NMAT.

Particularly relevant results have been shown by Zhang et al. (1997) and Mantua et al. (1997). Zhang et al. identify a decadal to multi decadal ENSO-like mode of sea surface temperature which has a distinctly different pattern in the east Pacific compared to the normal ENSO pattern and has more variability in the North west Pacific than does ENSO. Mantua et al. call a similar pattern the "Pacific Decadal Oscillation". We refer to these results later.

In this chapter we attempt to isolate the regions contributing most strongly to near bidecadal variability against the background of variations on El Niño-Southern Oscillation to century time scales. We demonstrate connections between ocean surface temperatures and tropical rainfall patterns on near bidecadal time-scales, and compare these relationships with already established links with ENSO on interannual time-scales. Particularly strong links appear between southern African rainfall and SST anomalies.

## 4.2   Data

Our main marine surface temperature data are SSTs from the Meteorological Office Historical Sea Surface Temperature (MOHSST6C) data set (Parker et al., 1995c). They are expressed as monthly anomalies on a 5° latitude x 5°longitude resolution from the 1961-1990 climatology of Parker et al. (1995b) and Rayner et al. (1996) and include the bias adjustments to pre-1942 SST data in Folland and Parker (1995). Missing monthly values with at least four spatially or two temporally adjacent neighbors are replaced by the average of these neighbors, as described by Parker et al. (1994). Outlying values, defined as deviating by more than 2.25° C from the average of these neighbors, are likewise replaced. As supporting evidence, we use NMAT anomalies relative to 1961-90 from the recent Meteorological Office Historical Night Marine Air Temperature (MOHMAT42) data set (updated from Parker et al., 1995a). Bias adjustments, smaller than those applied to SST, are also applied to older NMAT (Parker et al., 1995a).

The Global Sea Ice and Sea Surface Temperature Data Set (GISST) was not used. Earlier versions (Parker et al, 1995a) relax some regions to climatology. Later, better versions (Rayner et al., 1996) use eigenvector reconstruction methods which would impose some of the patterns investigated here. However, as described in Appendix 1, we have adapted a step used in Rayner et al. (1996) to create the eigenvector time series calculated in Section 4.3.4. This "enhancement" method gives almost unbiased estimates of the variance when marine data are incomplete. It boosts the variance of ENSO dominated EOF time series, particularly before 1950, in a similar way to that which occurs in the full GISST2 data.

Gridded precipitation data are updated from Hulme (1994). The southern African regional rainfall series used in Section 4.5 was derived from these data as follows. Gridded monthly values, at 2.5° latitude × 3.75° longitude resolution, were converted into seasonal anomalies for January-March, using 1951-1980 as the reference period. The anomalies were passed through a low-pass filter (Young et al., 1991) with half power at a period of 13.3 years to remove ENSO-related variability and to concentrate on the near bidecadal time scale and its longer term modulation. Unrotated correlation empirical orthogonal functions (EOFs) of Africa-wide rainfall (south of 25°N) were then calculated. The first EOF has large weights of one sign over southern Africa, including South Africa, Namibia, Lesotho, Swaziland, Botswana, Zimbabwe and extreme southern Mozambique. The corresponding time series correlates (r=0.90, significant at the 95% level) highly with an areal average of filtered rainfall over the broad southern African region. This time series was used as the 'observed' rainfall series in Section 4.5.

## 4.3    Low Frequency Variability in Marine Surface Temperatures

### 4.3.1    Global and Regional Time Series of SST and NMAT Anomalies

Fig. 4.1 shows global, hemispheric, and a selection of regional seasonal time series of SST (MOHSST6C) and NMAT anomalies for 1861-1996. Low-pass filtered series are superimposed using the method described in Section 4.2. These are much like the series shown in Bottomley et al. (1990). Bidecadal variability is particularly evident between 1870 and 1950 in the South Pacific but is less evident in the North Pacific. We choose to show marine data for the South Atlantic averaged over 0°-20°S in Fig. 4.1e because it best demonstrates the quasi-15-year fluctuation found by Venegas et al. (1996). It is especially evident after 1970 (in accord with the their findings and those of Chang et al. 1997), but also appears between 1890 and 1940. All the series in Fig. 4.1, especially those for the smaller regions, show reduced inter-seasonal coherence in the nineteenth century, and especially prior to 1875, when data were very sparse, as noted by Bottomley et al. (1990). Note the often high level of agreement between the SST and NMAT data down to the seasonal time scale in recent decades, with few exceptions.

### 4.3.2    Power spectra

Complete annual time-series of temperature anomalies were required to estimate power spectra. For the globe, hemispheres and ocean basins, the marine data yield complete annual series: for the smaller regions, Nino 1+2 area, (0°-10°S,

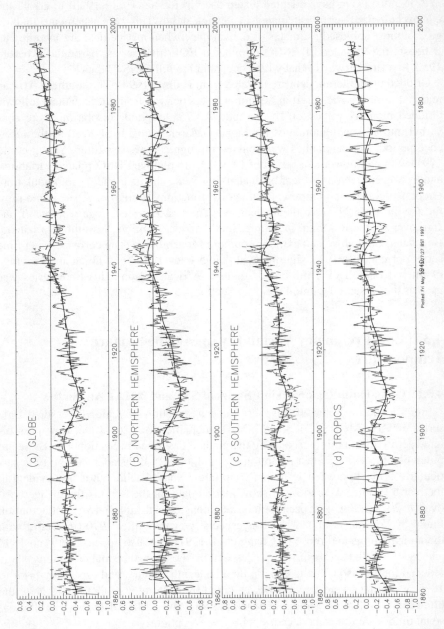

**Fig. 4.1** Seasonal (January to March, etc.) anomalies of SST (solid line) and NMAT (dashed line) relative to 1961-1990 climatology, 1861-1996. The smooth curves are filtered to pass >50% of the variance at periods >13.3 years. **a)** Globe: **b)** Northern Hemisphere; **c)** Southern Hemisphere; **d)** Tropics (20°N -20°S);

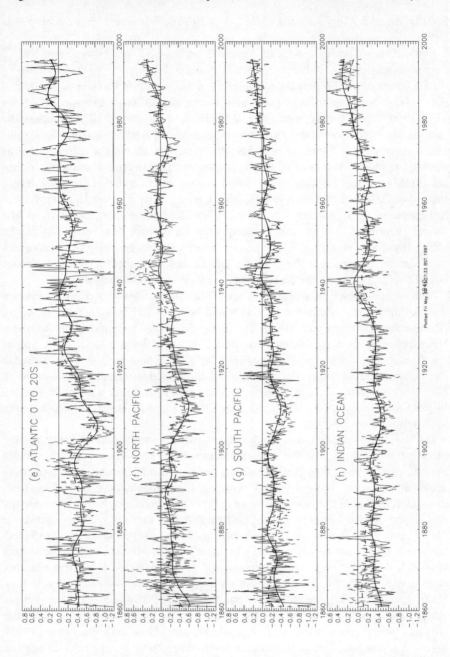

**Figure 4.1 (continued)   e)** Atlantic 0°-20°S; **f)** North Pacific (0°-65°N, 120°E to American coast); **g)** South Pacific (0°-50°S, 120°E to American coast); **h)** Indian Ocean north of 50°S.

80°-90°W) and Nino 3.4 area (5°N-5°S, 120°-170°W) linear interpolation was used.

Fig. 4.2 shows power spectra of detrended annual SST (thick lines) and NMAT anomalies (thin lines) for the globe, hemispheres and a range of regions. (The spectra are based on the Fourier transform of the autocorrelation function of detrended data, using about one third as many autocorrelations as data). The spectra are for 1876-1996 except for the west tropical Pacific and the Nino 1+2 area where the periods are 1898-1996 and 1906-1996 respectively. The 95% estimates of significance use a method due to Allen and Smith (1996) that allows for colored noise as an AR(1) process. Values of power significant at the 95% confidence level criteria are often very high because of the strong annual autocorrelation in many series; thus $r_1$ =0.70 for the Northern Hemisphere. Consequently few low frequency spectral estimates are significant. This problem is related to the short length of the historical record so that we cannot adequately distinguish between multidecadal variability that is part of an AR(1) process and variability resulting from processes in a particular frequency band. Where NMAT and SST spectra differ at periods beyond several decades, the SST spectral estimates are more reliable.

Many series show peaks near the main ENSO time scales of around 3.5-6 years that dominate the spectra for NINO3.4 and Nino 1+2 (Fig. 4.2n and Fig. 4.2o), (Rasmusson and Carpenter, 1982). The globe, Southern Hemisphere and Northern Hemisphere all show significant peaks near a period of 4 years, as do many Southern Hemisphere ocean regions (e.g.Fig. 4.2f, g, p). Some regions like the South Pacific show muted power because large warm and cold subregions associated with ENSO are prominent at the same time. The general warming and cooling of the Indian Ocean associated with ENSO shows up in Fig. 4.2i and Fig. 4.2m as peaks with periods near 3.5-5 years.

On longer time scales, quasi-bidecadal variations are evident, weakly for the globe but more strongly in the Southern Hemisphere (Fig. 4.2a,e,f), in accord with Folland et al. (1984), though the peaks are not significant. Interestingly the NMAT peaks are stronger, though NMAT data are less reliable than SST data. Century-scale variations are more dominant in both data sets, though even these are not significant. In the Northern Hemisphere (Fig. 4.2b), there is *minimum* power at 15-20 years in SST with a very weak maximum near 10 years. There is no bidecadal peak in the *annual average* SST of three main Northern Hemisphere oceans (Fig. 4.2c,d and l) though weak near decadal peaks are evident, but these are not statistically significant. The North Atlantic variations seen on near 10-year time scales appear to originate more from the tropics (0°-20°N, see also Chang et al., 1997) than from higher latitudes (>35°N, not shown). Although most of the latter area has similar phase in the decadal dipole found by Deser and Blackmon (1993), their result was based on winter data, so our findings based on annual data may not contradict theirs. The North Atlantic also shows the largest variation on a sub-century time scale. The western tropical Pacific (20°N-20°S, 120°E- 170°W, Fig. 4.2k) also shows near 10-year variations which almost reach 95% significance level for NMAT. In the Southern Hemisphere, a peak corresponding to bidecadal variations

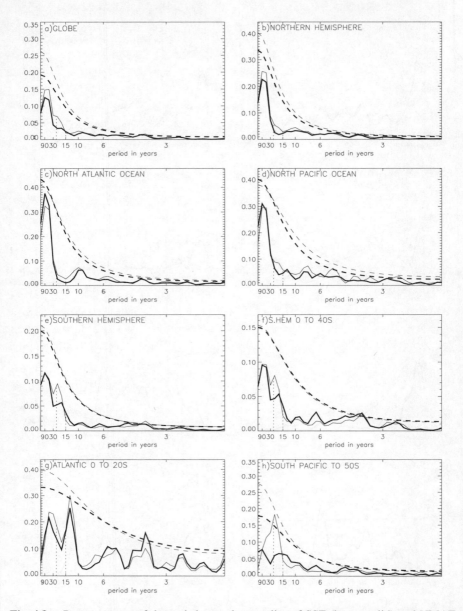

**Fig. 4.2**    Power spectra of detrended annual anomalies of SST (heavy solid) and NMAT (light solid), relative to 1961-1990 climatology. The analysis period is 1876-1996 unless otherwise specified. The dashed lines (thick, SST; thin, NMAT) are upper 95% confidence limits for Markov red noise. **a)** Globe: **b)** Northern Hemisphere; **c)** North Atlantic(0°-80°N, excluding Mediterranean but including Arctic to 70°E); **d)** North Pacific (0°-65°N, 120°E to American coast); **e)** Southern Hemisphere; **f)** Southern Hemisphere 0°-40°S; **g)** Atlantic 0°-20°S; **h)** South Pacific (0°-50°S, 120°E to American coast).

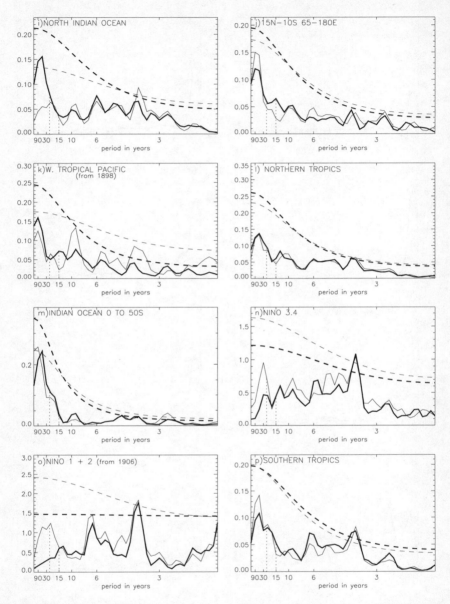

**Figure 4.2 (continued) i)** North Indian Ocean; **j)** Warm pool 15°N -10°S, 65°-180° E; **k)** Western tropical Pacific 20°N -20°S, 120° E-170° W, 1898-1996; **l)** Northern tropics (0°-20°N); **m)** Indian Ocean 0°-50°S; **n)** Nino 3.4 area (5°N -5°S, 120°-170° W); **o)** Nino 1 + 2 area (0°-10°S, 80°- 90°W, 1906-1996) **p)** Southern tropics (0°-20°S). Vertical dashed lines show periods of 15 and 24 years.

is strongest and nearly significant in the South Pacific in NMAT (Fig. 4.2h), but weaker in the southern Indian Ocean (Fig. 4.2m) and southern tropics (Fig. 4.2p), and not evident in the tropical South Atlantic (Fig. 4.2g). By contrast, in the tropical South Atlantic 12-14 year variations corresponding to those found by Venegas et al. (1996) and Chang et al. (1997) are prominent and just statistically significant for NMAT; similar results are found when the domain is extended to 30°S (not shown). There are no important spectral peaks near 20 years in the warm pool area 15°N-10°S, 65°-180°E (Fig. 4.2j), though SST has the hint of a peak near 15 years, or in the western tropical Pacific (Fig. 4.2k). In the Nino 1+2 area 0°-10°S, 80°W-90°W (Fig. 4.2o), there are non- significant multi decadal variations in NMAT, but not in SST: there is a weak peak near 14 years in both. The Nino 3.4 area, 5°N-5°S, 120°-170°W, shows statistically insignificant but moderately prominent 12-14 year variations and bidecadal variations around 25 years especially in NMAT. These are (Fig. 4.2n) weaker, as expected, than 4-6 year variations associated with El Niño, though the NMAT variation is not much weaker. Fig. 4.2n highlights the limitations of the estimates of statistical significance: even the well known longer time scale ENSO fluctuations around 5-7 years are significant.

We now investigate spatial patterns of near bidecadal variability in annual data.

### 4.3.3 Spatial patterns of near bidecadal ocean surface temperature variability

To estimate these patterns, we calculated global fields of low-frequency (LF) filtered SST and NMAT. First, seasonal (January-March etc.) mean anomalies for almost equal-area grid boxes (of dimension 10° latitude × 12° longitude at the equator) were formed by averaging all available monthly anomalies for 5° latitude × 5° longitude areas that contributed to the equal-area box. Anomalies for these 5° × 5° boxes were weighted according to the proportion of area covered within the equal-area box. An equal area box was used if there were at least 144 (40%) of seasons available between 1901 and 1990. Seasonal anomalies for equal area boxes were set to missing if there were fewer than 10 monthly anomalies available from their constituent 5° × 5° boxes. This criterion allows early, sparser, data to contribute as much to the analysis as possible without compromising its quality. Temporally complete LF data were then calculated using a low-pass filter with half-power at a period of 13.3 years. This time scale not only eliminates the usual ENSO time scales but it was also chosen to remove the near decadal time scales prominent in the North and tropical Atlantic which are beyond the scope of this study.

Fig. 4.3a depicts the standard deviation of LF annual SST anomalies for the epoch 1911- 1995 falling in the "near bidecadal" (18-30 years) range of periods, and Fig. 4.3b the standard deviation as a percentage of the standard deviation at all periods > 13 years. Fig. 4.3c,d are corresponding maps for the (slightly overlapping) "sub-bidecadal" 13-18 year range seen more in New Zealand surface air temperature.

Fig. 4.3a shows the highest values of 18-30 year standard deviation in the central Pacific from south of Alaska to at least 50°S with maximum (reliable) standard

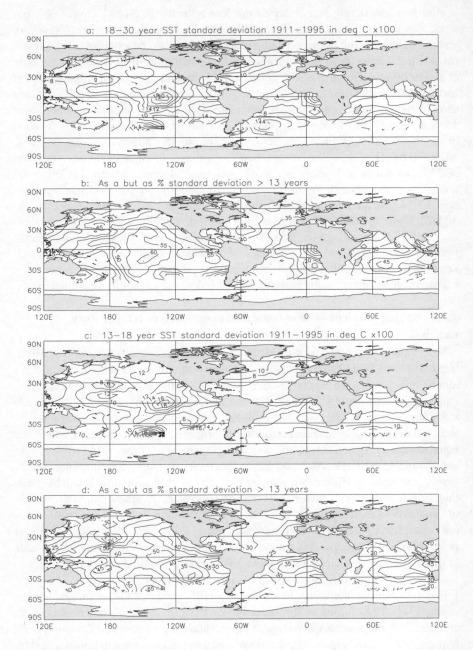

**Fig. 4.3  a)** Standard deviation of low-pass filtered SST anomalies for the period 1911-1995 falling in the bidecadal range (approximately 18-30 years). **b)** as a) but expressed as a percentage of the total low-pass (>13 years) variance. **c), d)** as a), b) but for range of periods prominent in New Zealand temperature (approximately 13-18 years).

deviations of SST just south of the equator around 120°-140°W. Values reach 0.2°C in the latter region, twice the mean global standard deviation of 0.1°C. Expressed as a percentage of local standard deviation on LF time scales >13 years, peak values are almost in the same region but extend south west to 160°W, 20°S, exceeding 60% around 15°S, 150°W. An area exceeding 55% occurs in the tropical eastern-most Pacific. Note that relatively large percentages extend across much of the tropical Pacific. In other oceans the maximum standard deviations are less and percentages only reach about 45% except perhaps for a small region in the Gulf of Guinea. The near decadal variability prominent in the tropical Atlantic is highly attenuated at these periods.

Fig. 4.3c shows a fairly similar picture for the 13-18 year time scale but with lower standard deviations (global mean 0.09°C). Expressed as a percentage of the total LF standard deviation (Fig. 4.3d) the peak values of >50% are less concentrated but still tend to extend across the tropical Pacific. Percentages near New Zealand are higher than for the 18-30 year interval but only reach 35-40% so it appears that the seat of the New Zealand 14-18 year temperature variations (presumably reflected in atmospheric circulation variations too) may lie over the tropical Pacific. The very large peak in absolute (Fig. 4.3c) and percentage standard deviation near 150°W, 50°S should be noted but requires the GISST data set for further analysis. This is a higher latitude region identified tentatively by Folland and Salinger (1995), and more definitely by Folland et al. (1997) using GISST2, as being quite strongly influenced by ENSO probably via a systematic modulation of the strength of the westerly winds.

### 4.3.4 Synthesis of variability: Empirical Orthogonal Function (EOF) analysis

We calculated LF global unrotated covariance all-seasons EOFs from the LF marine data and a corresponding set of high frequency (HF) EOFs (periods <13 years), all for 1911-1995. Many authors have calculated global SST EOFs, mostly over shorter periods e.g. recently Kawamura (1994). Our EOFs update those of Parker and Folland (1991) who did not distinguish between the HF and LF time scales and we use the EOF time series enhancement technique described in Appendix 1. Note that choice of the boundary of the LF time scale at 13.3 years (half power at this period) is not only compatible with the Southern African rainfall analyses but is deliberately designed to suppress near decadal variations prominent in the Atlantic which may confound the lower frequency LF results shown below. Interpolated values were excluded from fields input to the EOF analyses. We also excluded some coastal and landlocked sea areas in middle and high latitudes of the Northern Hemisphere, because the high variance of anomalies in these areas was found to have a disproportionate effect on the shape of the leading global EOFs. We chose covariance EOFs to highlight areas of strong absolute low frequency variability.

Fig. 4.4 depicts the first four LF EOFs of SST for 1911-1995 (Fig. 4.4a, d, f, i), along with their enhanced time series (Fig. 4.4c, e, g and j). LF EOF1 (45.5% of LF

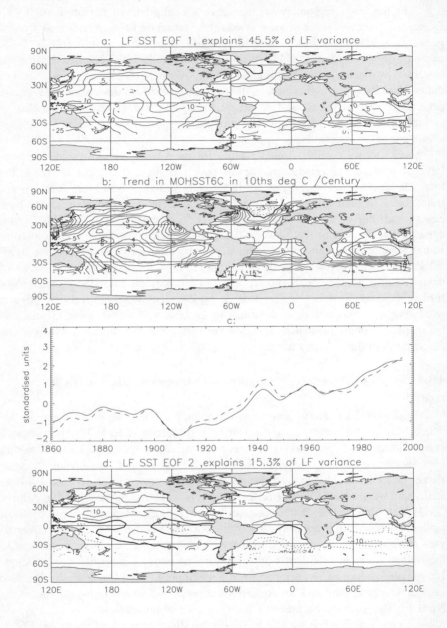

**Fig. 4.4**  a) Pattern of weights of LF SST EOF1, 1911-1995. b) Linear trends of SST, 1911-1995 c) Time-series of LF SST EOF1, 1861-1996. Global mean SST anomalies are shown as a dashed line. d), e) as a), c) but for LF SST EOF2. f), g) as a), c) but for LF SST EOF3 h) FFT power spectrum of LF EOF3. i), j), k) as f), g), h) but LF SST EOF4.

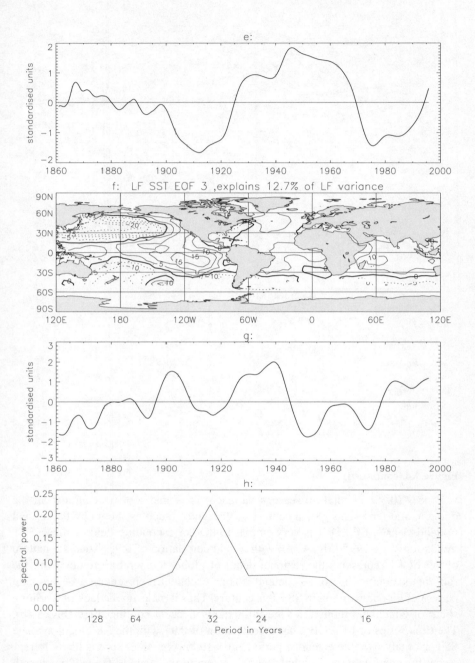

Figure 4.4 (continued)

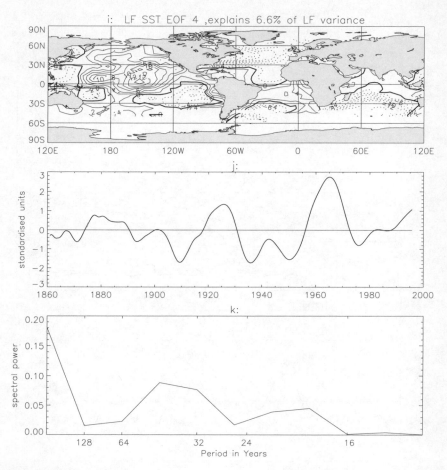

**Figure 4.4 (continued)**

variance, 10.3% of total all-seasons variance) is of one sign over almost all the domain, and represents global century-scale trends, much as shown by Parker and Folland (1991). LF EOF1 is very similar to the corresponding field of linear SST trends for 1911-1995 (Fig. 4.4b), with a field correlation of 0.93. Thus the pattern of LF EOF1 represents the regional detail of global temperature trends from the full data set very well so we contend it is an excellent representation of the main pattern of global warming in SST this century. Thus it picks up the lack of warming south of Greenland though not the full extent of recent cooling there (Fig. 4.4b). The time series of LF EOF 1 has a correlation of 0.97 with the LF global-average SST anomaly over the extended period 1861-1996 (Fig. 4.4c) so it is likely to represent the chief pattern of global marine temperature change in the late nineteenth century as well. The second LF EOF (15.3% of LF variance, 3.4% of total all seasons variance) represents the interhemispheric contrast associated by Folland et al.

(1986), Parker and Folland (1991) and Folland et al. (1991) with decadal episodes of Sahel wetness or drought. LF EOF2 contributes considerably to the multi decadal SST changes in the North Atlantic linked by Kushnir (1994) to regional atmospheric circulation changes. Its time series suggests variation on time scales of about 60-70 years. LF EOF2 is likely to be a manifestation of variations in the thermohaline circulation and temperature on the same time scale described by Schlesinger and Ramankutty (1994).

LF SST EOF3 (12.7% of LF variance, 2.9% of total variance, Fig. 4.4f) appears to be the focus of some bidecadal SST variability (Fig. 4.4g), though lower frequency variations dominate its spectrum (Fig. 4.4h). LF EOF3 has a pattern somewhat reminiscent of ENSO related SST patterns (see Fig. 4.7a). However it is not the same; the weights over the North west Pacific exceed in magnitude those of opposite sign over the tropical east Pacific. Nevertheless, the Indian Ocean, like ENSO, has weaker weights than in the tropical East Pacific but of the same sign. Another difference is the near zero weights over the easternmost Tropical Pacific. This helps to give the impression of an almost "horseshoe" shape of positive weights stretching from California through the equator to 15°S, 90°W seen also by Latif et al. (1997) in an LF principal oscillation pattern analysis and by Kleeman et al. (1996) for the "extended ENSO" period of 1990-1994 (their Fig. 1a). In the South Pacific again the pattern is like that associated with ENSO but the negative weights are relatively larger in magnitude than in Fig. 4.7a. Over the Atlantic the weights are small and positive and possibly not reliable. Recently Kachi and Nitta (1997) also show a broadly similar mode (their Fig. 9) using a singular value decomposition analysis. They analyse decadal variations in (i) December to February SST for 60°N to 40°S and (ii) decadal variations in 50°hPa height over the Northern Hemisphere, both from 1955 to 1995. The relative lack of variance over eastern most Pacific SSTs is again seen and many other SST features are fairly similar to our LF EOF3. Another less certain feature of our analysis is the extension of positive weights across the tropical West Pacific, again not seen on ENSO time scales. LF SST EOF3 is quite similar to LF SST EOF2 of an earlier unpublished analysis by Folland et al. (1993). The spectrum of LF SST EOF3 (Fig. 4.4h) shows a modest peak around 30 years.

The main features of LF SST EOF3 in the Pacific are borne out by LF NMAT EOF3 (Fig. 4.5a). Although the Indian Ocean centre is weaker in the NMAT analysis and the Atlantic has low weights of mixed sign, the correlation between the fields in Fig. 4.4f and Fig. 4.5a is 0.70, and the correlation between the time series in Fig. 4.4g and Fig. 4.5b is 0.77 over 1861-1996. Interestingly, the patterns in Fig. 4.4f and Fig. 4.5a, and the associated time series, seem consistent with the low frequency variations in Aleutian region winter mean sea level pressure and its relationships with surface temperature discussed by Trenberth and Hurrell (1994).

LF EOF3 is similar to the "Pacific Decadal Oscillation" of Mantua et al. (1997) and particularly to the first global EOF of unfiltered SST of Zhang et al. (1997). Zhang et al eliminated the effect of global warming by calculating an adjusted SST anomaly data set (based on MOHSST) expressed as the difference of their anoma-

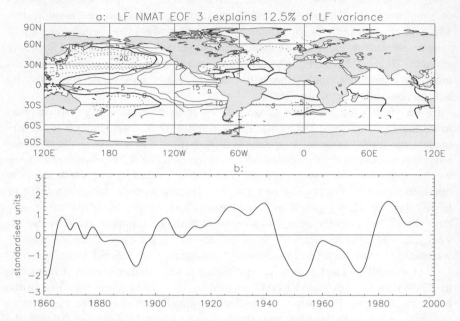

**Fig. 4.5**   a) Pattern of weights of LF EOF3 MOHMAT4N2 NMAT for 1911-1995. b) Time-series corresponding to a), 1861-1996.

lies from the global mean SST anomaly for the current year. They call such quantities "anomaly deviations". They carried out this procedure because they considered that forcing the remaining global EOFs to be orthogonal to a "global warming" EOF (our LF EOF1) was too strong a constraint. We appear to overcome this problem by filtering SST into separate LF and HF time scales. However Zhang et al. do not report our EOF2, though their analysis concentrates on ENSO-like patterns and they also use a smaller global domain (60°N-30°S). We consider the Zhang et al. "global" SST pattern (and their even more similar comparable SST pattern for 1950-93) is close to our LF EOF3 for 1911- 1995. Thus the time series in our Fig. 4.4g is quite like (but smoother than) the six year low pass filtered time series of their "global pattern" over this century. Our analysis method and results differ from theirs in several other respects. So this highlights the problem of the most incisive, or most parsimonious, methodology for studying large scale SST patterns.

Global LF SST EOF4 (Fig. 4.4i, 6.6% of LF variance, 1.6% of total variance) is included because of its quite strong relationship to LF southern African rainfall in Section 5, though we do not claim this pattern is as "real" globally as that of the first three LF SST EOFs. EOF4 has weights concentrated in the central Pacific, and weights of opposite sign in the north west Atlantic and South Indian Ocean. The time series shows a minimum around 1935-1950 and a maximum in the 1960s. Its spectrum (Fig 4k) shows comparable power across the resolvable low frequency spectrum.

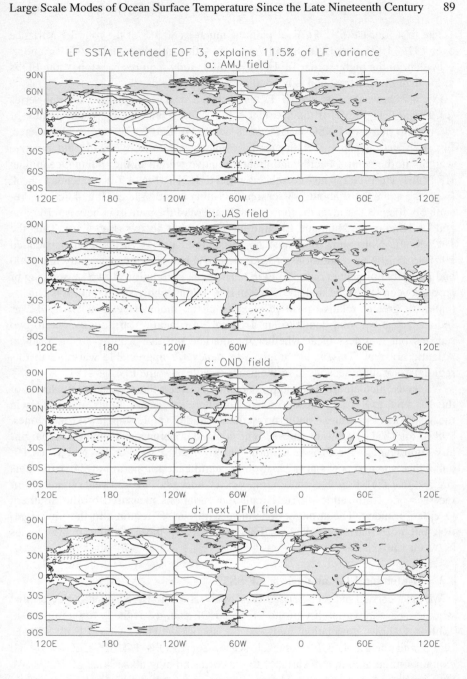

**Fig. 4.6**    Pattern of weights of extended LF EEOF3 for 1911-1995 for **a)** April-June **b)** July-September **c)** October-December **d)** for 1912-1996 for January-March.

The first four LF SST EOFs explain as much as 80.1% of the total LF variance over 1911- 1995, and 18.0% of all variance. Although a little bidecadal variability is evident in the higher order LF EOFs, a fuller study requires rotation of the EOFs to localize the patterns, and is beyond the scope of this paper.

Fig. 4.7 shows the first two HF EOFs (panels a, b) and their enhanced time series (c). HF EOF1 (12.6% of total variance, Fig. 4.7a) represents the usual picture of the mature phase of ENSO warm events, while HF EOF2 (3.5% of total variance, Fig. 4.7b) has negative weights in the tropical eastern-most Pacific but positive weights in the central Pacific and greater relative weight in the Indian Ocean than HF EOF1. HF EOF2 is lag- correlated with HF EOF1 (Fig. 4.7d) and tends to peak (with the same sign) about 2-3 seasons (six-nine months) later (Fig. 4.7c, d). Presumably it tends to reflect the more muted pattern of the transition between El Nino and La Nina or La Nina and El Nino but clearly not always (Fig. 4.7c). Thus the 1982-83 El Nino showed almost no projection onto HF EOF2. The extended 1991-1995 El Nino conditions projected clearly onto HF EOF1 but only late 1990 and mid 1995 projected onto HF EOF2. The peak correlation at two seasons lag of 0.56 between HF EOF1 and HF EOF2 in Fig. 4.7d is highly significant. It is consistent with greater relative positive weights in the Indian Ocean in HF EOF2 as Indian Ocean warming tends to lag Tropical East Pacific warming by one to two seasons but it is less clear whether the marked positive weights in the tropical Atlantic are real. This would imply a tendency for appreciable warming in this region two or three seasons after the peaks of at least some ENSOs.

We have calculated the squared coherence between the seasonal time series of the LF SST EOF3 and the normalized unfiltered seasonal Tahiti minus Darwin mean sea level pressure Southern Oscillation Index (SOI, Ropelewski and Jones, 1987), for 1871- 1995 (not shown). The squared coherence peaks at 0.47, marginally significant at the 95% level, at a period of 20 years. For LF NMAT EOF3, coherence squared peaks at 0.72, strongly significant at a period of 24 years. This suggest that the LF EOF 3 pattern imposes a "bidecadal" (20-30 year) signal on, or receives one from, atmospheric circulation variations measured by the Southern Oscillation index. The stronger coherence with NMAT requires further investigation but note that LF NMAT EOF3 has relatively more weight in the tropics between 150°-180°E.

### 4.3.5  Extended EOF analysis of LF SST

We have carried out an extended EOF analysis of the LF SST data to determine whether seasonal variability of the LF SST EOF patterns is likely to be important. Table 4.1 shows correlations between the spatial weights of the LF all seasons EOFS and the LF EEOFS for each season for EEOFs 1-4. LF EEOFs 1-4 all explain similar amounts of variance to the corresponding all seasons LF EOFs. We have used the "year" April to March to optimise the consistency of corresponding HF EEOFS (not discussed here) where we try to avoid the frequent transition from El Nino to La Nina or the reverse around April. The correlations are generally high, and in fact remarkably high for EOF4 and the corresponding seasonal EEOFS

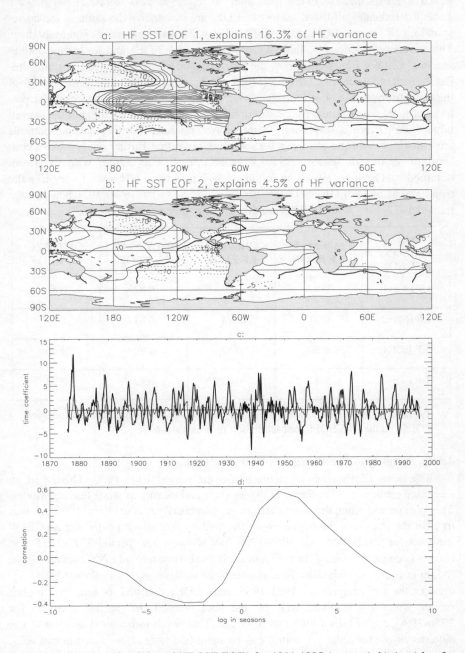

**Fig. 4.7**  a) Pattern of weights of HF SST EOF1 for 1911-1995 (see text). b) As a) but for HF SST EOF2. c) Time-series of HF EOF1 (heavy line) and HF EOF2 (light line) plotted seasonally. d) Lag correlations between HF EOF1 and HF EOF2. Positive lag implies EOF1 leads EOF2.

which suggests that LF EOF4 may after all reflect a "real" quasi-global pattern. Table 4.1 indicates all listed seasonal EEOFs are essentially the same as the corresponding all seasons EOFs. Beyond EOF4 (not shown) correlations suddenly drop. The lowest correlations in Table 4.1 are between LF EOF3 and its corresponding seasonal EEOFs. Fig. 4.6a-d show the four seasonal EEOFs. They are all quite similar to each other and to LF EOF3 (Fig. 4.4f). Scrutiny of Fig. 4.7a, least similar of these diagrams to Fig. 4.4f (correlation 0.57), shows detailed differences in the patterns of local weights, and regionally in the position of the zero line, but the overall large scale pattern is surprisingly similar. We conclude that the seasonal patterns are sufficiently similar allowing for sampling error that further analyses can concentrate on the all seasons SST LF EOFS1 to 4. This agrees with the result obtained by Graham (1994); he found an SST pattern in NH winter very like the tropical part of LF EOF3 between 20°N and 20°S, 110°E to 70°W, when he subtracted 1977-1982 winter SST from that of the colder period 1971-76.

| ALL SEASONs SST EOF | JAN-MAR LF EEOF | APR-JUNE LF EEOF | JULY-SEPT LF EEOF | OCT-DEC LF EEOF |
|---|---|---|---|---|
| LF EOF 1 | 0.87 | 0.77 | 0.82 | 0.83 |
| LF EOF 2 | 0.71 | 0.74 | 0.81 | 0.77 |
| LF EOF3 | 0.80 | 0.57 | 0.80 | 0.76 |
| LF EOF4 | 0.82 | 0.82 | 0.82 | 0.83 |

**Table 4.1** Correlations between the spatial weights of the low frequency, all seasons, EOFS and the LF EEOFS for each season for EEOFs 1-4

### 4.3.6 Stability of the all seasons EOF analyses.

All seasons EOFs were recalculated for the period 1901-1990. These had an increased tendency to similarity of eigenvalues and mixing of some leading modes. This is expected when the input data are geographically sparse (Kim, 1996), as was true for the Pacific at the beginning of the century. We also repeated the SST EOF analyses for 1921-1955 and 1956-1990 (not shown). The period 1921-1955 was chosen because it covered the epoch of weakened conventional ENSO relationships (Allan at al., 1996) whereas these connections were stronger in 1956-1990. However, in the LF analysis for 1921-1955, only EOF1 retained its integrity: higher modes were not resolved well by the short record. In the LF analysis for 1956-1990, even EOFs 1 and 2 were mixed. This result indicates that some of the patterns of interdecadal variability can be identified with about a century of data, but a full identification cannot be made. However, valuable insights can be gained through diagnosis of the processes at work through one or two cycles (e.g. Kleeman et al., 1996). A further general concern is that the weights of LF EOF2 and to a slightly lesser extent, EOF3, are large over many data sparse regions; this includes the tropical and South Pacific and other southern extratropical oceans. These caveats should be borne in mind in the remainder of the paper. On the other hand, HF SST EOF1 was well preserved in both periods, and HF EOF2 was also similar to 1911-1995 in the Pacific and Indian Ocean, but in the 1921- 1955 analysis there were differences in the tropical Atlantic where, as noted above, the rather strong weights in 1911-1995 need confirmation.

We also repeated the all seasons SST EOF analyses for 1911-1995 with the domain restricted to the Pacific. Because LF global EOF3 is largely a Pacific mode (Fig. 4.4f), the corresponding mode in the Pacific-only analysis was very similar (not shown) but became EOF2 for that ocean, explaining 17.2% of its LF variance, as against 12.7% of the global LF variance for LF SST EOF3.

## 4.4    Relationships between marine surface temperatures and tropic-wide rainfall on multi decadal time-scales.

We found above that LF EOF3 had a pattern quite like HF EOF1 in many respects. To determine if and how LF EOF3 influences LF rainfall, correlations were calculated between the time series of LF global SST EOF3 (Fig. 4.4g) and gridded LF seasonal worldwide (land only) rainfall for 1901-1995 for January-March and July-September. The tropical and subtropical ($40°N-40°S$) parts of these correlation patterns are  shown in Fig. 4.8a and b respectively. For comparison, we calculated EOF1 of unfiltered rainfall for $20°N-40°S$ in January-March and EOF1 of unfiltered rainfall for $40°N-20°S$ in July-September over the same period (Fig. 4.9a, d). The strong link of the former with ENSO is shown in Fig. 4.9b where the seasonal time series of HF SST EOF1 (which represents the mature phase of ENSO most typically around January) has a simultaneous correlation of 0.80 with the time series of January-March rainfall EOF1. Both time series were enhanced as described in the Appendix. The correlation in July- September is lower at 0.70 (Fig. 4.9e). Most interestingly, the two unfiltered rainfall EOF1s have pattern correlations of 0.61 and 0.46 respectively with the fields of Fig. 4.8a and Fig. 4.8b over 1901-1995. Thus moderately similar relationships between SST and rainfall with an ENSO-like global tropical pattern exist on both HF and LF time scales.

The patterns in Fig. 4.8b and Fig. 4.9d, particularly the weights of the same sign from India to the Caribbean are reminiscent of a boreal summer tropic-wide mode of precipitation found by Ward et al. (1994). Similar patterns are also found by correlating gridded rainfall with the time series of the equivalent LF Pacific-only SST EOF 2 (Section 4.3.4) or with the unfiltered SOI (not shown). Graham estimated the differences in rainfall in NH winter between 1977-82 and 1971-76 for the tropical Pacific. Graham's 4c is consistent with the implied pattern of rainfall differences between warmer and colder tropical Pacific phases of LF EOF3 shown in the same area of Fig. 4.8a. However we lack rainfall data in the tropical East Pacific east of $130°W$; here Graham showed an increase of rainfall just north of the equator

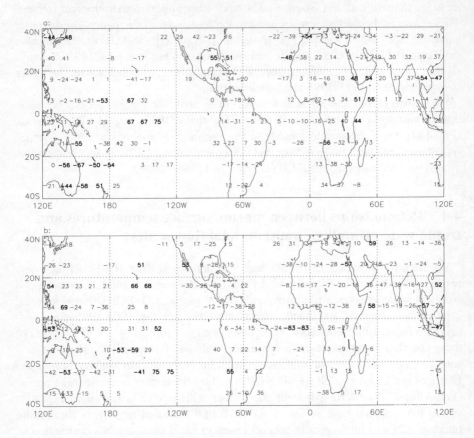

**Fig. 4.8**   Correlations between LF SST EOF3 time series in Fig. 4.4g and correspondingly LF gridded rainfall in **a)** January to March and **b)** July to September, 1901-1995. Bold values are significantly different from zero at the 95% level of confidence.

and a decrease just to its south. Nitta and Kachi (1994) have carried out analyses of decadal changes in annual rainfall in the tropical Pacific and related these to decadal changes of SST since 1955. Their annual results are not strictly comparable with our seasonal analysis but their finding of increased rainfall in the central tropical Pacific since the mid 1970s again agrees well with our Fig. 4.8 and Fig. 4.9.

Fig. 4.9c and Fig. 4.9f show power spectra associated with the unfiltered rainfall EOFs in Fig. 4.9a and b. Both spectra show some quasi-bidecadal variability more evident in January- March, though the bidecadal power is not statistically significant. This implies that some Southern Hemisphere tropical regions with rainfall maxima in this season are particularly likely to experience bidecadal modulation, perhaps those with larger weights in Fig. 4.9a and d. We now investigate LF southern African rainfall where strong negative weights can be seen in Fig. 4.9a. As explained in Section 4.2, LF southern African rainfall is represented by its first EOF.

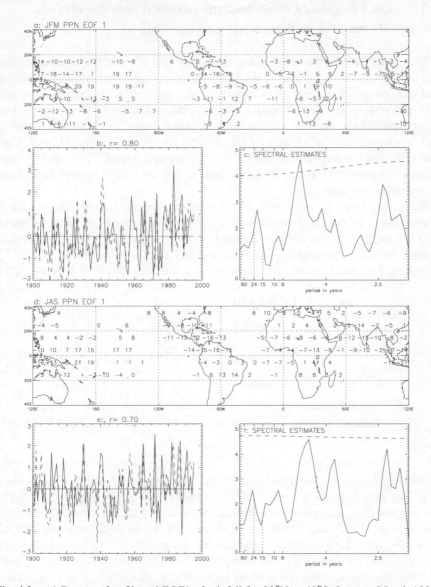

**Fig. 4.9** **a)** Pattern of unfiltered EOF1 of rainfall for 20°N to 40°S, January-March 1901-1995. **b)** time series of January-March unfiltered rainfall EOF1 (solid line) and HF SST EOF1 (dashed line), 1900-1995 **c)** FFT spectrum of January-March unfiltered rainfall EOF1 **d)** pattern of unfiltered rainfall EOF1 for 40°N to 20°S, July-September 1901- 1995. **e)** as b, but July-September. **f)** As c. but July-September. In the spectra, dashed lines are the upper 95% confidence limits. Rainfall EOFs were calculated using the same equal-area box scheme as for SST and NMAT, but, to qualify for analysis, an equal-area box required at least one constituent 5° × 5° box to have at least 50% of seasons with data in all 3 months. Vertical dashed lines show periods of 15 and 24 years.

## 4.5   Low Frequency relationships between marine surface temperatures and southern African rainfall

The importance of the quasi-bidecadal component of southern African rainfall has long been known (Tyson, et al., 1975) and is perhaps the best studied rainfall variation on this time scale. Although near bidecadal variability explains less than 30% of the total rainfall variability, the underlying oscillation has been used to predict the prolonged southern African drought of the 1980s (Tyson and Dyer, 1980), and to infer the climates of the Last Glacial Maximum (Cockcroft et al., 1987), the late Holocene (Tyson and Lindesay, 1992; Cohen and Tyson, 1995) and the next century (Tyson 1991). Physical mechanisms to explain this variability have, however, remained elusive.

Fig. 4.10a shows the time series of LF southern African January-March rainfall (1900- 1991) by the solid line. It shows strong variability on the quasi-bidecadal time scale (22 years is sometimes claimed but this is too precise) with a major peak around 1910, a minor peak around 1935, and major peaks near 1950 and 1975.

Table 4.2 shows stepwise multiple regression statistics for LF SST EOFS 2-4. F

| STEP | LF SST EOF | F | $R^2$ |
|:----:|:----------:|:----:|:----:|
| 1 | EOF4 | 48.81 | 0.35 |
| 2 | EOF3 | 77.94 | 0.64 |
| 3 | EOF2 | 116.66 | 0.80 |

**Table 4.2** Multiple Regression Statistics of Southern African Rainfall and SST EOFs.

is the F ratio significance statistic and $R^2$ the coefficient of determination for each step. The best fitting individual EOF is LF EOF4 with a correlation of 0.59. The southern African droughts of the 1980s, 1960s and 1920s, together with the wet spells of the 1970s and 1950s, are in phase with the modulation of SST by LF EOF4. A plot of LF EOF4 against the rainfall (not shown) indicates that EOF4 simulates the rainfall variability well except between 1935 and 1950 when it grossly overpredicts the amount. Addition of EOF3 largely corrects this and the further addition of EOF2 which varies on a time scale near 60 years improves the overall fit to give a correlation of 0.89. The dashed line in Fig. 4.10a shows the simulated Southern African rainfall based on these three LF SST EOFs. We judge these three LF SST EOFs to add statistically significant variance at the 95% confidence level according to the F ratio, based on estimates of the number of degrees of freedom in the LF time series calculated from their serial correlations. These three EOFs explain 34.6% of the LF SST variance in 1911-1995. Note that LF SST EOF1, global climate change, explains no significant rainfall variance even though it explains more LF SST variance (45.5%) than the three explanatory EOFS put together.

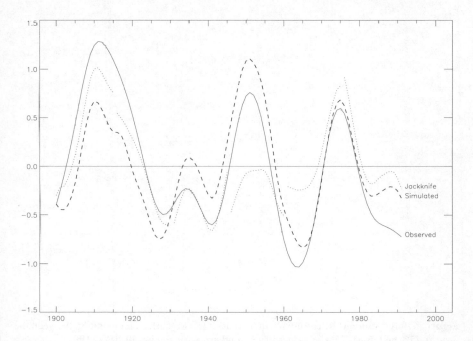

**Fig. 4.10 a)** (Solid line) Time series of LF rainfall for southern Africa, January-March, 1900-1991. (Dashed line) Simulated rainfall using a stepwise multiple regression of LF SST EOFs 2, 3 and 4. (Dotted line) Hindcast rainfall using a jackknife method in 15 year blocks (see text).

.We do not show maps of the local correlations of LF SST and Southern African rainfall because a provisional analysis indicates some strong temporal variations. Thus strong correlations exist with the Tropical East Pacific in 1900-1925 and 1956-1991 but in the intermediate 30 years (1926-1955) these correlations are weak. The strongest correlations then occur with LF SST around 10°-20°S, 120°W. Our EOF representation has somehow picked up these changes to a considerable extent. This needs more investigation. However we feel it useful to give an indication of the SST differences that are involved between wet and dry conditions. The absolute values of SST vary strongly through the century because of climate warming, affecting both wet and dry values in a broadly similar way and this can be aliased onto a small sample of extreme years sampled in LF mode. So Fig. 4.10b shows the differences in *linearly detrended* SST between the wettest 25% of LF Southern African rainfall values and the driest 25% of values. Detrending the rainfall makes little difference to the pattern. The regression equation will of course allow a variety of patterns to correspond to a given LF rainfall percentile depending on the relative strength of the constituent EOF time coefficients. Fig. 4.10a shows a

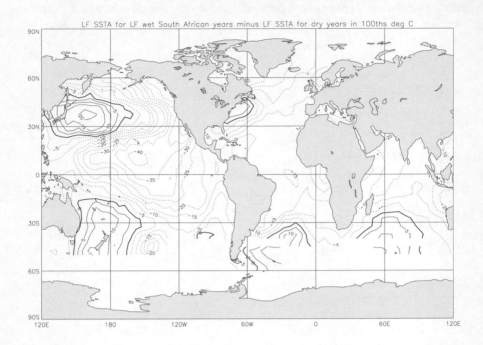

**Figure 4.10 (continued) b)** Difference in detrended LF sea surface temperature between the wettest 25% and driest 25% of values of LF southern African rainfall, 1900-1991.

broadly El Niño-like pattern but with only small differences in SST near the South American coast consistent with the pattern of, and strong effect of, LF EOF3. The largest differences in SST are displaced into the northern tropical Pacific around 160°W consistent with EOF4. Tests of the regression equation for 1900-1991 suggests that it places more weight on SST differences (of the same sign) south east of South Africa than indicated in Fig. 4.10b.

We have also attempted to create quasi-independent hindcasts of the rainfall, shown by the dotted line in Fig. 4.10a. This involves calculating SST EOFs over a period independent of the rainfall and forming multiple regression equations for these independent periods. A problem is that some of the lower LF SST EOFs of interest become confounded in some of these periods. This applies mainly to EOF4 which can be confounded with EOF5. So we have allowed up to 5 EOFs in the multiple regression formulae but have only retained those that are statistically significant. We have, experimentally, hindcast for 15 year periods at a time in this way. These were chosen to be 1901-1915, 1916-1930, 1931-1945, 1946-1960, 1961-1975, 1976-1990. In choosing these periods we are constrained by the need to keep as many years as possible for calculating LF SST EOFs The hindcasts are little different from the simulations (better before 1930), except that for 1946-60 which completely fails to pick up the increase in rainfall to the 1950 peak.

## 4.6    Discussion and conclusions.

There are divergent opinions regarding the importance of tropical and midlatitude processes in the interdecadal variations of oceanic and atmospheric climate in the Pacific sector. On the one hand, Graham (1994), suggests that changes in tropical Pacific SST and convection are important in forcing the observed decadal-scale variations in atmospheric circulation at midlatitudes, even though the midlatitude SST anomalies induced by the anomalous atmospheric circulation may provide further positive feedback (Lau, 1997). Knutson and Manabe (1997) find that low-latitude oceanic mechanisms drive both the conventional El Niño and bidecadal variations of tropical and subtropical Pacific SST in atmosphere-ocean model simulations. On the other hand, Latif and Barnett (1996) find strong evidence that the interdecadal variations of SST in midlatitudes are a result of ocean-atmosphere interactions involving the subtropical ocean gyre and the Aleutian low, without an important contribution from tropical SST anomalies. Our results lend support to the latter view to the extent that our first two LF SST EOFS do not include important variations in equatorial SST. However tropical SST anomalies on interannual time scales may provide a key triggering role through their influence on the extratropical atmospheric circulation (Trenberth and Hurrell, 1994). In addition, Gu and Philander (1997) have proposed that subduction and equatorward transport of subtropical Pacific surface waters may provide a mechanism for oceanic fluctuations on interdecadal time scales, along with interactions between the tropical Pacific Ocean and the tropical and extratropical atmosphere. Our LF EOF3 certainly indicates the possibility of a more active interaction between the extratropical and tropical ocean in the Pacific. On the interannual ENSO time scales, the extratropical parts of the SST patterns like Fig. 4.7a (HF EOF1) are often regarded as largely a response to tropical forcing. Relative to its equatorial weights, the extratropical weights of LF EOF3 are more than three times stronger than for HF EOF1. It is not completely clear whether this represents a more active tropical-extratropical Pacific Ocean interaction on bidecadal and longer time scales, or merely a coincidence of time scales of SST variation of reverse phase.

Another approach (Currie, 1996; White et al., 1997) has been to ascribe bidecadal climate variability to lunar tidal or "Hale Cycle" solar influences. A hypothesis involving the lunar 18-year tidal time scale or the 22-year Hale Cycle, however, does not explain the greater importance of bidecadal variability in the Pacific than in the Atlantic. Furthermore, if the dominant frequency bands of ocean-atmosphere interactions overlap with the frequencies of external (lunar or solar) forcings, the origins of bidecadal climate variations are likely to be misinterpreted in the absence of thorough, physically based, sensitivity studies. However solar influences may be enhanced through forced stratospheric ozone fluctuations (Haigh, 1996). So the energetics of hypothesized triggering by lunar tidal and solar influences still need to be assessed. White et al. (1997) claim that the magnitudes and phases of world-wide decadal and bidecadal variations in SST are consistent with those of 11-year cycle and Hale-Cycle solar forcing respectively, but they do not explain the greater bidecadal variability in the Pacific and greater decadal variability in the Atlantic, even though this is evident in their Fig. 6

It is likely that there are several "quasi bidecadal" phenomena. Identifying the several physical processes likely to exist on these time scales requires the use of coupled models, atmospheric models and observations. Our results indicate that more than one near bidecadal to multi decadal pattern of SST is likely to have a strong influence on the best known near bidecadal variations of climate, those of Southern African rainfall. Considering that rainfall tends to be strongly affected by chaotic internal atmospheric processes, we believe our results are noteworthy. Thus much more analysis of LF rainfall (e.g. Eastern Australia, Fig. 4.8; western USA, not shown) and temperature (e.g. New Zealand) in key regions influenced on decadal to multidecadal time scales is needed. We plan this in tandem with model simulations for the last century using the more realistic pre-1950 GISST data sets now available (e.g. Rayner et al, 1997 and personal communication) and new analyses of worldwide surface pressure since the 1870s (Basnett and Parker, 1997).

## 4.7   Appendix: Enhancement Of EOF Time Series For Missing Data

Consider a fixed spatial pattern of eigenvector weights $A_k$. Let there be m weights with the weight at a point $i$ being $A_{ki}$. If the field of data $X_{it}$ is complete, the true time coefficient is $Z_{kt}$ such that

$$Z_{kt} = \sum_{i=1}^{m} A_{ki} X_{it}. \qquad (4.1)$$

But if some data are missing, we cannot calculate $Z_{kt}$ as in equation (4.1) because we do not know all the $X_{it}$. The usual method is to ignore this problem and calculate $Z^{*}_{kt}$ from the available data $i^{*}$. Writing $Z'_{kt}$ as the contribution from the missing data that we wish to estimate then

$$Z_{kt} = Z'_{kt} + Z^{*}_{kt} \qquad (4.2)$$

where

$$Z^{*}_{kt} = \sum_{i^{*}} A_{ki^{*}} X_{i^{*}t} \qquad (4.3)$$

and

$$Z'_{kt} = \sum_{i'} A_{ki'} X_{i't} \qquad (4.4)$$

Because missing data points progressively increase in amount in earlier times, it

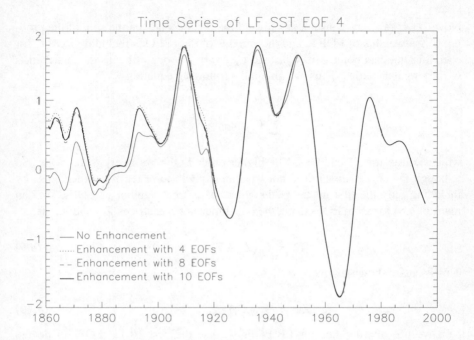

**Fig. 4.11** Time series of LF EOF4 before and after enhancement, 1861-1995.

would be expected from (4.1)a that the variance of an EOF time series $Z^*_{kt}$ would tend to reduce systematically, though details would depend on the distribution of missing data. We now make nearly unbiased estimates of the influence of missing data from the time series of the leading eigenvectors. We make these estimates from a function of the spatial weights at missing data points as follows. The actual but unknown value of a missing observation at time t is

$$X_{i't} = \sum_{k=1}^{m} A_{i'k} Z_{kt} \tag{4.5}$$

Substitute (4.5) into (4.4)

$$Z'_{kt} = \sum_{i'} A_{ki'} X_{i't} = \sum_{i'} A_{ki'} \sum_{j=1}^{m} A_{i'j} Z_{jt} \tag{4.6}$$

Substituting for $Z_{jt}$ from (4.2) with a small amount of manipulation gives

$$Z'_{kt} = \sum_{j=1}^{m} \left( \sum_{i'} A_{ki'} A_{i'j} \right) (Z'_{jt} + Z^*_{jt}) \qquad (4.7)$$

Now $\sum A_{ki'} A_{i'j}$ is a constant at a given time that depends on the sum of the product of the weights of EOF k, and the weights of the m EOFs including EOF k, for each missing data point. This constant, $C_{kjt}$, varies through time as the distribution of missing data varies. Thus, we obtain the following equation,

$$Z'_{kt} = \sum_{j=1}^{m} C_{kjt} (Z'_{jt} + Z^*_{jt}) \qquad (4.8)$$

which is true for all m EOFs. The higher order EOFs are likely to be noise. So although the above model does not have an explicit noise term we filter noise by including only the first m <M EOFs where M is total number of EOFs. We can rewrite (4.8) for all m EOFs using the one dimensional matrices $\mathbf{Z'}_t$ and $\mathbf{Z^*_t}$ as:

$$\mathbf{Z'}_t = \mathbf{C}_t (\mathbf{Z'}_t + \mathbf{Z^*_t}). \qquad (4.9)$$

This is easily shown to give:

$$\mathbf{Z'}_t = (\mathbf{I}_t - \mathbf{C}_t)^{-1} \mathbf{Z^*_t} \qquad (4.10)$$

here we use m=10. Thus for LF EOFs we use the first 10 LF EOFs to do the enhancement and for HF EOFs we use the first 10 HF EOFS. Fig. 4.11 shows time series of LF SST EOF4 before and after enhancement (m=10) and for m=4 and 8. The increase in signal magnitude before 1900, and a change in the First World War, is clear. An additional problem with LF EOFs is that the enhanced EOF time series is affected by the timing of data gaps. This creates a small but discernible amount of HF noise in the reconstructions. This is smoothed by reapplying the 13 year low pass filter to the initial version of the reconstructed time series.

**Acknowledgements**.The rainfall data set provided by M. Hulme (Climatic Research Unit, University of East Anglia), was developed with the support of the U.K. Department of the Environment under Contract No. PECD/7/10/198. This research was supported by the EC Environment Research Programme under contract EV5V-CT94-0538

# 5 The Indian Summer Monsoon and its Variability

JULIA SLINGO
*Centre for Global Atmospheric Modelling,*
*Department of Meteorology, University of Reading, Reading, UK.*

## 5.1 Introduction

The Indian Summer Monsoon is part of a larger scale circulation pattern (Fig. 5.1), known as the Asian Summer Monsoon, which develops in response to the large thermal gradients between the warm Asian continent to the north and the cooler Indian Ocean to the south. The strong south westerly flow in the lower troposphere (Fig. 5.1, upper panel) brings a substantial supply of moisture into India which is released as precipitation primarily along the Western Ghats of India and over the Bay of Bengal (Fig. 5.2). During June to September, the rainfall associated with the monsoon provides the main source of fresh water for millions of people in India. The influence of the monsoon also extends to many regions remote from India. For example, the arrival of dry summer weather over Turkey and surrounding areas can be related to the development of the Asian Summer Monsoon (Rodwell and Hoskins 1996).

Societies world-wide rely on a stable climate and nowhere more so than in those countries affected by the monsoon where the failure or even the delay of the monsoon can make all the difference between famine and plenty. Therefore the human implications of improved prediction for sub-seasonal, seasonal and climate timescales are enormous. The countries influenced by the monsoon have predominantly agrarian economies which are very sensitive to the weather and possible changes in the climate. The economies around the world are now so closely linked that the impact of a failed monsoon may be felt worldwide. The ability to understand and predict variations in seasonal monsoon circulations is thus of paramount importance not only for the countries directly affected by the monsoon but also for those remote from it. It is becoming increasingly evident that potential exists for providing seasonal predictions for countries influenced by the Asian Summer Monsoon. The benefits of seasonal and climate prediction for the societies and economies in tropical regions are enormous. Consequently considerable research has been conducted in recent years towards understanding the factors that give rise to the observed interannual variability of the monsoon and towards improving our capabilities to predict it.

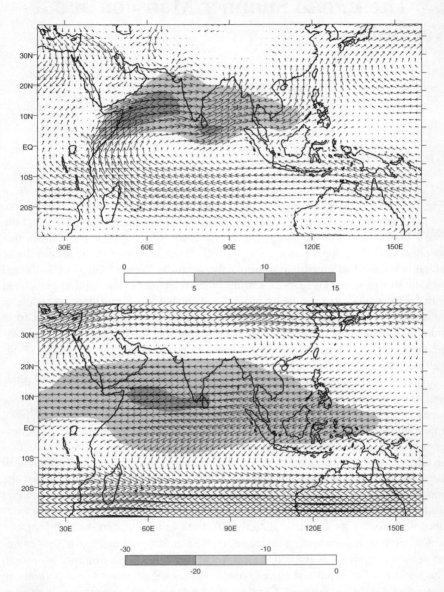

**Fig. 5.1** Climatologies of the seasonal mean winds (June - September) at 850hPa (upper panel) and 200hPa (lower panel). The data are from ECMWF Reanalyses and Analyses for 1979-95. In the upper panel westerlies are shaded; easterlies are shaded in the lower panel.

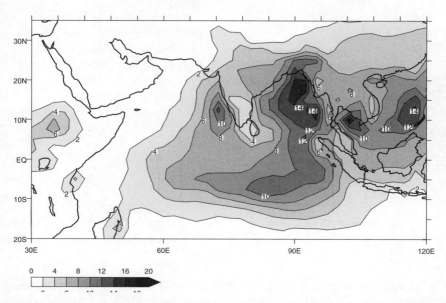

**Fig. 5.2** Climatology of the seasonal mean rainfall (June - September) from the data of Xie and Arkin (1996) for 1979-95. Contours are drawn every 2 mm/day

## 5.2    Mean behavior of the Indian Summer Monsoon

Since it depends primarily on the seasonal cycle in the solar heating of the Asian continent, the Indian Summer Monsoon shows remarkable reproducibility from year to year in its evolution, which can be characterized in terms of its onset, established and retreat phases (e.g. Soman and Krishna Kumar 1993). During May, the cross-equatorial  Somali jet develops in the equatorial western Indian Ocean and the lower tropospheric south westerly flow associated with the monsoon is established over South East Asia, with the development of a trough to the east of the Indian sub-continent. The onset of the Indian Summer Monsoon in terms of rainfall occurs over the Bay of Bengal in late May, and then over India around 1 June. Once the monsoon is established in June, the strong low level flow over India is maintained through July and August. In the upper troposphere (Fig. 5.1, lower panel), the subtropical westerly jet has retreated poleward and the monsoon anticyclone is established north of India with the strong monsoon easterlies over the equatorial Indian Ocean. During August and September the penetration of the south westerly monsoon flow into China ceases with the rapid retreat of the monsoon from the main Asian continent. The retreat of the monsoon from India itself occurs typically during September with the gradual equatorwards movement and deceleration of the low level westerly flow. The mean seasonal cycle in All India Rainfall  (Fig. 5.3) shows the rapid increase in rainfall during May and June with over half the sea-

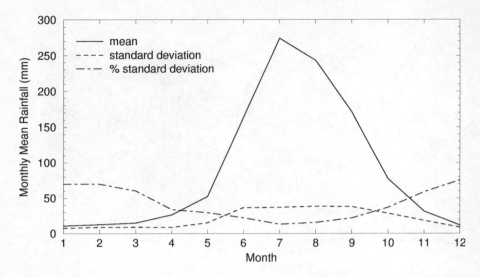

**Fig. 5.3** Seasonal variation of the monthly mean All India Rainfall (mm) and its interannual standard deviation. The standard deviation is expressed in terms of absolute rainfall amount and as a percentage of the monthly mean rainfall. All India Rainfall data from 1871-1994 are used from Parthasarathy et al. (1994,1995)

sonal mean (June-September) accumulation occurring during the established phase of the monsoon, i.e. July and August.

## 5.3 Interannual to decadal variability and its relationship with El Niño

The stability of the monsoon system from year to year is remarkable, as demonstrated in Fig. 5.3 where the interannual standard deviation is shown both in terms of actual rainfall amount and as a percentage of the monthly mean total rainfall. During the peak rainfall months of July and August, the standard deviation is less than 20% of the total. During the onset and retreat phases it rises to more than 30%, suggesting that the behavior of the monsoon may be more variable during the early and late parts of the season.

Although Fig. 5.3 shows that the interannual variations in All India Rainfall (AIR) are not large, nevertheless these can have profound social and economic consequences for the people of India (e.g. Pant and Rupa Kumar, 1997). Fig. 5.4(a) shows the All India Rainfall seasonal mean (June to September) anomalies for 1871 to 1994 based on the data provided by Parthasarathy et al. (1994, 1995). The long term mean is 853mm with a standard deviation of 84mm. Clearly there are

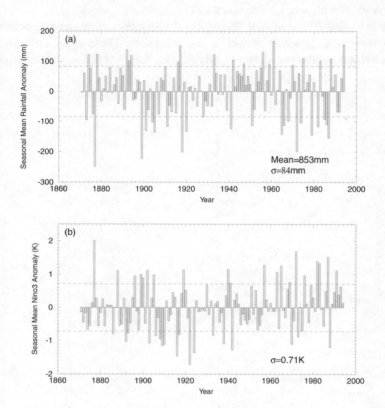

**Fig. 5.4** Bar charts of the seasonal mean (June - September) anomaly in (a) All Indian Rainfall (mm), and (b) Nino3 Sea Surface Temperature (SST; K). The dashed horizontal lines show +/- 1 standard deviation ($\sigma$). The All India Rainfall data are from Parthasarathy et al. (1994, 1995); the Nino3 SST anomalies are from Global sea-Ice and Sea Surface Temperature (GISST) dataset, Version 2.3 (Parker et al. 1995)

several years where the rainfall is above or below normal by more than one standard deviation, leading to widespread floods and droughts.

The possible influence of El Niño and the Southern Oscillation (ENSO) on Indian Summer Monsoon rainfall has long been recognized (e.g. Walker 1923, 1924), and has been studied extensively (e.g. Rasmusson and Carpenter 1983, Webster and Yang 1992). The record of seasonal mean (June to September) Niño3 Sea Surface Temperature (SST) anomalies for 1871 to 1994 is shown in Fig. 5.4b for comparison with the AIR. (The Niño3 region (50N-50S, 90W-150W) is frequently used to describe the state of El Niño). The correlation between the two time series is -0.57 confirming that El Niño significantly influences the Indian Summer Monsoon. In essence, drought years over India are often, but not exclusively, related to warm SST anomalies in the equatorial central and East Pacific (El Niño), and wet years with anomalously cold SSTs (La Niña).

This extended record of AIR (Fig. 5.4a) also displays multi-decadal behavior in which there is a clustering of wet or dry anomalies. These epochs of above and below normal rainfall are shown clearly in Fig. 5.5a where the timeseries of both

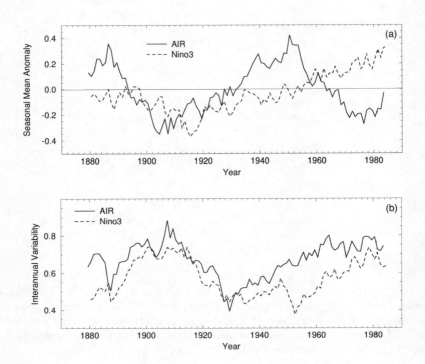

**Fig. 5.5** Low frequency, decadal timescale variations in (a) the seasonal mean hstav431_1981.medmex10.Z All India Rainfall and Niño3 SST anomalies, and (b) their interannual variability. The low frequency behavior has been identified by applying a 20-year running mean. This is applied in (a) to the time series of the seasonal mean anomalies shown in Fig. 5.4 and in (b) to the square root of the variance of the anomalies. In both (a) and (b) the All India Rainfall has been scaled by 0.01 to facilitate comparison with the Niño3 SST anomalies.

the AIR and Niño3 SST anomalies have been smoothed by a 20-year running mean to isolate low frequency behavior. It is clear from these results that the epochal behavior of the Indian Summer Monsoon rainfall is not associated with similar variations in El Niño, i.e. that dry epochs are not associated with a tendency for a clustering of warm El Niño events. Indeed the Niño3 record (Fig. 5.5a) shows mainly a century timescale warming which is not reflected in the AIR record.

In addition to wet and dry epochs, the AIR record (Fig. 5.4(a)) also shows more interannual variability in some decades than in others. For example, during the 1920's the AIR was very stable. Again the epochal behavior of the variability in AIR can be isolated by applying a 20-year running mean to the variance timeseries (Fig. 5.5b). Comparison of Fig. 5.5a and b suggest that during dry epochs the AIR is more variable than in wet epochs. The activity of El Niño also varies on the multidecadal timescale (Fig. 5.5b) similar to the AIR, with less variability between 1920 and 1960, suggesting that the level of interannual variability in AIR and Niño3 may be linked.

FromFig. 5.4 and Fig. 5.5 it can be concluded that teleconnections between El Niño and AIR operate primarily on the interannual timescale. In contrast, the epochal behavior of Indian Summer Monsoon rainfall, on which the interannual variability is superimposed, does not appear to be related to the behavior of El Niño. The mechanisms involved in this decadal variability are not understood, but the influence of decadal timescale fluctuations in SST, such as those described by Zhang et al. (1997) and which possibly occurred in the tropical Pacific Ocean during the 1990's, cannot be ruled out.

## 5.4   Understanding the factors that influence monsoon interannual variability

Although weak monsoons are frequently associated with El Niño, the correspondence between the strength of the rainfall anomaly and the strength of El Niño is relatively weak. The influence of factors other than El Niño is evident in the spectrum of AIR (Fig. 5.6) which, when compared with that for the Niño3 SST anomalies, shows that the Indian Summer Monsoon has more variability at higher frequencies associated with the quasi-biennial timescale of 2-3 years rather than at the lower frequencies (3-6 years) associated with El Niño. The dominance of the biennial timescale in AIR is well known (e.g. Annamalai 1995). Indeed the Tropical Biennial Oscillation (TBO) is gradually emerging as an important mode of interannual variability, with the coupled tropical ocean-atmosphere system in the Pacific displaying a marked quasi-biennial signal. There is no doubt that a strong link exists between the strength of the monsoon and the Pacific trades and this relationship has been used to explain the TBO (e.g. Meehl 1993). In addition, interactions with the extratropics may contribute significantly to biennial variability. A recent study by Tomita and Yasunari (1996) has pointed to the potential role played by the North East Winter Monsoon in setting up the TBO. They identified the

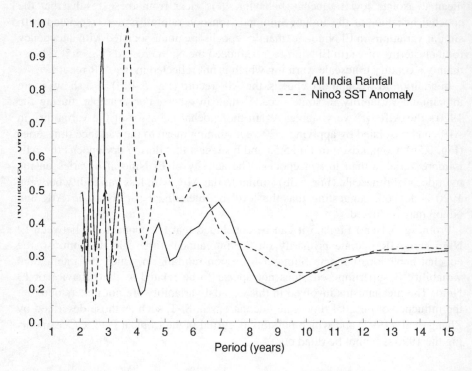

**Fig. 5.6**   Spectra of the seasonal mean (June-September) All India Rainfall and Nino3 SST anomalies for 1871-1994. The spectra have been smoothed with a Tukey window.

South China Sea as a key area through its response to the strength of the winter monsoon and its subsequent influence on the seasonal migration northwards of the West Pacific tropical convective maximum which heralds the onset of the Asian Summer Monsoon. Independently, Ju and Slingo (1995) and Soman and Slingo (1997) identified the South China Sea and the West Pacific as having a substantial influence on the date of onset and overall strength of the Indian Summer Monsoon.

   The teleconnection pattern between seasonal mean (June to September) AIR anomalies and tropical SSTs for the 45-year period 1949-1994 (Fig. 5.7a) is dominated by negative correlations in the East Pacific and positive correlations in the West Pacific. The positive correlations in the West Pacific tend to be associated with SST anomalies of the opposite sign which develop as a complementary pattern during the mature phase of El Niño. Again the teleconnection pattern shown in Fig. 5.7a confirms the important influence of El Niño on the Indian Summer Monsoon. The mechanisms through which this teleconnection operates has been the subject of considerable research. It is generally assumed that modification of the Walker circulation through the longitudinal displacement of the West Pacific tropical convective maximum from the warm pool into the central Pacific is the dominant process (e.g. Palmer et al. 1992). However Ju and Slingo (1995) suggested

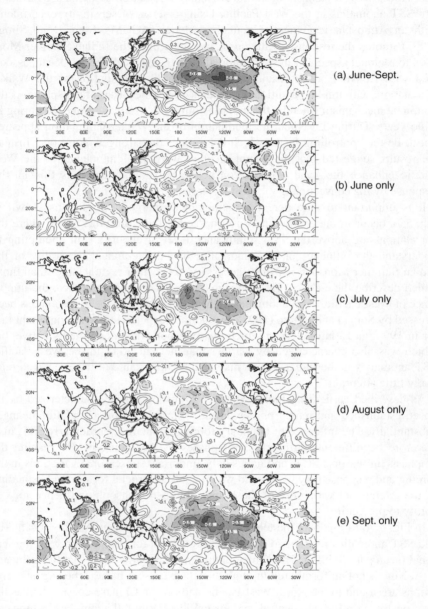

**Fig. 5.7** Zero lag correlations between AIR and SST anomalies for 1949-93 for (a) seasonal mean (June-September), (b) June only, (c) July only, (d) August only, and (e) September only. The SST data are from the GISST dataset, Version 1.1. The contour interval is 0.1 and correlations of less than -0.2 are shaded. No zero contour is drawn.

that the modulation of the West Pacific convection might be associated with either changes in the Walker circulation through the influence of the East Pacific SST anomalies, or with changes in the local Hadley circulation associated with the in situ SST anomalies in the West Pacific. Using a series of sensitivity experiments with an Atmospheric General Circulation Model (AGCM), Soman and Slingo (1997) studied the response of the onset and strength of the Indian Summer Monsoon to regional aspects of the SST anomalies in the tropical Pacific Ocean, associated with El Niño. Their results suggested that the modulation of the Walker Circulation, with implied additional subsidence over the eastern hemisphere, is the dominant mechanism whereby the Indian Summer Monsoon is weakened during El Niño years. During La Niña, the modulation of the Walker Circulation appeared not to be the controlling factor which determines the stronger monsoons. Instead the results suggested that the complementary warm SST anomalies in the West Pacific enhance the local tropical convective maximum, and it is this in situ response by the convection that leads to a stronger monsoon.

It is important to note here that, although the largest correlations shown in Fig. 5.7a occur in the central and Eastern Pacific, there are other areas where the correlations are above the 95% significance level of about 0.3. In attempting to understand the influence of SST anomalies in the interannual variability of the Indian Summer Monsoon it will be necessary to consider regions other than those which describe the classical El Niño signal (e.g. Niño3). For instance, the importance of West Pacific warm anomalies in the development of the monsoon has been stressed by Soman and Slingo (1997); indeed they appeared to be an important factor in 1994, an El Niño year in which the monsoon was unexpectedly active, but which was also characterized by warmer than normal SSTs in the West Pacific. SST anomalies in the Indian Ocean may also be important but have received relatively little attention.

As described earlier, the Indian Summer Monsoon can be characterized by its onset, established and retreat phases during which the global circulation changes substantially. The influence of boundary forcing anomalies, particularly SST, may therefore be different during different phases of the monsoon. By placing the emphasis on seasonal mean anomalies only, important subseasonal variations in the forcing and response may be masked which could be crucial for our understanding of the interannual variability of the Indian Summer Monsoon and hence for our ability to predict it.

In Fig. 5.7b to Fig. 5.7e the teleconnection patterns between monthly mean AIR and SST anomalies for June to September are shown for comparison with the seasonal mean pattern inFig. 5.7a. The monthly mean patterns show that the influence of El Niño is not uniformly felt through the whole season. In June the largest correlations are found in the North West Pacific and off the Chinese coast, whereas the relationship with El Niño is weak and insignificant. Rainfall anomalies in June can be influenced by the timing of the onset and it is interesting to note that Joseph et al. (1994) identified significant correlations between the date of onset and the SSTs over the North West Pacific. Ju and Slingo (1995) suggested that a shift in the latitudinal position of the West Pacific tropical convective maximum in response to the SST anomalies in the North West Pacific might be an important factor in explaining the teleconnection between the onset date and SSTs. Although no relationship has been found between the date of onset and the seasonal mean rainfall anomaly, nevertheless the ability to predict the timing as well as the overall strength of the monsoon is important. For example, in 1995 the monsoon onset over India was later than normal by about one week and its subsequent progression northwards was slower than usual. Combined with an unprecedented heatwave, this led to severe water and electricity shortages over much of India and considerable hardship for many people.

During July (Fig. 5.7c), the influence of El Niño is seen more strongly but it again appears to wane in August (Fig. 5.7d). In neither June nor August does the absolute correlation coefficient exceed the 95% significance level of 0.3 over substantial areas of the Pacific and Indian Oceans. Even in July the maximum absolute correlation barely exceeds 0.5, suggesting that the established phase of the Indian Summer Monsoon is less susceptible to the influence of SST anomalies. It is only in September that the strong correlation with El Niño is seen; indeed this month would appear therefore to dominate the seasonal mean teleconnection pattern (Fig. 5.7a). The importance of the retreat phase of the monsoon in determining the seasonal mean anomalies is well known (e.g. Rupa Kumar et al. 1992). The teleconnection patterns suggest that it is during this part of the season that the monsoon system is most fragile and likely to be perturbed by the remote effects of El Niño; the established monsoon, on the other hand, is relatively robust and not so easily perturbed by changes in the slowly varying boundary conditions.

As well as SST, other slowly varying boundary conditions, such as soil moisture, vegetation and Eurasian snow cover may influence the behavior of the monsoon. The Indian Summer Monsoon is primarily maintained by the land-sea temperature contrast, so that an anomalous increase in snow cover over the Tibetan plateau may reduce the land-sea temperature contrast and weaken the monsoon circulation. This was the basic mechanism proposed as early as last century (Blandford 1884) to explain the interannual variability of the monsoon. Observational studies have suggested a significant correlation between monsoon rainfall and Himalayan snow cover (e.g. Dickson 1984). More recent studies have raised the possibility that enhanced Eurasian snow amounts may themselves be part of a remote response to the warm phase of El Niño in which there is a systematic equatorwards shift in the Asian subtropical jet in the winter and spring preceding the monsoon (e.g. Webster and Yang 1992, Soman and Slingo 1997).

## 5.5    Subseasonal behaviour and its relevance to interannual variability

The teleconnection patterns shown in Fig. 5.7 suggest that the slowly varying boundary conditions associated with SST do not exert a constant influence on the monsoon. This is confirmed by considering the correlations between consecutive months of the monthly AIR anomalies (Table 5.1). Taking the complete record for 1871 to 1994, Table 5.1 shows that the correlations are very small between the months of June, July and August when the monsoon is in its established phase. This analysis can be repeated taking only those years when firstly, the forcing by El Niño is strong (defined in terms of the Niño3 SST anomaly exceeding 1 standard deviation), and secondly when there is an extreme monsoon (defined in terms of the AIR anomaly exceeding 1 standard deviation). Again the correlations between the rainfall anomalies in June and July are very small. In the case where the forcing by El Niño is strong the consistency between the months remains low suggesting that the influence of El Niño is not the dominant factor in determining the magnitude of the monthly mean anomalies. However, in extreme monsoon years, the correlation between August and September rainfall is high and probably significant. This implies that the latter part of the monsoon season, including the retreat phase, may be the most important for determining the seasonal mean rainfall anomaly in extreme years as suggested by Rupa Kumar et al. (1992). It is interesting to note that the correlations in Table 5.1 are nearly all positive, albeit at times small, suggesting that there is consistency between months in the sign, if not the magnitude, of the rainfall anomaly.

|  | (a)<br>All Years | (b) Strong<br>El Niño/La Niña<br>Years | (c)<br>Extreme Monsson<br>Years |
|---|---|---|---|
| May with June | 0.30 | 0.46 | 0.46 |
| June with July | 0.00 | 0.03 | 0.13 |
| July with August | 0.11 | 0.32 | 0.44 |
| August with September | 0.25 | 0.29 | 0.69 |

**Table 5.1** Correlations between monthly mean All India Rainfall anomalies for (a) all years (1871-1994), (b) 40 years with strong tropical Pacific forcing (El Niño or La Niña), and (c) 38 years with extreme seasonal mean monsoon rainfall. Note that only 17 years were common between (b) and (c).

The lack of correlation between the monthly mean rainfall anomalies during the established phase of the Indian Summer Monsoon is not surprising since the monsoon exhibits strong intraseasonal variability associated with a variety of phenomena. Monsoon depressions, which originate in the Bay of Bengal and move west/northwest, have a frequency of between 5 and 20 days, whilst north-south shifts in the location of the monsoon trough, described by Gadgil et al. (1992), have a lower frequency of between 30 and 60 days. The monsoon trough itself shows bi-weekly variations in intensity associated with westward moving synoptic scale disturbances (Krishnamurti and Bhalme, 1976). In addition, the rainfall over western India has been correlated with the strength of the low level Somali Jet (Findlater 1969) which itself displays considerable variability and may be linked to interactions with the mid-latitudes of the southern hemisphere (Rodwell 1997). During the established phase of the monsoon the circulation often undergoes significant variations with a pronounced northward excursion of the monsoon trough which brings the monsoon into an inactive or break phase over India.

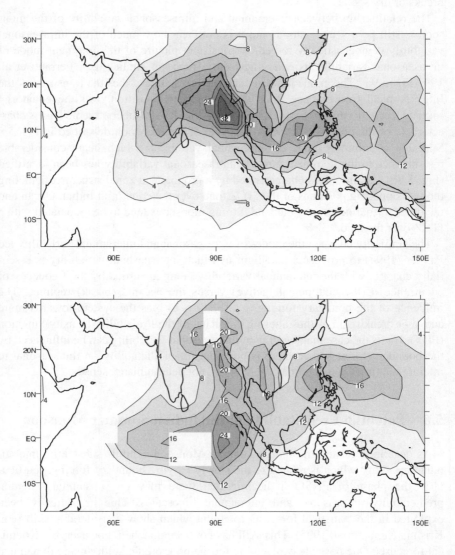

**Fig. 5.8** Composite precipitation distributions (mm/day) for active (upper panel) and break (lowe panel) phases of the Indian Summer Monsoon, created from pentad precipitation data for 1986-1991 provided by the Global Precipitation Index (GPI, Janowiak and Arkin 1991). The contour interval is 4 mm/day and white squares denote missing data.

Fig. 5.8 shows typical rainfall distributions associated with active and break phases of the Indian Summer Monsoon. The difference in rainfall over the Indian subcontinent is sufficiently large that these active/break cycles may affect the monthly or seasonal means for any particular year. The major difference between strong and weak monsoon seasons appears to be in the duration and intensity of the breaks or dry spells.

The relationship between interannual and intraseasonal variability of the monsoon is still poorly understood, and it is not clear how much of the intraseasonal variability is stochastically forced. A consistent picture of the dominant mode of intraseasonal variability is emerging from a range of studies (e.g. Ferranti et al. 1997;Fig. 5.9a . This mode has a zonal structure and describes the transition of the Inter Tropical Convergence Zone (ITCZ) from its position over the equatorial Indian Ocean, northwards over S.E. Asia and the Indian subcontinent. This is characteristic of the observed active/break cycle of the monsoon described in Fig. 5.8 Similarly a common mode of interannual variability which bears considerable resemblance to the dominant mode of intraseasonal variability has been identified (Fig. 5.9b). Again, this is characterized by a pronounced zonal structure, depicting the two opposing positions of the ITCZ, one over the equatorial Indian Ocean and the other further north over the land. Strong monsoons tend to be associated with a ITCZ over the land.

The similarity between the modes of intraseasonal and interannual variability led Palmer (1994) to propose a paradigm in which intraseasonal variability is essentially chaotic, with the interannual variability being governed by the frequency of occurrence of the continental (active) versus the oceanic (break) regimes. The influence of the boundary forcing (e.g. SST) is to bias the system towards more active or break regimes thus altering the shape of the Probability Density Function (PDF). If it is the case that the PDF of intraseasonal variability can be influenced by the boundary forcing, then there is potential for predictability on the seasonal to interannual timescale, but in a probabilistic, not deterministic, sense.

## 5.6    Potential Predictability of the Indian Summer Monsoon

As described earlier, the Indian Summer Monsoon exhibits substantial interannual variability which is related to the slowly varying boundary forcing, particularly the phase of ENSO, indicating that there may be substantial potential predictability at seasonal and interannual timescales. This has already been exploited in the statistical forecasts for India which show considerable skill (e.g. Krishna Kumar et al. 1995). This skill has not been matched, however, by dynamical forecasts. One possible explanation for the poor predictability using dynamical methods is that model errors still dominate. One of the future major challenges will be to understand why numerical models show so little skill and to identify those key physical processes that are crucial for improving model performance. This means developing our understanding of the mechanisms involved in the mean evo-

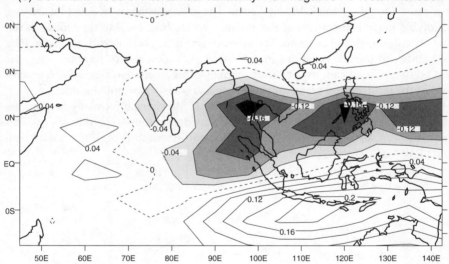

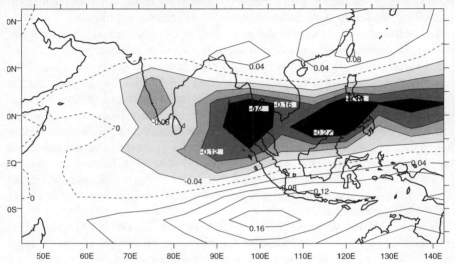

**Fig. 5.9** Dominant EOFs of (a) daily mean Outgoing Longwave Radiation (OLR; Wm$^{-2}$) for May to September, and (b) seasonal mean (May-September) OLR, from a 4-member ensemble of 45-year integrations with the Hadley Centre climate model (Version HADAM2a) forced with observed SSTs from GISST dataset, version 1.1. (a) describes the dominant interannual mode which explains 21% of the interannual variance. (b) describes the dominant intraseasonal mode which explains 7% of the intraseasonal variance. Negative (shaded) OLR anomalies correspond to enhanced convective activity.

lution of the monsoon and its spatial and temporal variability through diagnostic studies of observations and numerical weather prediction analyses and reanalyses, as well as experimentation with a hierarchy of models ranging from simple hypothesis testing models to state-of-the-art atmospheric and coupled models.

The potential predictability of the monsoon may also be limited by its high frequency, intraseasonal behavior. The relationship between interannual and intraseasonal variability of the monsoon is still poorly understood, and it is not clear how much of the intraseasonal variability is stochastically forced. The hope is that the low frequency boundary forcing influences the intraseasonal variability in a predictable way. Model results have shown that the dominant modes of interannual and intraseasonal variability do indeed have a similar structure, but no relationship has so far been found between the temporal behavior of the intraseasonal mode and the seasonal mean anomaly. A major goal is therefore to understand the mechanisms involved in monsoon intraseasonal variability, to improve its simulation and to determine how much intraseasonal variability limits the potential predictability at seasonal to interannual timescales.

In May 1994 the World Climate Research Programme (WCRP) organized an international conference on "Monsoon Variability and Prediction" in Trieste, Italy. The challenge to the modeler represented by monsoons could not be more succinctly put than in the introduction to the Conference Proceedings (WCRP 1994):

Understanding the physical processes that determine the monsoon phenomenon has proved to be a remarkable challenge. The processes are subtle and interactive, involving many components of the climate system, both local to the monsoon and remote from it. However, more than the pure scientific challenge of understanding, our ability to predict variations in monsoons, on timescales of days to years, has extraordinary implications for the economies of tropical countries, and represents one of the most fruitful areas where science and society can, and do, interact.

**Acknowledgments**: The author thanks Dr. H. Annamalai for many helpful discussions and for providing the observational datasets used in this study. This research was supported by the EC Environment Research Programme under contract EV5V-CT94-0538

# 6 Climate Variability in Northern Africa: Understanding Droughts in the Sahel and the Maghreb

M. Neil Ward
*IMGA - CNR, Via Gobetti 101, Bologna, Italy*
*and*
*Cooperative Institute for Mesoscale*
*Meteorological Studies, School of Meteorology,*
*The University of Oklahoma, Norman, Oklahoma*
*73019, USA.*

Peter J. Lamb, Diane H. Portis
*Cooperative Institute for Mesoscale*
*Meteorological Studies, School of Meteorology,*
*The University of Oklahoma, Norman, Oklahoma*
*73019, USA.*

Mostafa El Hamly, Rachid Sebbari
*Moroccan Direction de la Météorologie Nationale*
*(DMN), Casablanca, Morocco*
*and*
*Cooperative Institute for Mesoscale*
*Meteorological Studies, School of Meteorology,*
*The University of Oklahoma, Norman, Oklahoma*
*73019, USA*

## 6.1 Introduction

The climates of North Africa are still relatively poorly known and understood. Yet, just to the north and south of the Sahara desert are regions that are coming to be recognized as having climates that are among the most variable in the world. The variability is on timescales of interannual-to-decadal and beyond, with the decadal variability being among the most pronounced so far identified in the historical climate records of the 20th Century.

Africa is the largest tropical land mass with its east-west extent in the Northern Hemisphere being particularly impressive. The distance from Dakar in Senegal to Mogadishu in Somalia is about 6,000km. Continuing eastward for a similar distance would take you across the Indian Ocean and into Oceania. This size of the

North African sub-continent, along with the amplitude of its climate variability, make the region an important component of not only the climate system of surrounding regions but also probably of the global climate system itself. However, the most direct need to study North African climate is generated by the societies of North Africa which are still largely dependent on rain-fed agriculture, local surface water supplies and hydro-electric power, so that they are particularly vulnerable to climate fluctuations. Also, the human health and spread of disease in Africa is increasingly being related to climate variability. Applications of climate information and predictions will accordingly make a strong contribution to sustainable development in North Africa.

Globally, no resource can be considered to be more strategic than water. Its use is essential for every inhabitant and for a wide range of socioeconomic sector activities. It is vital for health, agriculture, industry, and hydropower. Water is also an integral part of the environment and is home for many forms of life, on which human well-being ultimately depends. Given the increasing demand for water, in terms of both quantity and range of use, precipitation variability is a more sensitive and important issue today than at any time in the past.

A basic problem in North Africa, as in many regions of the world, is inadequate knowledge of the natural and potential water resources and of the present and future water demand. The science of climate can assist by providing not only the precipitation average values but also their spatial and temporal distributions. Furthermore, such background information is also essential for the optimum application of any available precipitation climate predictions.

The North African region south of the Sahara desert is tropical to equatorial. The atmosphere at these latitudes is sensitive to small changes in surface boundary forcing (Charney and Shukla 1981), though the highest sensitivity may be confined to mainly equatorial latitudes (Stern and Miyakoda 1995). These findings are central to the postulated causes of interannual-to-decadal variability in tropical North Africa. Firstly, coupled ocean-atmosphere variations that lead to persistent sea-surface temperature (SST) anomalies have been shown to be important for tropical Africa (e.g., Rowell et al. 1995). The persistent SST anomalies are able to force atmospheric anomalies that extend across vast expanses of the tropical African continent. However, the size of the land mass suggests that changes in land surface characteristics could also represent a significant boundary forcing that could lead to persistent atmospheric anomalies at interannual and especially decadal timescales (Dirmeyer and Shukla 1996). Finally, for interannual variability, there is expected to be a component of variability that is generated by internal atmospheric dynamics (Palmer 1993). However, for longer timescales, our current understanding of internal atmospheric dynamics suggests little internal variability, such that distinct forcing mechanisms are needed to maintain decadal atmospheric anomalies.

The region north of the Sahara desert is subtropical and its climate is closely related to that of Europe. Northwestern subtropical Africa and much of Europe are influenced by the North Atlantic Oscillation (NAO, Walker and Bliss 1932; Lamb and Peppler 1987), which alters the strength of the mean westerly winds across the mid-latitude North Atlantic. The evidence for SST-forced variability in sub-tropical North Africa is weaker than in tropical Africa, though a role for North Atlantic coupled ocean-atmosphere processes and for remote SST forcing has been proposed (e.g., Grötzner et al. 1998). However, at the interannual timescale, internal atmospheric dynamics may here be a major generator of seasonal atmospheric anomalies (Palmer 1993). The source of the pronounced decadal variability in these regions is one of the most pressing questions in climate dynamics today. Specialists on water resources around the Mediterranean (both in North Africa and southern Europe) are becoming more and more convinced that the growing water scarcity and misuse of available water resources in the region are now major threats not only to sustainable agricultural development, which accounts for up to 80% of the water consumption, but also to the other water-dependent sectors mentioned earlier.

In addition to their shared NAO influence, the climates of North Africa and Europe are connected in other ways as well. It is likely that diabatic heating anomalies associated with convection over tropical North Africa induce, through atmospheric teleconnection processes, at least some atmospheric variability over Europe. A further common interest is the northward oceanic heat flux across the Equator in the Atlantic, which is at least in part associated with the global oceanic thermohaline circulation. One hypothesis suggests that variations in this heat flux generate large scale patterns of Atlantic SST anomalies which potentially can influence the climates of North Africa and Europe. Finally, the monsoon winds of tropical North Africa extend across the tropical Atlantic and clearly both respond to and effect changes in SST patterns and ocean circulation, with the potential for feedback on the climate of surrounding regions.

This chapter considers the nature and causes of the interannual-to-decadal variability exhibited by some of the climates of North Africa. Section 6.2 provides background on the data and methods used in the climate analyses. Section 6.3 considers the July-September summer monsoon season of tropical North Africa. Section 6.4 makes some observations on the transition seasons of tropical North Africa which, though less studied, bring two economically important rainy seasons to latitudes south of about $10^{\circ}$N. Section 6.5 focuses on sub-tropical northwestern Africa, and especially the Kingdom of Morocco. This region is particularly closely tied to the interannual and decadal variations of the NAO.

## 6.2    Discussion of Data and Methods

The analyses of observed interannual-to-decadal variability discussed in this chapter have used a variety of data sources. The sources are summarised here. Station reports of monthly precipitation have been processed by Hulme (1994) into a monthly grid-box format. In this chapter we show standardised area-average rainfall anomaly indices (e.g., Katz and Glantz 1986) that have been calculated from the grid-box values. For the specific studies on Morocco, monthly station precipitation totals were provided by the Moroccan Direction de la Météorologie Nationale (DMN), while for the NAO index, monthly sea-level pressure (SLP) values for Ponta Delgada (Azores) and Akureyri (Iceland) were provided by Rogers (1995, personal communication) for the 1874-95 period (values to April 1998 were obtained from the US National Climate Data Center (NCDC)).

Near-surface ocean-atmosphere parameters have been recorded by ships since the middle of the 19th Century, and these data are now digitised into computerised datasets. These include the U.K. Meteorological Office Historical SST dataset (MOHSST, Bottomley et al. 1990) and the Comprehensive Ocean-Atmosphere dataset (COADS, Woodruff et al. 1987). Though not used in this chapter, re-analysed atmospheric data, assimilating all available observations into a numerical model, are now beginning to provide a new tool with which to study interannual-to-decadal atmospheric variability in Africa. However, this will not completely remove problems associated with a lack of homogeneity in the data sources. Thus, there will still be a need for careful appraisal of decadal data sources (e.g., see Folland and Parker 1995).

To study decadal variability, it is often useful to filter climatic indices, separating the decadal variability (the low frequency, LF) from the sub-decadal (high frequency, HF) variability. The analyses reported in this chapter have used filters with a 50% cut-off amplitude set to about 10 years (see Section 6.7). The analyses are mainly performed using relatively simple correlation and Principal Component Analysis (Empirical Orthogonal Function, EOF - see Richman 1986; Jolliffe 1986) techniques. The more complex but powerful coupled pattern techniques of Singular Value Decomposition Analysis (SVDA, Bretherton et al. 1992) and Canonical Correlation Analysis (CCA, Barnett and Preisendorfer 1987) are also employed. Coupled pattern techniques are used to analyse time-sequences of fields for two different variables, seeking the coupled spatial modes with common temporal variation. For example, one set of fields may be for precipitation, while the other set of fields may be for SST. Each statistical tool should always be used along with other supporting techniques to ensure conclusions are not an artifact of any one technique. In this chapter, when the coupled tools are used for prediction experiments, all verifications are made using cross-validation, i.e., by successively removing the year to be predicted from the EOF and CCA analysis. The SVDAs are also cross-validated by forming SVDA time series values for year i by projecting data for year i onto SVDA modes that were calculated excluding year i (see Moron et al. 1998).

## 6.3    The Boreal summer season July-September: The Sahelian rainy season

On the northern side of the Sahara, the intensity of the summer subtropical high pressure system precludes almost any precipitation during July-September (JAS). At this time, to the south of the Sahara, the intertropical convergence zone (ITCZ) has reached its most northerly location. This is the single short rainy season for the

Sahelian region immediately to the south of the Sahara desert (Fig. 6.1). For regions further south (approximately 5°-10°N), boreal summer generally constitutes a relative break between the main rainy seasons of March-June and October-November/December. This is especially true for the southern parts of the Guinea Coast region marked on Fig. 6.1, since the ITCZ and associated rains are generally well to the north.

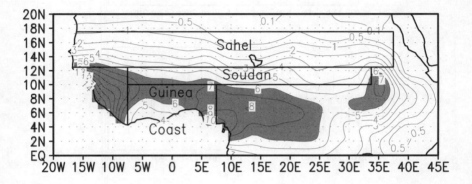

**Fig. 6.1**  July-September 1961-90 rainfall climatology (mm per day) for tropical North Africa, based on the Hulme (1994) dataset. Marked on the map are the regions for which standardised area-average rainfall indices were calculated. Contour interval is 1 mm per day, with extra contours at 0.1 and 0.5 mm per day. Region above 6 mm per day is shaded.

A number of studies have defined the large-scale rainfall anomaly regimes of tropical North Africa (Nicholson 1980; Janowiak 1988; Nicholson and Palao 1993). The Sahel, Soudan, and Guinea Coast regions in Fig. 6.1 were defined in Rowell et al. (1995) broadly following the above studies, but based additionally on a requirement for consistent large-scale associations with SST. Note that the southern region is referred to as the Guinea Coast region so as to keep terminology simple, even though in reality it extends a substantial distance inland.

The Sahel (Fig. 6.2a) and Soudan (not shown) standardised area-average rainfall indices both show strong decadal variability (Nicholson 1980, 1996; Lamb 1982). In contrast, the JAS rainfall in the Guinea Coast region has been relatively stationary over the period of historical record (Fig. 6.2b). The Sahel and Soudan indices reveal a relatively dry period early in this century, a period of maximum wetness through the 1950s, followed by an extended dry period after 1970, which has recently shown some amelioration. The contrast between the abundant Sahelian rains of the 1950s compared to the deficient rains of the 1970s and 1980s constitutes one of the most marked decadal fluctuations in climate this century. Several studies have documented the accompanying tropical Atlantic changes in surface atmospheric circulation and SST (Lamb 1978a,b; Lough 1986; Hastenrath 1984, 1990; Lamb and Peppler 1992), in particular noting a southward shift of the zone of

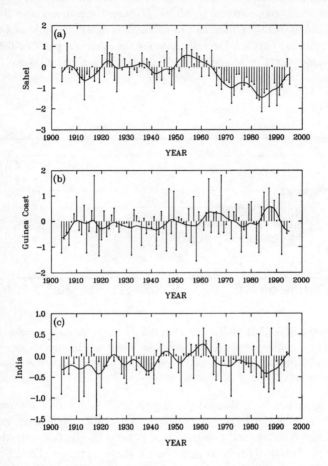

**Fig. 6.2**    Standardised July-September rainfall anomaly time series 1904-95. (a) Sahel, (b) Guinea Coast, (c) All-India. Bars give the raw individual seasonal values. The smooth line is a low-pass filter with 50% cut-off amplitude at 11.25 years. It is used to define the low-frequency (LF), while residuals from the LF for each individual season define the high-frequency (HF).

near-surface convergence and SST maximum during the Sahelian drought periods. Folland et al. (1986) showed that, in addition, a near global change in SST had accompanied the 1950s-to-1970s shift in climate in the Sahel, with the North Atlantic and North Pacific becoming colder and the South Atlantic and Indian Oceans becoming warmer in recent decades. The third EOF of global SST in Folland et al. (1991) captured especially well the global signature of the LF SST change (Fig. 6.3). Correlating the LF signal in the Sahelian index with LF SLP at each location over the global ocean shows large-scale SLP associations with the LF Sahelian rainfall changes (Fig. 6.4). The negative correlation with SLP in the northern Indian Ocean is consistent with an in-phase signature for the Sahelian and

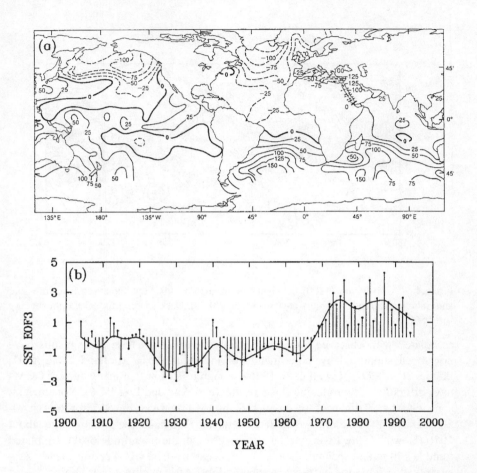

**Fig. 6.3** Third covariance eigenvector (EOF) of all-seasons sea-surface temperature anomalies for 1901-80. (a) Spatial pattern, weights are *1000. (b) July-September time-coefficients. (Taken from Folland et al. 1991).

Indian monsoons, which is supported by comparison of the LF rainfall indices (Fig. 6.2a,c). Increased cyclonicity near northwestern Europe is also suggested for the wetter Sahelian period (Fig. 6.4). Gray et al. (1997) report a range of other climate phenomena that shifted around 1970, including hurricane activity, which is directly linked to Sahelian precipitation (Landsea and Gray 1992; Goldenburg and Shapiro 1996). Gray et al. also tentatively interpret the above SST change around 1970 Fig. 6.3b) as being the result of a change in the global oceanic thermohaline circulation, though the cause of this SST change remains controversial.

A role for SST-forcing of decadal Sahelian rainfall was initially suggested by idealised General Circulation Model (GCM) experiments forced with all (Folland et al. 1986) and individual parts (Palmer 1986) of the global SST anomaly pattern

Correlation LF Sahel v LF SLP

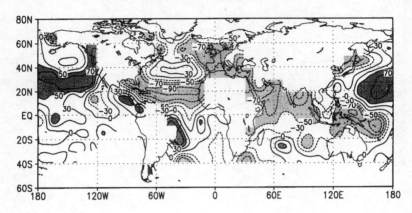

**Fig. 6.4** Correlation (*100) between low-frequency July-September sea level pressure and Sahelian rainfall. Contours at 0, +/-30, 50, 70, 90. Dark shading for >50, light shading for <-50.

associated with observed Sahelian rainfall fluctuations. These early results have now been supported by experiments for individual years (Folland et al. 1991; Palmer et al. 1992; Rowell et al. 1995), though it is clear that many current GCMs have difficulty in reproducing those results (e.g., Sud and Lau 1996). Nonetheless, some GCM experiments forced with the observed time-varying SST for the last 40-50 years have successfully simulated reduced Sahelian precipitation after about 1970 (Rowell 1996; Livezey et al. 1995). Though the magnitude of the simulated trend is still model specific, the overall message from the GCM results clearly suggests that SST-forcing has a role to play in West African climate variability.

An alternative hypothesis suggests changes in land-surface characteristics in North Africa can play a strong role in decadal climate variations. Following the work of Charney (1975) that considered the impact of changes in surface albedo on semi-arid climates, other effects of a modified land-surface have also been investigated, including changes in available moisture from the land-surface, altered land-surface roughness and changes in aerosol properties of the atmosphere, which may result from either changed land surface characteristics or changed wind patterns (reviewed in Nicholson 1988). Modifying the land-surface in GCMs that contain sophisticated land-atmosphere interaction schemes can clearly impact upon regional precipitation in tropical North Africa (Xue and Shukla 1993; Dirmeyer and Shukla 1996). Though the realism of the land-surface changes imposed is not yet firmly established, the results do suggest a possible role for the land-surface, even if the primary forcing turns out to be from large-scale SST changes associated with coupled ocean-atmosphere processes.

The SST patterns associated with sub-decadal Sahelian rainfall variability are quite distinct from the LF patterns discussed above (e.g., Rowell et al. 1995). A CCA analysis between the West African rainfall indices and global SST on the HF timescale reveals clearly the principal two modes of variation(Fig. 6.5, taken from Ward 1998). The two modes account for similar fractions of the total rainfall variance. The second CCA SST mode (Fig. 6.5b)  is mainly confined to the equatorial

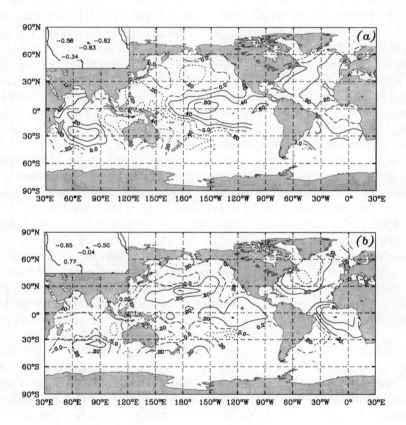

**Fig. 6.5**    Canonical correlation analysis between July-September (JAS) global SST and the four tropical North African rainfall indices (regions marked on Fig. 6.1, but additionally here with the Sahel divided into an eastern and a western region, see Ward 1998). (a) Loadings for SST mode 1 and, in the top left insert, rainfall mode 1. (b) Same as (a) but for mode 2.

and southern tropical Atlantic. Its warm phase is associated with more rain on the Guinea Coast and less rain in the Sahelian strip, with the Soudan strip being a zone of transition. Over the years 1949-88, a JAS SST index for $10°N-10°S$, $20°E-10°W$ correlates at r=-0.53 with Sahelian rainfall and r=0.68 with Guinea Coast rainfall. Janicot (1992) and Fontaine and Bigot (1993) show the tropical Atlantic atmospheric patterns for this mode, while Ward (1998) shows the patterns after removing the effects of decadal fluctuations from the results.

The first CCA SST mode (Fig. 6.5a) is clearly an expression of El Niño / Southern Oscillation (ENSO). Semazzi et al. (1988) first proposed that ENSO warm events were associated with less rain in the Sahel, based on an analysis of the years 1970-84. However, to demonstrate this result over many years, it is best to remove the decadal variability in the indices before calculating the correlation. For example, over 1950-88, the correlation between the HF Sahel index and the Niño3 SST index is -0.55, while for the Soudan region, the correlation is -0.58. The association can also be seen in the composite results of Nicholson and Kim (1997) and in the first HF SVDA mode between precipitation over North Africa-Eurasia and the tropical Pacific SST (Fig. 6.6, taken from Ward 1998). The SST mode (Fig. 6.6a) is clearly the ENSO signature. The corresponding precipitation mode is coherent between India and the eastern Sahel and ultimately to the western coast of Africa. A belt of out-of-phase rainfall anomalies is present through the Mediterranean, with an in-phase zone encompassing the UK and Scandinavia. This pattern is also supported by SLP analyses (Ward 1992); a connection between Sahelian rainfall and UK summers was first proposed in Folland et al. (1988). Furthermore, Walker and Bliss (1932) included the eastern Mediterranean in their original Southern Oscillation index and recently Rodwell and Hoskins (1996) have shown an inverse dynamical link between the intensity of the Indian monsoon and vertical motion in the eastern Mediterranean. Thus, it appears that the Indian monsoon, Sahelian rainfall, the Mediterranean atmosphere, the Mediterranean SST and mid-latitude European circulation may be interlinked during boreal summer. This interconnection is clearly often connected to tropical Pacific SST (Fig. 6.6), but analysis also shows that it can operate as well independent of ENSO (Ward 1998). The interactions appear difficult for GCMs to capture and the ability of GCMs to simulate the observed interannual Sahelian rainfall varies greatly and is generally rather poor (Sud and Lau 1996). However, the JAS Guinea Coast rainfall anomalies appear easier to simulate, perhaps because the main SST forcing is relatively local in the equatorial Atlantic, so that the GCM does not need to represent a web of remote teleconnections in order to achieve simulation skill.

## 6.4    Transition seasons in tropical North Africa

During the transition seasons, the ITCZ passes northward through the Guinea Coast region in March-June and southward during October-December. These are the two main rainy (Fig. 6.7a) and agricultural seasons for this area. Analysis of

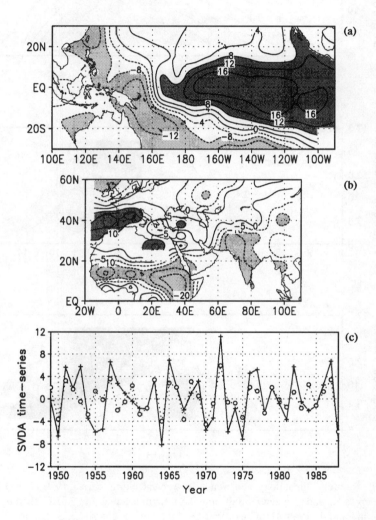

**Fig. 6.6**    Singular value decomposition analysis (SVDA) between JAS high-frequency tropical Pacific SST and JAS rainfall in the Asia-Africa-Europe sector, 1949-88. (a) SST mode 1. Pattern shown is the first singular vector with weights *100. Dark shading >8, light shading <-8. (b) Same as (a) but for rainfall mode 1, with dark shading >10, light shading <-10. (c) Cross-validated mode 1 time series for SST (solid with crosses) and rainfall (dashed with circles). The correlation is 0.76.

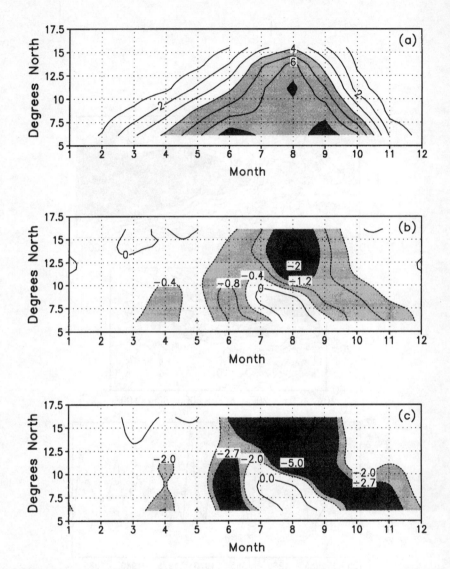

**Fig. 6.7** Rainfall climatology and trends in tropical North Africa 1950-87. (a) Month-latitude cross-section of mean rainfall (mm/day). Light shading 4 to 7, dark shading >7. (b) Change in rainfall 1969-87 minus 1950-68 (mm/day). Light shading -0.4 to -1.2, dark shading <-1.2. (c) t-value for the change in rainfall. Light shading -2.0 to -2.7, dark shading <-2.7. Approximate two-tail significance levels are: +/-2.0 = 5%, +/-2.7 = 1% (no attempt to account for serial correlation in degrees of freedom). All values are averaged over longitudes from 37.5°E to the western coast of Africa. Grid-boxes in West Africa excluded from the Soudan and Guinea Coast indices (Fig. 6.1) are also excluded here, since they are generally mountainous and with poor data coverage

individual months shows that the region of rainfall decline in recent decades over tropical North Africa tends to migrate meridionally with the annual cycle (Fig. 6.7b,c). In the transition season October-December, and also in June before the Sahelian rainy season, significant declines have occurred south from the Sahel to the belt near 5-10°N, including the Guinea Coast region. Ward (1994) shows that there have also been accompanying changes in the tropical Atlantic ocean-atmosphere system. However, attribution of cause and effect is complicated because it is possible that SST forcing and land-surface forcing could both create such a pattern of climate variation. For example, Dirmeyer and Shukla (1996) reproduce a rainfall change similar to Fig. 6.7b in GCM experiments forced with changed land-surface characteristics.

The interannual variability of the transition seasons has received less attention than that for boreal summer. One of the few relevant studies is Moron et al. (1995) who could find little evidence for an ENSO impact on the first rainy season (March-May, MAM) but some evidence for a warm equatorial Atlantic being associated with less rain in the Guinea Coast region in MAM, opposite to the correlation found for JAS, but physically plausible since the ITCZ is further south in MAM, and a warmer equatorial Atlantic could force the ITCZ anomalously south with more precipitation over the Gulf of Guinea ocean areas, at the expense of the land areas. Encouraging skill for this season has also been found in GCM experiments forced with observed SST through the period 1961-94 (Moron et al. 1998). The model mode of rainfall variation was an opposition between rainfall anomalies over the ocean and rainfall anomalies over the land. Out-of-phase variations between land and ocean in the transition seasons of West Africa can also be seen in the observed outgoing longwave radiation record (not shown), which acts as a proxy for tropical rainfall. As noted above, one way to trigger the change in preferred location for convection between ocean and land is to change the SST in the equatorial Atlantic. Mutai et al. (1998) show that an SST EOF measuring a warmer than normal equatorial and tropical South Atlantic in September is accompanied by reduced precipitation in West Africa in October-December (OND). The OND Guinea Coast rainy season also showed a tendency for less rain in warm event ENSO years (Moron et al., 1995). Further quantification and experimentation is clearly required on the transition seasons' characteristics, but the above initial findings of decadal and interannual variability suggest further research should yield results with useful applications.

The early season (June) and late season (October) rains of the Sahel, once the decadal variability is removed, have no correlation with the main July-September rains. In fact, the June and October HF Sahelian rains are positively correlated with warm phase ENSO events (Ward 1994, 1998), in contrast to the main JAS rains which tends to have reduced rainfall in ENSO warm phase.

## 6.5   Sub-tropical Northwestern Africa

The precipitation season for subtropical northwestern Africa is approximately
October-April (Fig. 6.8). During the height of the rainy season (November-Febru-
ary) most precipitation is generated as a result of depressions being steered southward during mid-latitude atmospheric blocking episodes. These episodes are
closely linked to the North Atlantic Oscillation (NAO, Walker and Bliss 1932;
Barnston and Livezey 1987; Lamb and Peppler 1987, 1991). The NAO is one of the
major atmospheric modes of variation at monthly, seasonal, interannual and decadal timescales. A simple index is the SLP difference between the Azores and Iceland, calculated here as the normalized SLP difference between Ponta Delgada
(Azores) and Akureyri (Iceland) (raw pressure data provided by Rogers, 1995, personal communication). A positive NAO phase implies strengthened westerly winds
across the mid-latitude North Atlantic, leading to mild and relatively wet winters in
northern Europe (van Loon and Rogers 1978) and anomalously dry conditions for
the Iberian peninsula (Zorita et al. 1992) and subtropical northwestern Africa,
especially in the region of Morocco (Lamb and Peppler 1987; Lamb et al. 1997a).

Since the late 1970s, there has been a tendency for the winter NAO index to be in
positive phase (Hurrell, 1995), and for precipitation in sub-tropical northwestern
Africa and the Iberian peninsula to decline. The winter of 1995-96 saw a dramatic
return to a strong negative seasonal winter NAO value, while 1996-97 and 1997-98
winters have been close to the long term normal, or slightly negative. It is too early
to tell whether this marks an end to the generally positive phase on the NAO since
the 1970s, but an ability to make such a statement would be of great practical value,
and is a motivation for ongoing research.

As a basis for the comprehensive investigation of Moroccan precipitation variability, the regionalisation in Fig. 6.9 was developed using various forms of unrotated and rotated Principal Component (see Richman 1986) and Cluster Analyses
applied to October-April station precipitation totals for 1932-95. Standardised seasonal precipitation indices for the five regions are shown in   Fig. 6.10, along with
the (October-April) NAO index for the same period. The standardised precipitation
anomaly time series for each region shows strong interannual variability and, especially since the middle of the century, strong multi-decadal variability. The correlations entered on Fig. 6.10 show that the October-April precipitation totals for
regions 1 (northwestern) and 2 (southwestern) have the strongest negative associations with the NAO when all timescales are considered (i.e., using the raw data)
and also when HF timescales are considered alone. On a monthly basis, the strongest associations are found in January and February (not shown), when up to 50%
of the interannual variability of the precipitation in northwestern Morocco is tied to
the NAO. From the standpoint of operational seasonal prediction, there is, however,
also an urgent need to better understand that part of the variance not connected to
the large-scale climate system behaviour represented by the NAO.

Furthermore, the decadal NAO change may have rather different regional atmospheric anomaly patterns associated with it, compared to the interannual NAO vari-

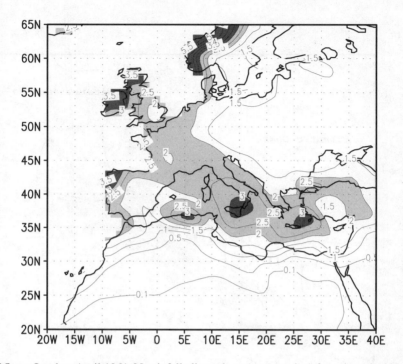

**Fig. 6.8**    October-April 1961-90 rainfall climatology (mm per day) for subtropical North Africa and Europe, based on the Hulme (1994) dataset. Contour interval is 0.5 mm per day, with extra contour at 0.1 mm per day. Light shading for regions between 2 and 3 mm per day, dark shading for >3 mm per day. Data have been interpolated across the Mediterranean Sea using available grid-boxes with island stations. Note that detailed orographic influences on precipitation, such as maxima over the mountains of Morocco, cannot be resolved by this dataset (raw resolution 2.5° lat x 3.75° long).

ations. For example Fig. 6.10 shows that at the LF timescale, the strong negative correlation between the NAO and seasonal precipitation totals extends eastward to the Atlas Mountains (Region 3) and the northeastern region (Region 5), whereas at the sub-decadal timescale (HF) there is little or no association with the northeastern region. The difference between the LF and HF correlations is also especially large for the Atlas Mountains region (Region 3) where the LF correlation is much stronger than the HF correlation, and the decline in rainfall since the late-1970s is most pronounced, representing a major loss of the entire country's water resource. The decline has been especially characteristic of December, an important "dam-filling" month.

   The decadal NAO swing appears to have a different spatial expression in the different calendar months. Fig. 6.11a,b show the composite SLP and precipitation pattern for December over the North Atlantic, subtropical North Africa, Western Europe and extending eastward as far as 50°E. The recent period (1979-93), rela-

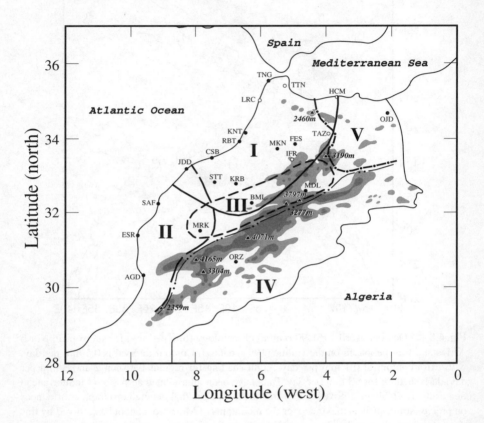

**Fig. 6.9**    Regionalisation of October-April Moroccan precipitation 1932-95. Based on 17 stations with data from 1932 to present (solid dots) and 5 additional stations with data from 1964 to present (open circles). A station may belong to more than one region. Shading indicates Moroccan topography: light shading for 1500-2000m, dark shading for >2000m

tive to the preceding 1961-78 period, is clearly marked by a strengthened meridional pressure gradient across the Atlantic and extending into Europe, with the rise in pressure in the southern pole of the NAO extending eastward into Morocco, the rest of the Maghreb and southern Iberia. This is the month of strongest precipitation decline in Morocco. In contrast, Fig. 6.11c-d shows the counterpart pressure and precipitation differences for February. The rainfall in Morocco is less affected, though along with most of Europe, it still experienced a decline. The reason for the smaller Moroccan precipitation decrease in February appears to be the location of the southern pole of positive pressure anomalies, which are now further north and east than in December. These results point to the need to study the decadal variations of the NAO by calendar month, as done in the rotated Principal Component

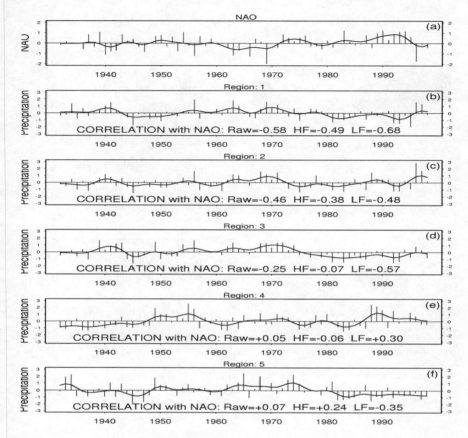

**Fig. 6.10**   Mean standardised anomaly time-series for North Atlantic Oscillation (NAO) (a) and Moroccan rainfall regions 1 to 5 (b-f) for the October-April season (based on Fig. 6.9). Bars give the raw individual seasonal values. The smooth line is a Lanczos low-pass filter with 50% cut-off amplitude set to 10 years (number of weights=11, Lanczos sigma factor=2). Panels (b)-(f) indicate the correlation between the unfiltered (Raw) rainfall index and the unfiltered (Raw) NAO, followed by the corresponding correlations for the high frequency (HF) rainfall and NAO (the residuals from the smooth lines) and for the low frequency (the correlation of the rainfall and NAO low-pass smooth lines).

analysis of Barnston and Livezey (1987), and to make further interpretations of the NAO in terms of the orientation of storm tracks and upper tropospheric jet streams, in order to shed more light on the mechanisms of the precipitation variability.

Given the strong decadal variability of the seasonal NAO values (Fig. 6.10), it is at first sight puzzling why the NAO should have such a low one month lag serial correlation (r=0.13, January 1874 to April 1998), contrasting strongly with the Southern Oscillation Index (r=0.63, Trenberth 1984). One factor contributing to this autocorrelation difference is likely the higher internal atmospheric variability

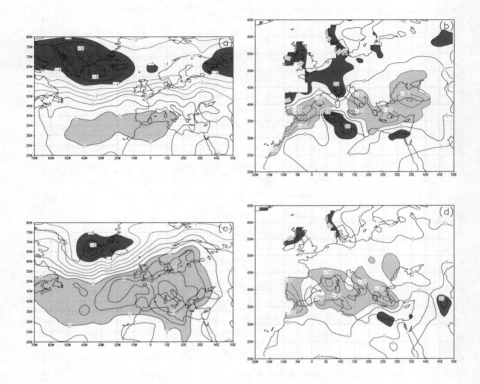

**Fig. 6.11** (a) Composite difference (mb) for December SLP 1979-93 minus 1961-78. Dark shading for <-4mb, light shading for >2mb. Contour interval is 1mb. (b) Same as (a) but for precipitation (mm), with light shading for <-10mm, dark shading for >10mm. Contour interval is 10mm. (c) Same as (a) but for February SLP 1980-94 minus 1962-79 (shifted by one year from (a) to cover the same winter seasons), (d) Same as (c) but for precipitation.

at the monthly timescale for the mid-latitude Atlantic compared to the tropical Pacific. However, the reconciliation of the low NAO monthly autocorrelation with the strong decadal variability may lie in a tendency for the NAO to internally oscillate from July to April. This involves the NAO moving from a positive phase in July/August, to a strong negative phase in October/November, and back to a strong positive phase in January that tends to persist through to March. Conversely, the approximate inverse sequence culminates in negative January NAO values. This feature emerges as the first unrotated EOF of the NAO's monthly evolution from July to April. The loadings of that EOF (Fig. 6.12a) describe the most characteris-

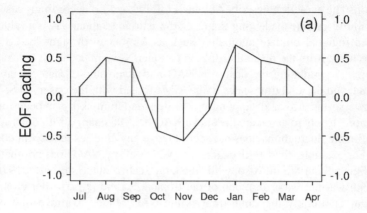

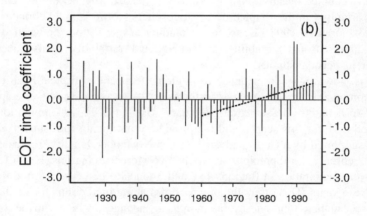

**Fig. 6.12**    First correlation eigenvector (EOF1) of NAO July to April monthly evolution, for the years 1922-1995. (a) Component loadings (weights), (b) Time-coefficients of the EOF in each year. Dashed line shows linear regression trend from 1960-1995.

tic NAO evolution, while its time coefficients (Fig. 6.12b) measure the extent to which this evolution occurred in each year during 1922-95. During the post-1960 period of overall rise in the October-April NAO values (Fig. 6.10), the EOF time coefficients (Fig. 6.12b) show a clear trend toward the phase of NAO evolution with positive NAO values in January. Indeed, six of the seven largest positive January NAO values since 1932 (not shown) were the result of the seasonal evolution mode depicted in Fig. 6.12a. We thus suggest that the recent multi-decadal uptrend of the seasonal winter NAO value has likely resulted in part from an increasing frequency of high positive mid-winter values produced by the above internal NAO oscillation. The evolution described by the EOF Fig. 6.12a) accordingly also offers the possibility of foreshadowing some of the extreme January NAO values, and associated regional climate anomalies such as Moroccan drought, based on the NAO evolution from the preceding July to October/November.

Attempts at simulating the observed NAO and associated climate variability at interannual and decadal timescales with GCMs forced with the observed SST variations have so far proved disappointing. Little simulation skill has been reported. Furthermore, trying to associate the above NAO decadal change with other low-frequency climate fluctuations does not seem to give any clues as to associated mechanisms. For example, the LF (decadal) component of the NAO does not match that of the LF Sahelian rainfall. Also, while the large scale climate shift after 1976 associated with warmer water in much of the tropical Pacific (e.g., Miller et al. 1994) does occur at roughly the same time as the NAO tended to move into positive phase, the overall correlation of tropical Pacific LF SST and the LF NAO is very low (for Niño4, r=0.20; for Niño3, r=0.35). The ultimate climate system source of the decadal variability in the NAO remains a mystery. It is possible that atmospheric GCMs forced with observed SST are not suitable for studying this problem, and a fully coupled ocean-atmosphere model is needed to generate decadal atmospheric variability in the coupled ocean-atmosphere North Atlantic climate system (e.g., Grötzner et al. 1998). Better understanding of interactions between synoptic systems, intraseasonal variability, and interannual variability is also needed to underpin the decadal studies.

There is, however, good evidence for a global (especially tropical Pacific) SST-forced component to sub-tropical northwestern Africa late season precipitation (March-April, and to a lesser extent, February). Skill in ECHAM4 simulations with prescribed SST for this season and region is very promising (not shown) and the result is supported by a CCA analysis between November-January tropical Pacific SST and February-April precipitation in the five Moroccan regions (Fig. 6.13). The late season precipitation in Regions 1, 2 and 3 tends to be below normal in warm event ENSO years. This offers the prospect of making predictions for the late season Moroccan precipitation several months in advance, which can be valuable information for agricultural and water resource planning.

## 6.6    Discussion

This chapter has described some of the known features of North African climate variability during this Century. The research has formed the basis of experimental prediction systems (e.g., Ward et al. 1993; Lamb et al. 1997b) and underpinned the First Forum on Seasonal Climate Prediction and its Application to Early Warning Systems for Food Security in West Africa, held in May 1998 in Abidjan, Côte D'Ivoire. While the basic mechanisms of SST-impact on the overlying atmosphere are coming to be understood, much remains to be revealed by further research

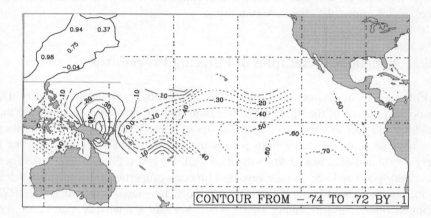

**Fig. 6.13** Canonical correlation analysis for 1951-95 between HF tropical Pacific SST in November-January and the five HF Moroccan rainfall indices (Fig. 6.9) in February-April. Loadings for SST mode 1 and, in the top left insert, rainfall mode 1. Cross validated correlation skill scores for (the unfiltered) region 1 using 3 modes is 0.43, and for region's 2 through 5 it is 0.42, 0.43, -0.14, 0.37. Except for Region 4, the skill levels are statistically significant at 1%.

directed towards North African climate. Precipitation is clearly the most important socio-economic climate variable for North Africa, and it is therefore natural that most research has focused on this variable. However, most of the effort to date has employed a monthly or coarser temporal resolution. For many agricultural applications, it is necessary to study the daily timescale, and consider such features as rainy season onset and dry spell lengths. Furthermore, to better understand the mechanisms of African climate variability will require observational data bases beyond those of precipitation. In this regard, the reanalysis datasets offer the potential to study the three dimensional structure of the observed African atmosphere. However, it may be necessary to assess the minimum data input to make the reanalyses reliable at given spatial and temporal scales over Africa.

There is also a need to improve the performance of GCMs over North Africa. Developments such as improved parametrizations of convection over large continental land masses, and better treatments of the radiative fluxes over dusty semiarid regions, should enable GCMs to become more useful tools with which to assess the relative roles of different mechanisms of climate variability in North Africa. There will also be a need to address the teleconnection mechanisms. For example, we need to know whether the impact of SST is felt through changing the probability of different internal intraseasonal modes of North African variability, as is currently being investigated for the Indian monsoon (Slingo, this volume). Finally, it is necessary to better resolve and link spatial scales in numerical models.

One approach to this is to nest mesoscale models inside GCMs. This should enable interactions between the planetary scale and meso-scale (such as lake effects) to be better represented, and lead to more useful applications-oriented statements about interannual-to-decadal climate variability.

## 6.7    Appendix - Examples of Filtering Procedures

Filtering climate indices is an important part of studying decadal variability. The response function of the filter should be established prior to analysis to ensure that the filter satisfactorily isolates the time scales of interest. Some less-effective filters can have negative side lobes in their response function, and a cut-off slope that is very gentle and does not partition the data well. However, the two filters used here have very good and very similar response function properties. For the Sub-Saharan work, the filter used is the recursively estimated Integrated Random Walk (IRW) Kalman filter (Ng and Young 1990). The filter has a relatively sharp and very smooth cut-off with no negative side lobes, and does not require assumptions of stationarity. It is a very flexible filter, whose 50% cut-off amplitude is controlled by one parameter, the noise variance ratio (NVR). Visual inspection of many NVR choices indicated that a value of 0.1, corresponding to a 50% amplitude cut-off at 11.25 years, partitioned the decadal variance from the sub-decadal variance very well for the indices in this climate problem.

Similar comments apply to the filter used for the Moroccan work, the symmetric Low-Pass Lanczos Filter (Duchon 1979). Duchon showed that the simplicity of calculating the corresponding weights and the adequate frequency response function make Lanczos filtering an attractive filtering method. The principal feature of the Lanczos filter is to use its "sigma" factors that significantly reduce the amplitude of the Gibbs oscillation. The only two inputs for this filter method are the number of weights and the cutoff frequency. Another advantage of this filter is that the cutoff frequency can be controlled independently of the number of weights. In comparison with other filters, Duchon (1979) demonstrated that the Lanczos approach to the design of the filter is quite easy and yields the good response characteristics noted above. To isolate the multi-decadal variations and trends of the NAO and precipitation indices, we set the number of weights to 11 and the cutoff frequency to 0.1, since for time series with one value per year, the 0.1 cutoff frequency corresponds to the period of 10 years. Also, we set the Lanczos sigma factor to 2, since it yielded a better response function. To keep the filter stable at the ends of the series, we extended the series at both ends with the average of the adjacent 10 values before applying the filter.

*Acknowledgments*. The research reported in this chapter was supported by the U.S. Agency for International Development ( Grant 608-G-00-94-00007), the Moroccan Direction de la Météorologie Nationale, NOAA Office of Global Programs (Grant NA 67RJ0150) and the European Union project DICE, EV5V-CT94-0538.

# 7 SST-Forced Experiments and Climate Variability

ANTONIO NAVARRA

*IMGA-CNR, Via Gobetti 101, 40138 Bologna, Italy*

## 7.1 Introduction

The extension of the simulation range to several months and then years in the early 1970s caused a re-assessment of the role of the Sea Surface Temperature (SST) in the regulation of the climatic variability. Namias (1959, 1963) proposed that the low-frequency variability over North America might be connected with the SST variations over the equatorial and tropical Pacific Ocean. An early attempt to produce long-range forecasts based on SST was attempted by Adem and Jacob (1968).

A more complete argument describing the SST influence on the general circulation of the atmosphere was proposed by Bjerknes (1966, 1969). The argument, originally intended as a theory for Pacific Blocking, was based on the effect of SST to shift the mid-latitude jet latitudinally, causing the transition from a "high-index" circulation to a "low index" one. A positive SST in the central Pacific would have caused a strengthening of the Hadley circulation and an acceleration of the jet, zonalizing the flow and decreasing the chances for blocking.

These theoretical arguments justified the investigation of the effect of SST with numerical models of the general circulation. Rowntree (1972, 1976a, 1976b) pioneered the early numerical studies devoted to the investigation of the capability of SST anomalies to generate persistent atmospheric anomalies at the monthly and seasonal time scales. This early study revealed the strong effect that anomalies of SST have on precipitation in the tropics.

A few years later the perception of the role of the SST became more prominent. The existence of coherent patterns, organized in patches of negative and positive anomalies at monthly and seasonal time scales, was demonstrated by Wallace and Gutzler (1981). They discovered a number of patterns that could be roughly divided into dipolar patterns composed of two center of actions, and more complicated patterns composed of several positive and negative centers, that could extend around the globe, connecting areas very far one from the other. Such patterns have been noticed before, for instance by Walker and Bliss in their papers (Walker 1923,1934; Walker and Bliss, 1932), but Wallace and Gutzler were able to show their existence systematically. The importance of the patterns was further enhanced by the result obtained shortly afterward by Horel and Wallace (1981) who argued very convincingly about the possibility that such patterns could be linked to the SST. The theoretical framework had already been laid down by Hoskins and

Karoly (1981) who interpreted the teleconnection patterns as stationary Rossby waves forced by anomalous diabatic heating generated by the anomalous SST.

Rasmusson and Carpenter (1982) discussed in detail the existence of a strongly coupled phenomenon involving atmospheric and marine anomalies with a coherent time scale of the order of several years. They were able to synthesize in a unified picture the Southern Oscillation (known since the result of Walker and Bliss, 1932) and the appearance of the ElNino current, known since the previous century as a local oceanographic event in the eastern Pacific.

The realization that powerful ocean-atmosphere interactions have a major role in shaping climate variability propelled to centerstage the usage of coupled models. Coupled general circulation models were therefore necessary not only for understanding climate on centennial and longer time scales, but also on the interannual, even seasonal time scale. However, though coupled models are feasible and manageable, their results are often as difficult to interpret as the observations of the real world. Furthermore they are still plagued by large systematic errors whose full impact on the quality of simulations of the variability is not fully understood. A hierarchy of approaches, based on a variety of design choices, is necessary to attack the problems from many angles.

Fig. 7.1 shows a conceptual scheme of possible experiments of climate simulations. The top cartoon (panel a), shows a simulation where an atmosphere-only model is forced by fixed SST. Possible choices for the fixed SST are the annual mean, equinoctial condition or a specific month, January or July. This approach economically produces large amounts of statistics, but at the cost of a strong distortion of the climate variability by eliminating the seasonal cycle. In the case of experiments investigating the variability of ENSO it is particularly unsatisfactory because of the strong phase-locking between ENSO and the seasonal cycle. Panel (b) shows an improved approach using seasonally varying SST, though averaged to obtain a climatologically repeating seasonal cycle. Panel (c) shows instead an experimental set-up where observed SST are used every month, a set-up similar to what has been used in the AMIP experiment (Gates, 1993). Panel (e) shows a coupled model, with SST and surface forcing now free to vary according to their respective interactions dynamics.

Method (a) and (b) are effective at producing good statistics economically. Samples of monthly or seasonal means anomalies of sufficient size can be produced and the signal from the SST can be analyzed very efficiently. On the other hand, method (e) is still very complicated to understand even if simpler versions of coupled models have also been introduced by describing the ocean as a single mixed layer, neglecting horizontal advection, but introducing a finite heat capacity (Manabe and Stouffer, 1996; Manabe and Broccoli, 1985, among others). Method (c), both realistic and assuming a "perfect" ocean model, is therefore a good way to gain insight into the workings of the interacting atmosphere-ocean system, but the length of the experiments is limited by the length of the available observed record, about 100 years. As a consequence, the size of the sample is correspondingly limited.

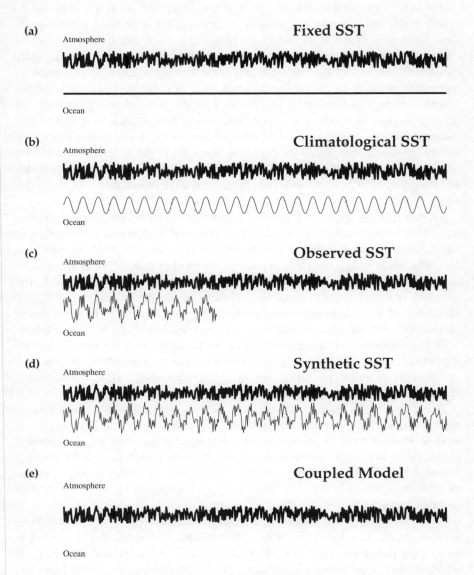

**Fig. 7.1** Schematic of the possible types of numerical experiment with climate models with prescribed SST (panels a-d) and fully coupled models (bottom). The atmosphere variability is indicated with the top fast-varying curve, whereas the ocean is indicated by the relatively slower one. Possibilities include fixed, or perpetual SST (panel a), climatological repeating SST (panel b), observed SST (panel c), statistically reproduced SST (panel d). Except the arrangement in panel (d), the others combinations have been heavily investigated, often including multiple experiments (ensemble experiments).

A possible way out is to expand the sample size by introducing extra experiments with the same SST. The slowly varying boundary forcing caused by SST has been shown to be a measurable climatic signal above the natural variability of the atmosphere itself. The detection of the signal is, however, far from being a trivial problem. The signal of the SST, though measurable, is often small compared to the intrinsic variability of the atmosphere, e.g. in the midlatitudes. In order to partially eliminate the effect of the quasi-random components Leith (1974) suggested an ensemble approach. In climatological applications an ensemble approach implies almost universally a number of simulations with the same boundary forcings, but different initial conditions. The natural instability of the system is then sufficient to generate different realizations of the climate variability. A recent analysis of the benefit of the approach can be found in Barnett (1995) and more details are described by Royer (1991), Palmer (1993), Stern and Miyakoda (1995), Harzallah and Sadourny (1995). These series of experiments have shown that a single experiment, i.e. a single realization of a simulation, is of limited use.

Another possible approach is to extend the sample size retaining the realistic features of the observed SST record, has been proposed by Navarra et al. (1998). In this case, a statistical model is constructed based on the observed record of the SST. The model is then used to generate time series of synthetic SST of an arbitrary length that can be used in simulations experiments, as in panel (d), Fig. 7.1. The approach is broadening the ensemble technique to include the case in which the variability of the SST in also considered. Ideally, experiments with ensembles of atmospheric-only simulations with the same prescribed synthetic SST are needed.

This approach has the advantage to allow the investigation of the spectral cascading between time scales, for instance between the interannual and the decadal time scales. This experimental set-up has not been investigated at this time, but it appears to be a very promising one.

The effects of the interannual variations of the ocean and atmospheric forcing have been shown to be within the reach of the simulation capabilities of numerical models. Philander and Siegel (1985) first demonstrated the possibility that oceanic numerical models could simulate such interactions with an ocean model properly forced. Lau (1985) then showed that the interannual variability of the Southern Oscillation Index (SOI) could be simulated very realistically by a GCM forced by monthly varying SST.

Since then the effects of SST on the general circulation has been the target of several studies. Lau and Nath (1990; 1994) extended the tropical SST forcing to include the global and mid-latitude SST. Other studies on the interannual time scale effects of SST were performed by Latif et al. (1990), Konig et al. (1990), Stern and Miyakoda (1995), Zwiers (1996), Harzallah and Sadourny (1995), Kumar and Hoerling (1995); Brankovic et al. (1995). The role of boundary forcings, dominated by the SST, was found to be so important that an international experiment was set up to compare the response of a very large number of atmospheric GCMs, the AMIP intercomparison project (Gates, 1993). The decadal effects of SST using multi-decadal simulations (20 years or more) or observations has been investigated by Lau and Nath (1994), Smith (1994, 1995), Kitoh (1991a, 1991b), Hastenrath (1990), Folland et al. (1986, 1991), Graham (1994), Graham et al. (1994), Rowell et al. (1995), Rowell (1997), Harzallah et al. (1996), Moron et al. (1998)

The ability of prescribed SST experiments with observed data to describe accurately the variability of the mid/latitudes has been recently been questioned by several studies. The obvious difficulty of neglecting the damping atmospheric mechanisms on the SST seems to be particularly damaging in the case of the air-sea interaction in the midlatitudes (Saravanan and Mc Willians, 1997). The hypothesis is that the midlatitude SST may cause the atmosphere to adjust, rather than create a specific response as in the tropics (Bladè, 1997; Barsugli and Battisti, 1998).

## 7.2    Ensembles of Simulations

### 7.2.1    The need for ensembles

Theory and numerical experiments strongly indicate that SST can influence the atmosphere mainly through its impact on the convective activity. The convective zones are modulated by the SST and in its turn the convection partially controls the regions of release of latent heat. The release of latent heat stretches vorticity tubes generating atmospheric motions. Linear dynamical studies have clarified the mechanism by which the release of latent heat, modulated by the SST, can affect the atmospheric circulation. A steady heat source generates a local response (Gill, 1980) and a remote response (Hoskins and Karoly, 1981). The local equatorial stationary response is composed of a mixture of Kelvin and Rossby waves, but the Rossby waves can penetrate into the mid-latitudes, affecting the entire globe, whereas the Kelvin waves are confined to the equatorial area.

Dynamical studies, the results of which are substantially unmodified by inclusion of the effects of nonlinearity and transients, provided a strong indication that the effect of SST might be qualitatively different in the tropics and in the mid-latitude. In the midlatitudes (see article by Frankignoul in this book) the existence of a strong interaction between SST and the atmosphere is not yet established, whereas in the tropics the evidence is now well documented. The tropical atmosphere and the tropical surface ocean are in fact part of the same physical system.

The turbulent character of the climate system is revealed by the absence of a clear, linear, one-to-one correspondence between SST anomalies and atmospheric anomalies. In general, the same or similar SST anomalies can be dynamically consistent with a variety of atmospheric anomalies. In other terms, several atmospheric states are possible for a single SST state, even in the tropics.

This conclusion cannot be demonstrated, but it can be formulated as a conjecture derived from a careful consideration of the observations. The combined study of atmospheric and oceanic observations shows that the variability of the ocean-atmosphere system is very high. The typical month-to-month situations tend to never repeat themselves and in practice the same exact situation never returns.The examination of monthly anomalies, i.e. deviations from the monthly varying climatology, either of SST or atmospheric fields show readily this difficulty. Early attempts to exploit possible repeating patterns for predictions purposes ("the analogue method") have all been unsatisfactory.

Numerical experiments illustrate this point with great clarity. In the following we will be showing some selected result from the analysis of simulations with the GCM ECHAM4 at resolution T30 with 19 vertical levels. The boundary conditions are SST prescribed from the GISST data set (Parker et al, 1995). The experiments consist of two ensembles. The first were performed under the European Program DICE and covered the period 1961-1994, the second was performed in collaboration with ENEL, the italian electric utility, for the period 1979-1994. The first ensemble was composed of three members, a bare minimum, and the second of 25 members. The long experiments are described in detail in Moron et al. (1998).

Fig. 7.2 illustrate the typical evolution in time of the average precipitation over North America from the second set of experiments. The experiments have been performed with a General Circulation Model using different initial condition, but with the same monthly varying SST for each member of the set. The ensemble is therefore sampling the possible states that are consistent with the prescribed SST. The time evolution of each member, represented by the thin lines, does not seem to have any coherence with the others. Clearly, for a given SST state there is a great number of possible atmospheric states at the monthly time scales. The situation is not very different over Europe (Fig. 7.3), where the rain evolution is as chaotic as in North America.

The tropical situation is very different (Fig. 7.4). The experiments are very consistent and the lines are piling on top of each other. In the tropics, the number of possible atmospheric states for a given SST is more limited and the variations of SST largely control the occurrence and distribution of rain on long time scales.

These experiments show that the effect of SST is superposed on variability from other sources in the atmosphere-ocean system. In the tropics, the SST are the dominant factor and the SST variability dominates the overall variability at monthly and seasonal time scales. In the midlatitudes, the synoptic variability due to baroclinic instability at scales longer than a week dominates the effect of the SST, though it is possible that SST may affect just the synoptic activity itself. In general, it is possible to state that the detection of the impact of the tropical SST in the mid-latitudes is made more difficult by the small magnitude of the signal. The impact of the mid-latitude SST in even more complicated to assess, since as we have seen inother chapters of the book, it is not even clear if a coupled mechanism is or not active.

Dynamical mechanism through which the SST could exert its influence have not been firmly established yet, and probably several mechanisms may be responsible, one at a time, or concurrently. A direct emission of planetary Rossby waves has been proposed by Hoskins and Karoly (1981). SST anomalies generate anomalous convection and anomalous release of latent heat that is transformed via a Gill (1980) process in upper level vorticity forcing. The subtropical centers of vorticity interact with the mid-latitude mean-flow to generate a stationary wavetrain of Rossby waves that arches spectacularly across the hemisphere. Another possibility

## NORTH–WEST AMERICA AREA tprec

### Deviations from Climatology

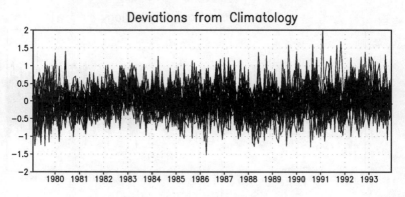

### Deviations from Ensemble Mean

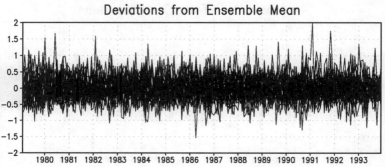

### Area Averages

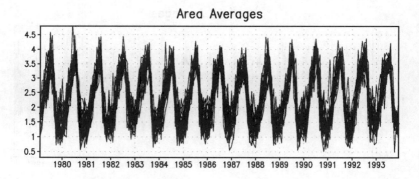

**Fig. 7.2** Area average of precipitation for the North-West American region for an ensemble of 25 experiments for the period 1979-1994 forced by AMIP SST. All values are in mm/day. The bottom panel contains the original values, the middle panel the deviation from the ensemble mean whereas the top panel show anomalies, i.e. deviations from the monthly climatology.

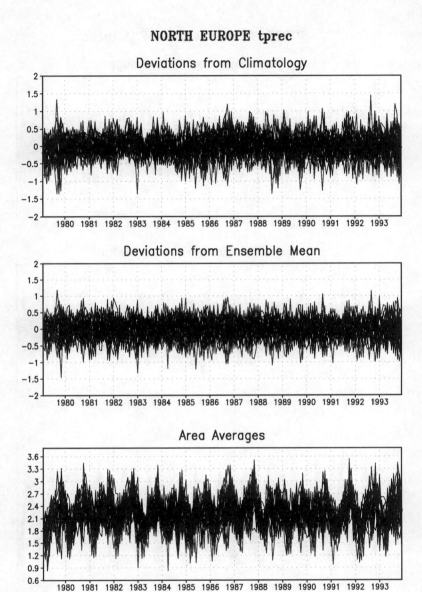

**Fig. 7.3**  As in Fig. 7.2., but for the North European region.

**EQUATORIAL PACIFIC tprec**

### Deviations from Climatology

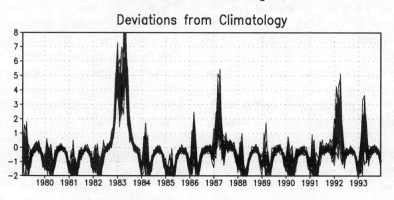

### Deviations from Ensemble Mean

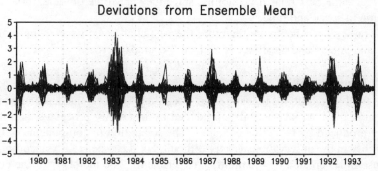

### Area Averages

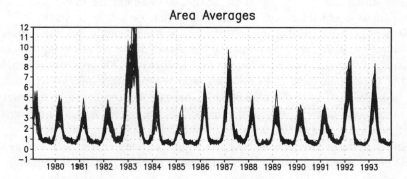

**Fig. 7.4**  As in Fig. 7.2 but for the equatorial Pacific region from 5N-5S and 150W-80W

is that the SST acts by modifying the probability distribution of the climate states, without actually modifying the main equilibria (Palmer, 1993). The mean flow is affected and as a consequence the storm tracks are modified because the transient are developing differently and following an altered track.

The conclusive evidence in favor of one scenario compared to the other is slim and probably they are both active. Whatever the cause the signal from the SST is small and swamped by the large variability from other causes. The only possibility for detection is to enhance the SST signal either by artificially increasing its magnitude (i.e, by using artificially large anomalies) or by improving the signal to noise ratio.

Though the first alternative is suitable for process studies and theoretical investigations, it is not particularly useful for applications in which realism is the primary concern. In this case, multiple numerical experiments have proven very effective to enhance the chances of detecting the signal by improving the signal to noise ratio of the response. In the case of the 25 members shown earlier, the experiments would differ for the part of the variability not linked or caused to SST, but they will share the same SST-related variability, since they are forced by the same SST distribution in space and time.

Ensemble experiments are therefore the best chance we have at the moment to study and understand the effect of SST on the atmosphere.

## 7.2.2  Reproducibility

The total variability of the atmosphere can therefore be considered as formed by two parts, the internal variability produced by instability processes, and the variability produced by the SST anomalies. The separation in components of the total variability in the observations is very difficult, but it can be readily accomplished in the ensembles created by the multiple experiments. The members of the ensemble usually differ only for the initial conditions. A small change in the initial state is sufficient to launch the evolution of the atmosphere on a different track.

It is to be expected that in each member of the ensemble the instabilities will develop quite differently, the times and the phases of the growth being affected by the small differences in the evolution of the simulations. This part of the variability will be uncorrelated and an external observer without other information could not tell from the analysis of this kind of variability that the runs belong to an ensemble. From the point of view of the instabilities, the experiments will not have anything in common.

However, the experiments do have something in common. The SST forcing is the same across the ensemble and we might hope that the portion of variability that depends on SST should be simulated similarly in each member. If any variability is in common in the runs, than it must be caused by the SST, the only factor that is common to all members of the ensemble. A genuine SST effect should be seen in all experiments.

In principle, the situation might be as complex as in the case of instabilities. The SST response could be composed of several modes and each experiment might

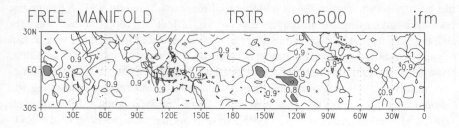

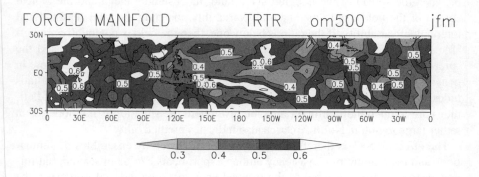

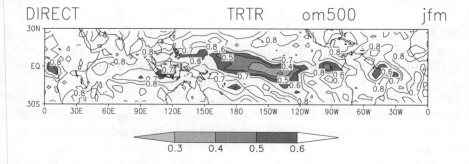

**Fig. 7.5** Reproducibilty ratio of seasonal means according to Stern and Miyakoda (1995) for vertical velocity at 500mb in JFM. The index is normalized between 1.0 and 0.0, low values of the index indicate a high reproducibility. The bottom panel shows the index for the original precipitation field, showing high reproducibility in the west-central Pacific. Using the Forced Manifold is possible to separate the field in a highly reproducible part (middle panel) and in a chaotic pert (top panel). Values smaller then 0.6 are shaded.

pick one of those according to the small errors typical of that experiment, resulting in a semi-chaotic situation not different from the previous case. Such result would not be very promising for the development of forecasts based on SST. A promising result for forecasting would require that the SST effect be "reproducible", namely that the SST must affect the atmosphere in each experiment in the same way. It is a very remarkable fact that the atmosphere-ocean system behaves exactly like that, showing a substantial amount of reproducibility, though varying in space and time.

A number of investigators has defined quantitative indices for measuring the amount of reproducibility in an ensemble. Stern and Miyakoda (1995), Rowell (1997), Harzallah and Sadourny (1995), have proposed that the ratio of the SST-derived variability and of the rest of the variability is varying in space and time. Fig. 7.5 (bottom panel) shows the reproducibility index of Stern and Miyakoda computed for the ensemble 1961-1994. This index essentially measures the ratio between the spread of the integrations around the ensemble mean and the typical spread of the field due to the "natural" variability, meaning the total seasonal and interannual variability. If the ratio is below one, then the field is potentially predictable in that area, because the realizations tend to agree with each other and the spread is less than the typical interannual variability. It is possible to see how in general the tropics are characterized by lower values of the ratio than the mid-latitudes, indicating a substantial potential predictability for the tropics. The midlatitudes are a much more difficult area, but even here the situation is not hopeless: a rather large region in North America also indicates predictability.

The effect of SST is therefore consistently reproduced by ensembles of simulations and the amount of consistency among experiments can be measured, but this conclusion is mostly limited to the tropical area and some special regions in the midlatitudes.

### 7.2.3 Manifolds

It would be interesting to be able to divide the fields themselves in a portion that is linked to SST forcing and another one that is independent. If such a separation were possible, it would make possible a detailed dynamical analysis of the processes that are active in each subspace. Navarra and Ward (1998) indicate a possible way to identify, within some statistical confidence, the subspace of the anomaly field that is covarying with the SST and the subspace that is independent from the SST variability.

The subspace of the states depending on the SST, denoted the "Forced Manifold", can be identified using simple techniques of analysis of the covariance. For instance, in the case of Fig. 7.5, an SVD was performed taking as left fields the Z500 field and as right field the SST. The Forced Manifold can be reconstructed by projecting on the Z500 part of the SVD modes. Because of the natural degeneracy introduced by the ensemble the reconstructed field will not be the entire time series of Z500 but a subspace. The actual calculation would involve rather elaborated estimates of the reliability of each mode, since only reliable modes can be used for the reconstruction procedure. The determination of the Forced Manifold is however probabilistic since it cannot be determined with an uncertainty that is smaller than the uncertainty on the SVD modes themselves.

The orthogonal complement of the Forced Manifold, by definition, is the Free Manifold. The Free Manifold is populated with modes that do not covary with the SST, removing the effects of ocean forcing.

The removal of the free part of the variability has a large impact on the reproducibility. Fig. 7.5, middle and top panels, show the reproducibility index for the Forced and Free manifold. It is possible to see that the reproducibility for the Forced manifold is much higher, as it was to be expected, whereas the Free Manifold is characterized by almost chaotic behavior. The Forced and Free Manifold appear to be useful concepts to understand the interannual and decadal variability and may represent a valid alternative to more conventional filtering.

## 7.3   Evaluation of performance: skill and reproducibility

### 7.3.1   Information and ensemble

The whole rationale behind performing an ensemble simulation is to enhance the signal to noise ratio for the SST induced signal; or, in other words, the elimination of the chaotic component of the variability. The basic methods to achieve the elimination of the internal component is to average the simulations across the ensemble dimension. The SST signal will be maximized in the ensemble mean and the effect of the chaotic component minimized. Fluctuations that do not depend on the SST will be compensated in the mean. The interannual variability could then be investigated by performing EOF on the time mean. Such was the approach followed, for instance, by Harzallah and Sadourny (1995). Unbiased estimates of the EOF of the ensemble mean can be obtained using the approach by Venzke et al. (1998)

The ensemble mean is effective but it represents an overkill. The SST signal is recovered, but the rich information deriving from the multiple experiments has been lost. In terms of variability, the mean reflects the SST signal, but it also contains contributions from the autocovariance of each member.

We can gain a little insight into this issue by considering the ensemble mean from a different point of view. Assume you have an ensemble made up of three members organized in a matrix $X = (x_1, x_2, x_3)$, where the fields are represented as vectors $(x)$. Defining the SST signal as the part of the variability that is common to all members it is then natural to consider the covariance matrix $XX^T$ between the members,

$$XX^T = \begin{bmatrix} X_{11} & X_{12} & X_{13} \\ X_{21} & X_{22} & X_{23} \\ X_{31} & X_{32} & X_{33} \end{bmatrix} . \qquad (7.1)$$

It is shown in Ward and Navarra (1997) that the EOF of the correlation matrix (1) coincide with the EOF of the ensemble mean, but the inspection of the structure of (1) reveals that there is a contribution from the autocovariance of the run themselves, the diagonal terms in (1). This procedure is then biased by the autocovariance of each member.

A way to reduce the influence of the autocovariance of each member has been proposed by Ward and Navarra, (1997) who developed an SVD-based procedure that yields the mode that maximizes covariation among the members of the ensemble. The algorithm yields a single sequence of modes that maximizes the common variation among all members of the ensemble. It is possible in this way to dissect the spatial and temporal characteristics of the SST induced variability. Reproducible situations should have a large fraction of the variance explained by the common model-model mode (MM), whereas situations that are characterized by a large spread should have a small fraction of variance in the MM mode.

Ward and Navarra (1997) proposed to monitor also the amount of covariation between the model simulations and the observations. The monitoring could also be realized via an SVD, with observations on the left field and the simulations on the right field. The fraction of total model variability explained by such modes covarying with the observations is a very powerful indicator of the most "optimistic" performance indicator. A "good" (reproducible) season or area should have a substantial portion of the variance covarying with observations, whereas "bad" (chaotic) periods should have a very small fraction of total variance explained by the observations-model mode (OM)

## 7.4   Skill and Reproducibility for 1961-1994

The ideas discussed above have been applied to a set of three simulations for the period 1961-1994. The experiments have been performed using prescribed SST updated every month obtained from the data set GISST 2.2(see article by Parker in the same volume). The experiments and the results shown here have been described in detail by Moron et al.(1998). The objective of the experiment was to investigate the capability of the model to simulate climate variability at long time scales. The model used in those simulations is the ECHAM4 model (Roeckner et al., 1995, 1996), developed at the MPI, in the version set at resolution T30 that was developed jointly with IMGA. The model uses advanced parametrizations and numerical formulations, and it has been used in scenario calculation for the greenhouse effect. The members of the ensemble were generated by repeating the simulations changing slightly the initial conditions for each case. The development of instabilities originating from the differences in the initial conditions leads very soon to substantial divergences in the evolution of the runs.

The experiments can be used to investigate the capability of the model to reproduce interannual variability and interdecadal variability. As we have discussed in the previous sections, this type of simulation cannot be assessed using the same criteria that are used in short-term forecasts. The chaotic nature of the atmosphere prevents a point-by-point comparison in space and time between simulations and observations, the objective is rather to reproduce realistically the statistics of the variability. Fig. 7.6 shows the evolution of indices of tropical precipitation defined as the average over the area, for the regions of the Sahel, India and the Brazilian Nordeste in the boreal summer. It is possible to note the regional differences in the capability of the model to follow the variations of the precipitation over the tropics. Even if the tropical atmosphere is more reproducible than the middle latitudes, nevertheless there are large differences in the quality of the simulations, in different tropical regions.

The quality of the simulations can be assessed by the degree of divergence between the ensemble members (spread) and by the agreement with the observed variations (skill). The best result is obtained in the Brazilian Nordeste. The correlation between the ensemble mean and the observation is 0.74. The discrepancies between the members are small and the ensemble mean (solid thick line) follows the observations very closely (dashed thick line). In the Indian region the performance is less satisfactory and the correlation drops to 0.28. The individual members (thin solid line) show a larger spread and the fit is not as accurate as before. The Sahel region yields the worst result with a correlation of 0.15 and there is only a faint reminiscence of the long downward trend in the precipitation for the Sahel in the decades since 1960.

Spread and skill of the simulations depend not only on the location, they also depend on time. The calculation of the correlation for the same three indices of Fig. 7.6, but for the period 1979-1988 yields similar values for the Nordeste and the Sahel, but significantly improves the score for the India region, reaching 0.58. This indicates that the model tends to capture the teleconnections from the Pacific to the India region mainly after 1980.

The analysis of the indices of the precipitation based on spatial averaging is not, however, allowing for the possibility of a systematic difference in the representation of the teleconnections pattern by the model. In the real world, a certain SST pattern is linked to a characteristic pattern of teleconnection in the atmosphere. For instance, equatorial SSTs in the Pacific are linked to the PNA in the midlatitudes and in the tropics they are connected with precipitation patterns in the Indian region. The simulations can fail drastically in a more gentle manner. In the first case, they do not reproduce the entire teleconnection mechanism and therefore that particular teleconnection is missing from the variability of that model. In the second, that is more frequent, the model exhibits sensitivity to the SST and therefore a teleconnection relation exists, but it is distorted, modified in space, time, its probability distribution different from reality. In this case the method discussed in 7.3 can be useful to identify the capability of the model to reproduce reality.

The following picture (Fig. 7.7) illustrates how the method by Ward and Navarra (1997) can be used in this context. The method is applied to the precipitation field in the tropics. Panel (b) show the first mode resulting from the application of the SVD analysis to the ensemble members. The mode can be interpreted as the con-

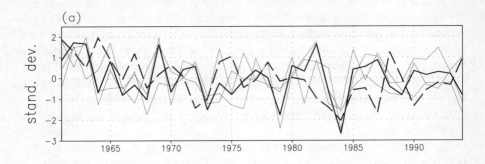

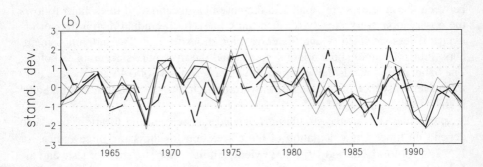

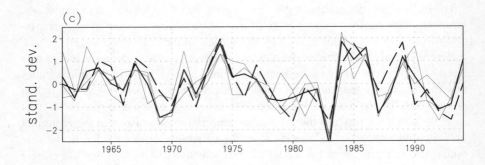

**Fig. 7.6** Standardized rainfall anomaly indices for (a) Sahel, (b) India and (c) Nordeste, following the definitions in Sperber and Palmer (1996). Thin solid lines show the three GCM simulations, thick solid is the ensemble mean, thick dashed is the observed.

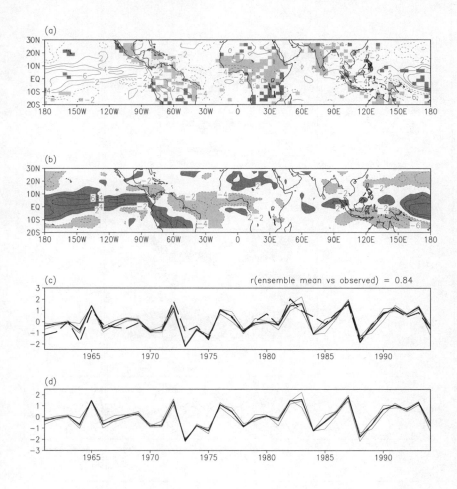

**Fig. 7.7**    Singular Value Decomposition analyses (SVDAs) for JJAS. (a) SVDA between GCM simulated and observed precipitation (OM analysis). Observed singular vector 1 (OMo) indicated by shading, dark shading >2, light shading <-2. GCM singular vector one (OMm) indicated by contours at +/-2, 4, 6. (b) SVDA between the three GCM simulations (MM analysis). Contours at +/-2, 4, 6, dark shading >2, light shading <-2. (c) SVDA time series from (a). Thin solid shows the three GCM simulations, thick solid is the ensemble mean, thick dashed is the observed. (d) SVDA time series from (b). Thin solid shows the three GCM simulations, thick solid is the ensemble mean.

sensus signal between the ensemble members. The pattern corresponds to the dominant mode of variability that the members have in common. The time evolution of the coefficient corresponding to the members is shown in Panel (d). It is possible to see a high degree of temporal consistency. It must be noted that the SVD method

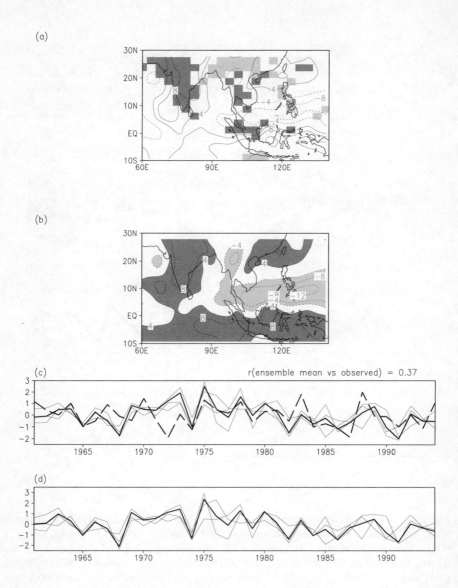

**Fig. 7.8** Same as Fig. 7.7 but for the South East Asia monsoon (SEAM) in JJAS. Dark shading for >4, light shading for <-4 and contours at +/-4, 8, 12.

does not require any relation on the temporal behavior of the coefficients. The large coherence among the members is real and it is not an artifact of the method.

This analysis identifies the mode that is explaining most of the cross-covariance among the ensemble, but it cannot say anything about the correspondence of the common mode with the real world. The simulations could be such that the entire bundle of the ensemble is deviating from reality, creating good agreement between the members, but a poor correspondence with the real world.

In order to check this last point, we can still make use of the SVD approach, to attempt to assess the amount of covariability between the ensemble and the observations. It is a very generous test towards the simulations, since we try to detect any kind of correspondence. In this case the SVD must be built using as left field the observations and as right field the simulations. The result is Panel (a) in Fig. 7.7, that shows the distribution of the first mode of covariability. The contours are the model part, whereas the shading indicates the observation part, in this case land precipitation observations from Hulme (1991). The time distribution of the coefficient is in Panel (c). It is possible to observe that the there is some correspondence in space between the patterns, indicating some skill of the model to reproduce the major mode of variability. There is also a good correlation in time, indicating a high reproducibility and a good confidence in the reproduction of the model. The confidence in the simulations is also increased by the similarity between the model part of the mode covarying with observations, panel (a), and the common mode, panel (b). The similarities indicate that whenever the members of the ensemble agree, they tend to agree because a realistic mechanism is reproduced correctly by the model.

The analysis of the global tropics is encouraging, but there are considerable regional differences. Fig. 7.8 shows the same analysis applied to the South East Asian monsoon. The skill and reproducibility of the model are considerably inferior. The model has some difficulty in following the interannual variability of the monsoon and the time correlation over the entire period is 0.37, even if the same correlation, restricted to the period after 1980, increases somewhat. The combined usage of the SVD analysis allows however a detailed investigation of the performance of the model.

## 7.5   Longer time scales

The previous discussion was mainly centered on interannual variability, but it would be interesting to assess the capability of the model to reproduce longer time scale variations. This is a very difficult problem that involves delicate problems of detection and statistical reliability. Though the duration of the experiments is barely enough to allow a discussion, the situation is not without hope, because the experiments are carrying the signature of longer time scale trends. Fig. 7.9 shows the time evolution of precipitation averaged in two areas in Europe, a North European area and the Mediterranean basin. The observation (bottom panel) show clearly a distinct trend of diminishing precipitation in the Mediterranean and increasing precipitation in Northern Europe.

The simulations (top panel) are more difficult to interpret. There is no apparent trend, even in the ensemble mean, that is reproduced here for simplicity. The appar-

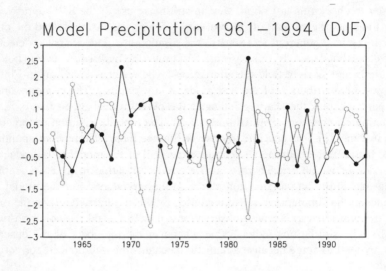

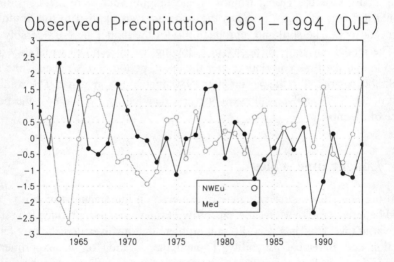

**Fig. 7.9**   Index of precipitation in DJF between 1961-1980 and 1980-1994 for model (top) and observations (bottom) for areas in North Europe (NWEu) and the Mediterranean (Med).

Model

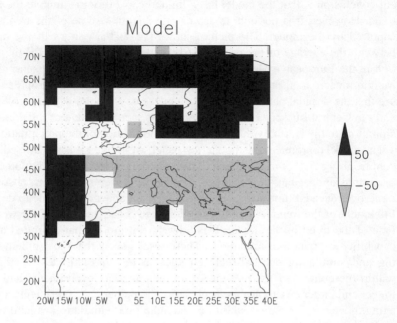

Observations

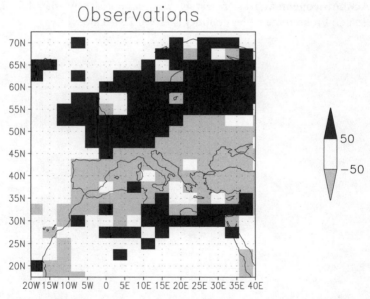

**Fig. 7.10** Precipitation differences in DJF between the average for the 1961-1980 period and 1980-1994 period. Model data (top) and observations (bottom). Units are mm.

ent conclusion is that the model has a limited capability to simulate the long term trend. However, it is possible to see in Fig. 7.10 that some of the trend has been captured by the model. The picture shows the spatial pattern of the differences between the average of the period 1961-1980 minus 1980-1994 for the precipitation in the European - Mediterranean region. The observations display the drying Mediterranean area, the progressively more humid Northern Europe and we can see that the decadal pattern has the appearance of a dipole, with a zero line apparently in Central Europe. The recent period has seen a shifting of the dominant precipitation to the North, with a consequent modification of the precipitation pattern.

The model reproduces the dipole (top panel) well. Even if is not immediately evident from Fig. 7.10, the amplitude of the pattern is overestimated, but otherwise the geographical distribution is realistic. The model is therefore capturing some of the trend over decadal time scales, even if it does not appear to be able to describe the full extent of the trend. A word of caution is in order, because of the weak amplitude of the trend in the model. It is however intriguing that a limited amount of capability exists in the simulations. These results show that the problem of detecting and simulating decadal-scale changes in the mid-latitudes is still difficult within the context of the present set-up of numerical experiments and probably longer and larger ensembles are necessary. The numerical experiments are encouraging, however, for the possibility of extending our simulation capability and our understanding to longer time scales than interannual.

**Acknowledgement.**This research was supported by the EC Environment Research Programme under contract EV5V-CT94-0538

# 8 Impact of Ensemble Size on the Assessment of Model Climate Signal

Z.X. Li

*Laboratoire de Météorologie Dynamique du CNRS, Ecole Normale Supérieure, 24 rue Lhomond, 75231 Paris cedex 05, France*

## 8.1 Introduction

Climate simulations realized with an atmospheric General Circulation Model (GCM) are mainly a boundary-forced problem, the forcing being sea surface temperature (SST) and sea ice distribution. However, recent results obtained by, among others, Harzallah and Sadourny (1995), Stern and Miyakoda (1995), and Barnett (1995) demonstrate that model internal variability can be large and it obscures sometimes the good detection of climate signal. Although the cause of this internal variability is not fully determined, it seems that ensemble approach can overcome the difficulty, since we can then enhance the signal/noise ratio. In the scientific community, ensemble approach is more and more employed although the need on computing resource is largely increased. One important question is thus the following: what is the minimum number of runs necessary to overcome the model internal variability and to study, with confidence, the model's response to external forcing ? It is clear that the answer depends upon the signal to study and the method used to detect the signal.

For a climate problem, both signal and noise levels are unknown a priori. However when a large number of simulations are realized, we can make statistical test to check if a climate signal is significant. An alternative method is the Monte-Carlo technique, i.e., we increase systematically the ensemble size and search for the point where the result becomes stable. This saturation point is precisely the optimum number of realizations to assess the signal. In this study, we present some results of the Monte-Carlo approach. The studied signal is the interannual scale signal of El Niño / South Oscillation (ENSO), but the methodology can be easily extended to decadal variability. Section 8.2 describes the model used and the climate signal to study. Section 8.3 presents the impact of ensemble size on the assessment of climate signal.

## 8.2 Model and climate signal

The model used in this study (LMD-Z, version 1) is derived from the LMD standard GCM (Sadourny and Laval 1984). But the code has been rewritten with more physical consistency, and organized in a more flexible manner for utilization. In the standard model used in LMD, the meridional discretization was regular in sines of the latitude. This made the model to have a poor resolution for high latitudes and thus the simulated atmospheric circulation for these regions tended to loss some reality. That is why the model's grid points in the present study are chosen to be regularly distributed in the longitude-latitude coordinates with resolution of $5° \times 4°$. The model has 11 vertical layers, unevenly spaced to allow a better resolution in the boundary layer. The time step is 6 minutes, however, the contributions of physical parameterizations are evaluated only every 30 minutes.

The model was integrated from January 1979 to December 1994. We used the observed monthly SST and sea-ice distributions based on the COLA-CPC analyses (Reynolds, 1988) as the boundary condition of the model. The ensemble of our simulations contains 8 members. The initial state for the first run was from a 6-month simulation forced by the climatological SST, while the initial state for each of the subsequent realizations were assigned with the same values as those at the last time step of the preceding integration. The present study hence entails altogether $16 \times 8 = 128$ years of model integration. Fig. 8.1 shows the SST anomalies of the Niño-3 region (5S-5N and 92W-150W) which provides a good proxy (Niño-3 SST index) for El Niño events. Note that a low-pass filter has been applied before drawing Fig. 8.1 in order to keep only the variability with time period longer than 8 months. Several warm and cold episodes occurred over the experiment period. We can note the El Niño events of 1982/1983, 1987, 1991/1992 and the most pronounced La Niña event of 1988.

The climate signal that we would like to study is the teleconnection between the tropical Pacific SST and the Northern Hemisphere winter circulation. The latter can be well represented by the 500-hPa geopotential height. We will show, in the rest of the section, that this signal is well captured in the ensemble mean of our simulations. Comparison will be performed against NCEP re-analysis.

The teleconnection detection method is the SVD (Singular-Value-Decomposition) analysis, which is a powerful technique for identifying the dominant coupling modes between two data fields (Bretherton et al. 1992 ; Wallace et al. 1992 ; Lau and Nath 1994, among others). Note that the SVD technique is here applied to the cross-correlation matrix, i.e., the anomaly fields have been normalized by the local standard deviations in prior to the SVD analysis. Furthermore, in order to take into account the spatial inhomogeneity of the grid, the anomaly fields are also normalized by cos $\phi$, where $\phi$ is the latitude of the grid.

The left three panels of Fig. 8.2 show the leading SVD mode obtained by using the ensemble-mean signal of the model. The middle three panels are from the NCEP re-analysis of the same period (1979-94). We should emphasize that the NCEP data have been objectively interpolated to the model's grid before the statistics is applied, so that the analysis was done in the same manner for both model and NCEP data. The top and middle contour panels give the heterogeneous correlation maps for the height and SST anomalies respectively, and the bottom panels their time coefficients for the leading mode. For the tropical Pacific SST, it is a typical El Niño structure for both observation and simulation. The resemblance between the

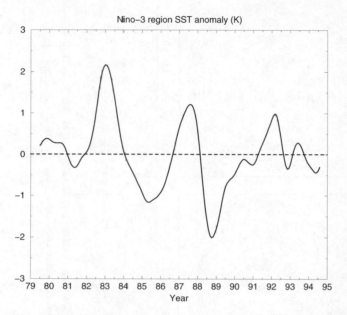

**Fig. 8.1**   Filtered sea-surface-temperature anomalies (K) of the Niño3 region (equatorial Pacific: 5S-5N and 92W-150W). The labels of abscissa are plotted at the beginning of each year.

model and the re-analysis is also good for the 500-hPa geopotential height, both of which show the same zonally-coherent structure in the tropics and several centers of action in the extratropics. The dominant structure is the PNA pattern, but other teleconnection patterns are also superimposed on this leading structure. Several discrepancies should be noted: the negative centers over North of Japan and North Atlantic are absent in simulation. Compared to the observation, a pair of centers over East and central Asia is eastward shifted to China and a negative center of the PNA over the Aleutian islands is shifted westward.

To make sure that the obtained SVD structures are not statistically spurious results, we also made EOF analysis separately for each of the two fields (tropical SST and 500-hPa geopotential height). The leading EOF structures (not shown here) are very close to the leading SVD structures.

## 8.3   Impact of ensemble size

We now perform, once again, the SVD as in Section 8.2. But instead of using the ensemble-mean data, we combined the eight simulations (16 years each) together to yield a long single simulation of 128 years. So the total signal is now analyzed.

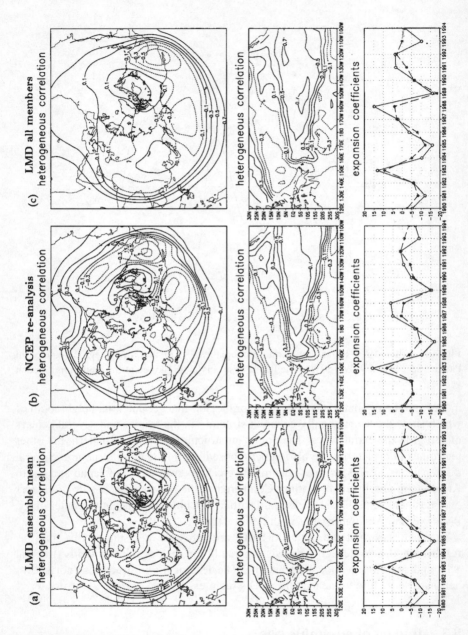

**Fig. 8.2** Leading SVD modes for respectively the model's ensemble mean (left three panels), NCEP re-analysis (middle three panels) and all ensemble members (right three panels). The top contour maps are for the Northern Hemisphere 500-hPa geopotential height, and middle ones the tropical Pacific sea surface temperature. The bottom curves show the expanded time coefficients.

The leading SVD structure is shown in Fig. 8.2c. The comparison with Fig. 8.2a reveals that the spatial structures are roughly the same, but the correlation is generally weaker. The comparison with NCEP re-analysis (Fig. 8.2b) is less satisfactory in Fig. 8.2c than in Fig. 8.2a. It is clear that the internal model variability modifies the results quantitatively, but still the results are qualitatively similar and remain robust.

We now evaluate the impact of ensemble size. Our approach is a Monte-Carlo method, already used by Cheng et al. (1995) and Richman (1986) for a similar purpose. We repeat the SVD analysis to subsets of the eight simulations. We can construct eight classes of the subsets based on the eight individual runs. For the subset $n$, the number of possible combinations is $C_8^n$. For convenience, the structure of the subset 8 (it is the total ensemble) will be referred to as master structure (Fig. 8.2c). The structures issued from other subsets will be called perturbed structures. We can now define a congruence coefficient to measure the resemblance between the master structure and the perturbed ones:

$$g = \frac{\sum (m_i p_i)}{\sqrt{\sum (p_i^2) \sum (m_i^2)}} \qquad (8.1)$$

where $m_i$ and $p_i$ denote elements of the respective (master or perturbed) structure vectors. The form of this coefficient is similar to a correlation coefficient, as the possible values range from $+1$ (perfect agreement) through $0$ (completely no relationship) to $-1$ (perfect agreement if sign reversed). As stated in Richman (1986), the congruence coefficient is a better measure than the pattern correlation coefficient in the context of quantitatively assessing the goodness-of-match of loading vectors since the former measures the similarity between two patterns without removing their respective means. However the congruence coefficient is biased towards a higher value than the correlation coefficient. According to Richman (1986) and the references therein, it is reasonable to attach the following goodness-of-match labels to specific ranges of absolute values: 0.98 to 1 (excellent match), 0.92 to 0.98 (good match), 0.82 to 0.92 (borderline match), less than 0.82 (poor match).

Table 8.1 shows the results of our calculation. We can see that, for a single realization, five out of the eight cases are under the borderline. However for a combination of five simulations, all the congruence coefficients are in the good or excellent range. We could conclude that at least 5 or 6 simulations are necessary to obtain a robust assessment of the teleconnection between the tropical SST anomalies and the 500-hPa height anomalies over the entire Northern Hemisphere. The time period of our study (1979-94) includes roughly three ENSO cycles, our conclusion may be then translated into another form, i.e., at least fifteen ENSO cycles are necessary for a robust identification of associated teleconnection patterns in the Northern Hemisphere. Here we used implicitly the assumption that the climate system is an ergodic one, without any significant lower-frequency modulation of the ENSO variability.

|          | Excellent | Good      | Borderline | Poor      | Total |
|----------|-----------|-----------|------------|-----------|-------|
| Subset 1 |           |           | 3 (37%)    | 5 (63%)   | 8     |
| Subset 2 |           | 8 (29%)   | 18 (64%)   | 2 (7%)    | 28    |
| Subset 3 |           | 40 (71%)  | 16 (29%)   |           | 56    |
| Subset 4 | 9 (13%)   | 56 (80%)  | 5 (7%)     |           | 70    |
| Subset 5 | 23 (41%)  | 33 (59%)  |            |           | 56    |
| Subset 6 | 21 (75%)  | 7(25%)    |            |           | 28    |
| Subset 7 | 8 (100%)  |           |            |           | 8     |

**Table 8.1** Number of combinations whose congruence coefficient is in the specific match range : excellent, good, borderline and poor. The percentages in parenthesis indicate the ratios over the total number of combinations which is shown in the last column.

To demonstrate that the minimum number of realizations is signal-dependent, we repeated the same analysis as above, but this time it was based on the relation between tropical Pacific SST and 200-hPa velocity potential between 45S and 45N. We find no single case in which the congruence coefficient falls below the borderline (Table 8.2). A robust detection of the ENSO signal in the velocity potential can be obtained based on just three (or even two) realizations. This result is consistent with the strong coupling in the tropics between the SST and atmospheric convective anomalies associated with ENSO.

## 8.4   Conclusion

By using an atmospheric GCM forced through the observed SST from January 1979 to December 1994, we have completed an ensemble experiment with eight members. Our results show that a single model simulation is inadequate and often misleading for studying the interannual climate variability, since the internal model variability (climate noise) is in general larger than the forced variability (climate signal), except for the tropical strip. In general, the length of simulation and the number of ensemble depend upon two factors : the signal-to-noise ratio and the degrees of freedom contained in the experiment. If a signal is absent, it can not be detected with a simulation of even infinite length. On the other hand, if the number of degrees of freedom (the number of ENSO cycles in our case) is small, the signal's statistic significance is not warranted.

By using a Monte-Carlo approach, we evaluated the influence of the internal model variability on the external climate signal and showed that fifteen ENSO cycles are the minimum length of simulation to assess the teleconnection between the tropical Pacific SST and the 500-hPa circulation for the whole Northern Hemisphere. However, this minimum length can be reduced if we are only interested in tropical signals. The Monte-Carlo approach used here can be generalized to other tests of climate reproducibility.

| | Excellent | Good | Borderline | Poor | Total |
|---|---|---|---|---|---|
| Subset 1 | 1 (13%) | 4 (50%) | 3 (37%) | | 8 |
| Subset 2 | 13 (46%) | 14 (50%) | 1 (4%) | | 28 |
| Subset 3 | 44 (79%) | 12 (21%) | | | 56 |
| Subset 4 | 65 (93%) | 5 (7%) | | | 70 |
| Subset 5 | 56 (100%) | | | | 56 |
| Subset 6 | 28 (100%) | | | | 28 |
| Subset 7 | 8 (100%) | | | | 8 |

**Table 8.2** Same as in Table 8.1, but for the 200-hPa velocity potential over the region of 45S to 45N instead of the 500-hPa geopotential height over the whole Northern Hemisphere.

**Acknowledgments.** This research was supported by the EC Environment Research Programme under contract EV5V-CT94-0538 and the French national program on climate dynamics (PNEDC).

# 9 Interannual and Decadal Variability in the Tropical Pacific

SIMON F. B.TETT, MICHAEL K. DAVEY AND SARAH INESON
*Hadley Centre,*
*U.K. Meteorological. Office,*
*London Road, Bracknell, Berkshire, RG12 2SY, UK*

## 9.1 Introduction

Differences in regional climate from one year to the next, such as seasonal rainfall variations, can have large direct and indirect effects on society. The largest such interannual variability occurs in the tropical Pacific, where strong ocean-atmosphere interaction associated with the warm seas generates substantial anomalies. These El Niño Southern Oscillation (ENSO) events have a global impact, and in many regions interannual climate anomalies are significantly associated with the cycle of these events. Glantz et al. (1991) provide a wide-ranging review of ENSO and its impacts. Background information and an atlas of ENSO impacts since 1871 can be found in Allan et al. (1996).

Understanding the ENSO phenomenon was a major goal of the Tropical Ocean Global Atmosphere (TOGA) programme. During TOGA (1985-1994), substantial advances were made through the use of enhanced observational networks and the application of a range of models. (For reviews, see TOGA, 1997).

The most complex climate models consist of coupled atmosphere and ocean general circulation models (AOGCMs). However simple coupled models, using reduced physics atmospheres or statistical atmospheres coupled to shallow water type ocean models are quite successful in simulating and predicting the timing and magnitude of significant SST anomalies in the equatorial east Pacific (Zebiak and Cane, 1987, Barnston et al., 1994). It is hoped that the continuing development of fully coupled AOGCMs will eventually lead to better and more comprehensive predictions, including the impacts on the extra-tropics from changes in the position of the warm pool and associated convective centers.

As well as interannual changes, there are climate fluctuations on decadal and longer scales (Parker et al., 1994) with evidence of decadal variations in ENSO (Diaz and Markgraf, 1992; Balmaseda et al., 1995; Gu and Philander, 1995; Wang and Ropelewski, 1995; Wang, 1995), and of extratropical variations (Mann and Park, 1994; Kushnir, 1994). Such variations also involve ocean-atmosphere interaction, and coupled models have provided insight into some likely mechanisms (Latif, 1996; Latif and Barnett, 1994, 1996; Delworth, 1996; Chang et al., 1997).

On interdecadal timescales, AOGCMs are used to make predictions about changes in the climate of the Earth as greenhouse gases such as $CO_2$ increase. Several modelling centers have made long integrations using coupled models to examine the transient effects of increases in such gases (Washington and Meehl, 1989; manabe et al., 1991; Murphy and Mitchell, 1995). More recently experiments have been run with such models in an attempt to simulate past climate change (Mitchell et al., 1995). Recent work in the detection of anthropologically caused climate change relies on model estimates of natural variability on interdecadal and greater timescales (e.g. Santer et al., 1996; Tett et al., 1996)). Confidence in the results of these coupled models can be increased if the processes such as ENSO occurring in the observed coupled system are properly reproduced by these models.

This chapter contains examples of the analysis and assessment of interannual and decadal variability in three AOGCMs which have been developed at the Hadley Centre, United Kingdom Meteorological Office (UKMO). The models, referred to as HADCM, HADCM2 and TOGAGCM, are described in section 9.3 while the datasets used to compare the models and observations are described in section 9.2 HADCM and TOGAGCM use the same AGCM, but have different ocean components (global and regional Pacific respectively).

In section 9.4 the interannual tropical Pacific behavior of HADCM and TOGAGCM is described and compared with observed behavior, concentrating on sea surface temperature (SST) and upper ocean vertically averaged temperature (VAT) fields. In section 9.5 brief examples of results from a 1300 year integration of a more recent AOGCM developed at the Hadley Centre, HADCM2, are described.

## 9.2    Data sources

In the study reported in this chapter we use several datasets:

1) The Earth Radiation Budget Experiment (ERBE) (Harrison et al., 1990) data provides observed top of the atmosphere outgoing longwave radiation (OLR) and shortwave albedo for the 1985—89 period. In the tropics OLR is a good proxy for convective activity (Zhang, 1993). In regions of high OLR the shortwave albedo gives an indication that low cloud is present.

2) For SST, we used the UKMO monthly Global Ice and Sea Surface Temperature (GISST) dataset (Parker et al., 1993), which is a development of a previous UKMO global sea surface temperature data-set described in the atlas by Bottomley et al. (1990). We use the period 1962 to 1993 from this data-set.

3) A Pacific Ocean analysis (assimilating ocean surface and sub-surface temperature observations into an OGCM (Leetmaa and Ji, 1989) from the US National Meteorological Center was used to provide an estimated ocean temperature dataset from the surface to a depth of 360.0 meters. The so-called RA3 dataset (Ji et al., 1995; Smith and Chelliah, 1995; Ji and Smith, 1995) for the 126 month period from July of 1982 to December of 1992 was used.

**4)** The final dataset we use is a long term record of the Southern Oscillation Index (SOI) (Parker, 1983) covering the period from 1866 to 1991.


## 9.3  Model Description

In this section we provide some details of the three related AOGCMs (HADCM, HADCM2 and TOGAGCM).

The global AGCM used in all the coupled integrations described below has resolution of 2.5° latitude by 3.75° longitude with 19 levels on a hybrid co-ordinate vertical grid. A split-implicit integration scheme is used allowing a timestep of 30 minutes. A comprehensive physics package is used, including a stability dependent cloudy boundary layer scheme (Smith, 1990), a land surface hydrology scheme (Gregory and Smith, 1990; Dolman and Gregory, 1992) and a radiation scheme with interactive optical properties (Slingo, 1989). The mass flux convection scheme used to represent shallow, deep and mid-level convection (Gregory and Rowntree, 1990) has been modified by the inclusion of deep convective downdrafts. A more comprehensive overview of the atmospheric model can be found in Cullen (1993).

The OGCMs are variations of the basic Bryan/Cox finite difference model (Cox, 1984). A mixed layer model, based on that of Kraus and Turner (1967), is embedded in the ocean models and the K-theory mixing scheme of Pacanowski and Philander (1981) is used to parameterize vertical turbulent mixing.

For HADCM and HADCM2 the ocean has 20 vertical levels, with 7 levels covering the top 113 meters. The depths and thicknesses of these levels are shown in Table 9.1. For HADCM and HADCM2 the horizontal OGCM resolution of 2.5° latitude by 3.75° longitude, matching the AGCM horizontal resolution is relatively coarse to allow long integrations.

The HADCM2 AOGCM is a later development of the HADCM AOGCM. The major differences between the two include the treatment of ice, fixing an error in the location of the stress due to gravity wave breaking and correcting the Levitus salinities near Antarctica for summer bias used to derive the flux corrections. Further description of HADCM2 and its applications can be found in Johns et al. (1997), Tett et al. (1997), Mitchell et al. (1995), Hewitt and Mitchell (1996) including a 1300 year integration in which forcing was kept constant. We analyse time-series of tropical Pacific temperature and Southern Oscillation from this simulation.

The ocean model for TOGAGCM is limited to the tropical Pacific region between 30°N and 30°S, but has higher and variable horizontal resolution to represent equatorial processes more accurately. The horizontal grid has a meridional grid spacing of 1/3° at the equator, increasing to 1° at the northern and southern boundaries, and zonal grid spacing of 1.5° decreasing to 0.5° near the eastern and western boundaries. The vertical grid uses the upper 16 levels shown in Table 9.1. The barotropic mode is omitted and there is no variation in model bottom topography. Temperature and salinity at open northern and southern boundaries are kept

| Model Layer | Thickness | Depth of mid point |
|:-----------:|:---------:|:------------------:|
| 1 | 10.0 | 5.0 |
| 2 | 10.0 | 15.0 |
| 3 | 10.0 | 25.0 |
| 4 | 10.2 | 35.1 |
| 5 | 15.3 | 47.9 |
| 6 | 23.0 | 67.0 |
| 7 | 34.5 | 95.8 |
| 8 | 51.8 | 138.9 |
| 9 | 77.8 | 203.7 |
| 10 | 116.8 | 301.0 |
| 11 | 175.3 | 447.1 |
| 12 | 263.2 | 665.3 |
| 13 | 395.3 | 995.6 |
| 14 | 616.0 | 1501.1 |
| 15 | 615.0 | 2116.0 |
| 16 | 615.0 | 2731.0 |
| 17 | 615.0 | 3347.0 |
| 18 | 616.0 | 3962.0 |
| 19 | 615.0 | 4577.0 |
| 20 | 616.0 | 5193.0 |

**Table 9.1** Ocean model level depths (m)

close to their seasonally varying climatological values by Haney forcing to Levitus (1982) values. Outside the ocean model domain SST is kept fixed to the observed climatological seasonal cycle. The seasonal cycle of TOGAGCM model is described in Mechoso et al. (1995), and further details of its interannual behavior can be found in Ineson and Davey (1997). The results discussed here are based on a 25 year integration of this model.

In each case the ocean models are coupled to the atmosphere model by the same procedure: heat and fresh water fluxes and windstress are accumulated by the atmospheric model and passed to the ocean every day. In the ocean model daily average values of SST are computed and passed to the atmosphere model once a day.

Flux adjustment is used in HADCM and HADCM2 to control climate drift. There is no flux adjustment in TOGAGCM, but the SST is constrained to observed climatology outside the ocean model domain.

## 9.4    Climatology and Interannual Variability

In this section the behavior of HADCM and TOGAGCM in integrations of 20 and 25 years respectively is compared with observed behavior, to illustrate assessments of their skill at reproducing tropical climatology and variability.

### 9.4.1  Climatology

The observations show three convective centers (Fig. 9.1, a-b) whose positions change throughout the seasonal cycle: one centered over Indonesia which extends west into the Indian ocean and east into the west Pacific, another centered over Africa and the third over central America/South America. As noted by Meehl (1987) the movement throughout the seasonal cycle is a north to south and a west to east movement. Other important features are the warm SST pool in the west Pacific (Fig. 9.1, a-b), with temperatures greater than 29°C, and the region of cooler waters in the equatorial east Pacific. The feature of note in the cross section of near-surface ocean temperature is the strong thermocline (Fig. 9.1, c), shallowest in the east Pacific and deepest in the west Pacific.

The SST pattern in HADCM (Fig. 9.2, a-b) is quite similar to that observed (recall that HADCM has flux adjustment), with the model qualitatively capturing the large-scale features of the tropical SST in both July and January. There are several differences: during July the model warm pool is smaller than observed and mainly consists of a tongue of warm water (greater than 29°C) extending to the east of New Guinea with small regions of warm water near the Philippines. Compare this to the observations for July which show a large region of warm waters to the east of the Philippines. In the Indian ocean the waters around India are cooler than observed during July and January.

If we examine the regions of low OLR (Fig. 9.2, a-b), corresponding to regions of high convective activity, we find that the low OLR values associated with the Indian monsoon have too small an extent and the convection over the north west equatorial Pacific is too strong. During January HADCM suppresses convection in the Indonesian region and strengthens it over both the Indian and west Pacific Oceans. The model is generating convection over the Andes mountains; something not observed in nature. The other convective centers, over Africa and South America are smaller than the comparable centers in the observations. These errors in the placement of low OLR values are similar to those of an AMIP integration (not shown) strongly suggesting that the errors are due to problems in the atmospheric component of the model.

HADCM captures the major upper ocean features (Fig. 9.2, c); the isotherms can be seen rising across the Pacific with slopes similar to those observed, as we travel from west to east. In the east Pacific the thermocline is quite strong, but in the west and central Pacific the thermocline is substantially more diffuse than observed. Furthermore, in the east Pacific the subsurface temperatures below the thermocline are too warm, probably due to the excessive diffusion of heat through the model thermocline.

Simon F. B.Tett, Michael K. Davey and Sarah Ineson

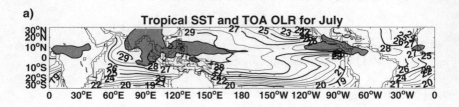

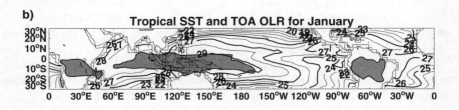

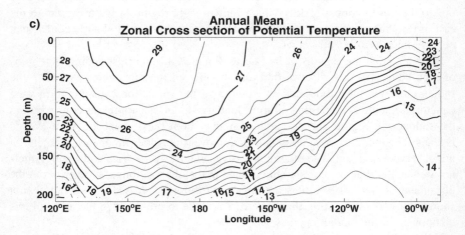

**Fig. 9.1** Observed tropical climatology. Plots (a) and (b) show the observed sea surface temperature in the tropics for July and January respectively using GISST data for 1985—89 for which period ERBE data is available Plot (c) shows an equatorial section of annual mean ocean temperature in the Pacific (averaged from 2.5N to 2.5S) using Levitus (1982) data. A contour interval of 1°C is used in all plots. Below 26°C every 5th contour is bold, above 26°C every odd contour is bold. Also shown in plots (a) and (b) by stippling are regions where the outgoing longwave radiation at the top of the atmosphere is less than 220 W/m² taken from ERBE data.

a)
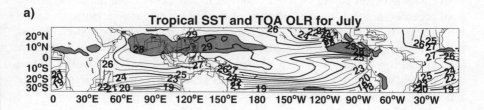

b)

c)
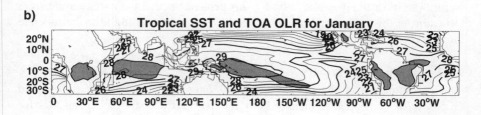

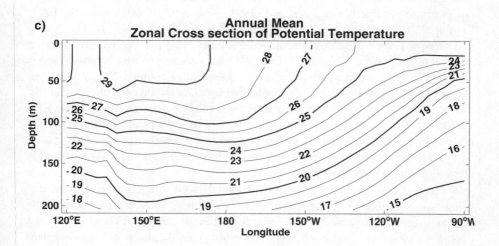

**Fig. 9.2** HADCM tropical climatology. The plots, contours and shading are the same as in Fig. 9.1 except that they are taken from the HADCM experiment.

The equivalent plots for TOGAGCM are shown Fig. 9.3. Only the SSTs from the active Pacific ocean model region are shown. In the west Pacific the model warm pool is cooler than the observed warm pool, with only very small regions where the temperature is greater than 29°C. The warm water in the west Pacific stretches too far to the east. Furthermore the model is failing to capture the cold pool in the east Pacific. During January TOGAGCM generates a closed region of cooler waters near the equator. The TOGAGCM equatorial Pacific zonal temperature gradient is much weaker than observed.

In TOGAGCM the thermocline is also too flat. This is consistent with the weak zonal winds (not shown) in this model. In common with other CGCMs without flux adjustment, SST quickly rises in the east Pacific after coupling the ocean to the atmosphere. In this model, this leads to weakening of equatorial winds and a drift to the model climatology shown in Fig. 9.3. (See Ineson and Davey,1997, for further details.)

Like HADCM, the simulated subsurface temperatures in the east Pacific are too warm. The most likely explanation for this behavior is that the surface waters are too warm in this region and more heat is diffusing down through the thermocline into the cooler waters below.

### 9.4.2 Inter-annual variance

The simulation of the observed tropical inter-annual variability by coupled GCMs is of considerable interest both for use in seasonal forecasting and for improving models used for climate prediction. Many modelling groups have described the ENSO-like behavior of their coupled AOGCMs (Philander et al., 1989; Lau et al., 1992; Meehl, 1990; Nagai et al., 1992; Neelin et al., 1992; Philander et al., 1992; Sperber and Hameed, 1991; Tett, 1995; Robertson et al., 1995; Latif et al., 1993b; Latif et al., 1993a; Terray et al., 1995).

There are several different mechanisms causing ENSO like behavior in different coupled models. see e.g. Neelin et al. (1994) for a review. The two principal modes which occur are a propagating (normally westwards) coupled SST mode, and another in which SST variability is controlled by eastward propagating changes in the thermocline depth with atmosphere/ocean feedbacks acting to amplify the SST anomalies. The second mechanism is believed to be responsible for the observed ENSO phenomenon (Philander, 1990). In most low resolution models the coupled SST mode is the primary mechanism for SST variability; however some models (Tett, 1995) have both mechanisms active.

In Neelin et al. (1992) many coupled models were intercompared, primarily using time-longitude plots of equatorial SST anomalies. These figures can be difficult to interpret, in particular determining dominant frequencies and regions of maximum variability is difficult. These time-longitude diagrams only present data from the near equatorial region and do not encourage a statistical comparison, focusing attention on the detailed time behavior of the individual models. In this subsection we carry out an intercomparison by first examining total power in different frequency bands. This techniques focuses attention on the statistical proper-

a)
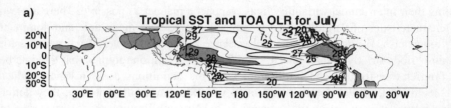

b)
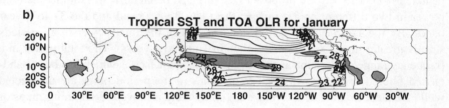

c)
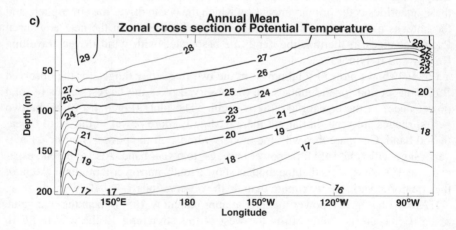

**Fig. 9.3** TOGAGCM tropical climatology. The plots, contours and shading are the same as in Fig. 9.1 except that they are taken from the TOGAGCM simulation.

ties of the model variance. We also compare timeseries of NINO3 SST to verify the usefulness of the technique.

In order to successfully intercompare models we need a method of characterizing their inter-annual variability in as compact a manner as possible. There are two characteristics of variability that we consider important; magnitude and dominant frequencies. We define inter-annual variance as the total power of a timeseries for all periods strictly greater than 12 months. This definition requires that the data be Fourier transformed and is different from the definitions used in other studies. Meehl et al. (1994) define inter-annual variability in a different way; they define inter-annual variance as the variance of a 12 month running mean timeseries of monthly anomalies. A disadvantage of using a 12 month running mean is that it has effects on frequencies greater than 1 year. Furthermore in our method no removal of the seasonal cycle is required.

The observed major timescales of ENSO are the biennial timescale (Meehl, 1993; Rasmusson et al., 1990) and the 3 to 6 year timescale (Rasmusson and Carpenter, 1982). The way we propose to capture this behavior is to examine the total power in two different period ranges; the 1+ to 3 year band and the 3+ to 6 year band. (By the 1+ to 3 year period range we mean the total power in all periods strictly greater than 1 year and less than or equal to 3 years.) The total power in a frequency range is the same as the variance of the timeseries which has been bandpassed filtered in such a way that frequencies in the period range are unaffected while frequencies outside the range are completely removed. Examining maps of total power in these period ranges enables us to determine the major characteristics of the model interannual variability. We consider SST and the upper ocean vertically averaged temperature (VAT) from the models and observations. The first of these quantities is the primary means by which the ocean drives the atmosphere and the latter is a diagnostic for changes in thermocline depth. In the near equatorial Pacific changes in thermocline depth are associated with wind driven travelling Rossby and Kelvin waves.

Fig. 9.4 shows plots of total power in the two frequency ranges for the observed SST data-set and for the models. In the observed plots the largest values of total power occur in the equatorial eastern Pacific. The plots (Fig. 9.4 a-b) of the two period ranges are very similar but there is slightly more power in the 3+ to 6 year period band. Compare that with the models; almost all the power is in the 1+ to 3 year range with relatively little power in the 3+ to 6 year band. Also apparent, especially in TOGAGCM, is that the models show a much narrower latitudinal extent of the equatorial variance maximum than do the observations.

HADCM has a SST power maxima at approximately 165°W peaking at a value of $0.3°C^2$. There is a long tail to the west of this stretching all the way to South America. The peak value ($0.4°C^2$) for this model is comparable to the observed 1+ to 3 year total power but the position is too far to the west.

TOGAGCM shows maximum values along the equator, with a large longitudinal extent and peak values ($> 0.8°C^2$) that are larger than those observed, at least in the 1+ to 3 year range.

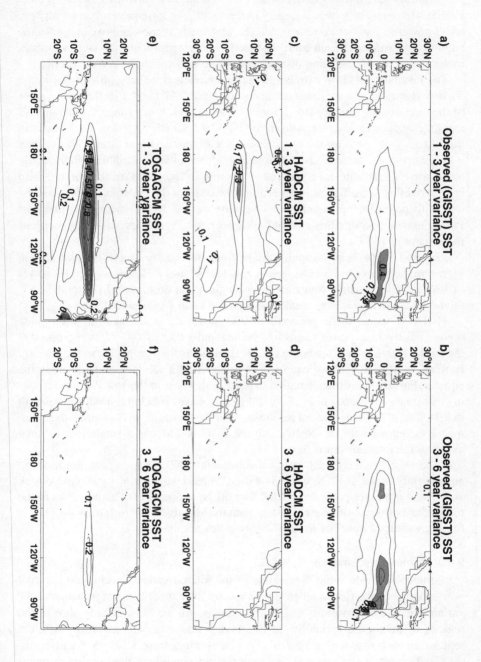

**Fig. 9.4** SST filtered variance. Panels (a), (c), and (e) show the total power in the 1+ to 3 year period range for observations, the HADCM and the TOGAGCM models respectively. Panels (b), (d), and (f) show the corresponding total power in the 3+ to 6 year period range. A contour interval of 0.1 $K^2$ is used in all plots with shading for values above 0.3 $K^2$.

To summarize, both models produce substantial SST variability in the tropical Pacific. However the power is largely in the 1+ to 3 year frequency band with very small power in the 3+ to 6 year band. The observations show approximately equal amounts of variance in each band. The models do not reproduce the large variances in the eastern Pacific, tending instead to generate maxima in the central Pacific.

The results for VAT are similar, in that the models have very little power in the 3+ to 6 year range, so we focus on the 1+ to 3 year VAT band. Fig. 9.5 shows plots of the inter-annual variance (total power for all periods greater than 1 year) in VAT from the surface to 360 meters from the NMC RA3 analysis and of total power in the 1+ to 3 year range for the two models. Recall that the observations are for only a 11 year period. The analysis shows clear evidence of large variability in the equatorial wave guide with an equatorial maximum in the east Pacific. There is also high variability spreading polewards along the coast of the Americas; this is consistent with poleward propagation of coastally trapped Kelvin waves. In the west Pacific maxima occur either side of the equator: at 8°S and, less distinctly, north of the equator at 10°N.

Both models also have variability peaks along the equator, suggesting that Kelvin waves are being excited by changes in zonal wind stress. However the total power in VAT is approximately four times weaker than that observed. (Like model SST, model VAT variance is even weaker beyond the 1+ to 3 year period range.)

HADCM seems to be generating VAT variability in the central wave guide region, but the total power seems to decline rapidly east of 140°W, suggesting that the model Kelvin waves are being rapidly damped as they propagate to the east. By contrast, in TOGAGCM we can see a clear signal of a variance maxima along the equator. Further, the variance region expands polewards in the east Pacific suggesting that poleward propagation of coastal Kelvin waves is being reproduced. (Recall that TOGAGCM has enhanced resolution near the coasts as well as along the equator.) By contrast with the NMC analysis, TOGAGCM has a smaller meridional extent of the total power maxima.

In the west Pacific, HADCM has a substantial VAT variance peak north of the equator only, at about 12°N, while TOGAGCM has a strong zonally extensive peak at 5°S and a weaker peak along 4°N. The off-equatorial west Pacific peaks in the models and observations are consistent with the idea that the wind stress variance is forcing westward traveling Rossby waves in this region.

### 9.4.3  Timeseries Analysis

In order to provide some verification of the filtered variance technique we shall now examine timeseries and their power spectra from the regions of maximum SST variance in the observations and in the models. As we have already shown, the regions of maximum variability in the models and observations are in different regions. In each case we have calculated area averages from 5°N to 5°S with a longitudinal extent of 15° centered on a region corresponding to the location of maximum variability. (105°W-90°W for observations, 165°W-150°W for HADCM, 150°W-135°W for TOGAGCM.) Fig. 9.6 shows plots of the area averaged monthly

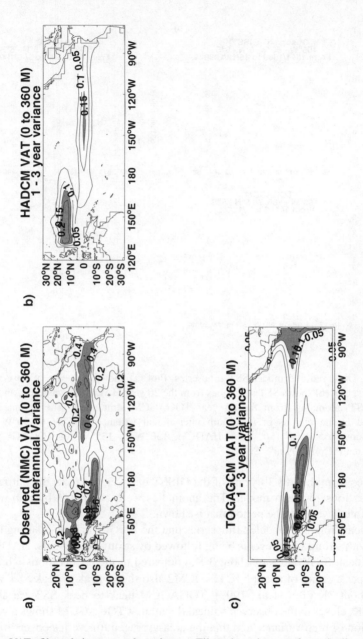

**Fig. 9.5** VAT filtered interannual variance. Filtered variance of vertically averaged temperature (VAT) from the surface to a depth of 360 meters is shown. Panel (a) is from 11 years of NMC RA3 analysis data. A contour interval of $0.2K^2$ is used, with stippling where values are greater than $0.6K^2$. Panels (b) and (c) contain similar plots for the HADCM and TOGAGCM models, but with a contour interval of $0.05 K^2$ and stippling for values greater than $0.15 K^2$.

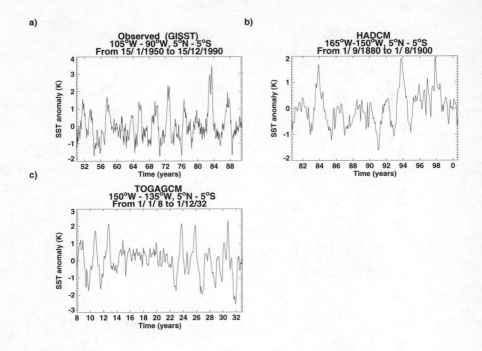

**Fig. 9.6**   Observed and model SST timeseries. Plot (a) shows 40 years of observed SST anomalies, plot (b) shows SST anomalies from the 20 year HADCM simulation, and plot (c) shows SST anomalies from the 25 year TOGAGCM simulation. The anomalies were calculated in the regions of maximum inter-annual variance: 105°W to 90°W for the observations, 165°W to 150°W for HADCM, 150°W to 135°W for TOGAGCM. Each region extends from 5°N to 5°S.

anomalies from the last 30 years of the GISST data-set and from the entire period of integrations for each model. The anomalies were computed by removing the mean annual cycle for the period that is shown.

It is clear from the modeled timeseries that the AOGCMs have a strong biennial mode, with strong warm events being followed by strong cold events. HADCM has peak to peak SST values of about 3 K, compared to the observations which have peak to peak values of about 5 K. HADCM2 also shows peak to peak SST variability of about 3K (Tett et al., 1997). TOGAGCM peak to peak SST variability is about 4 K, closer to the observed values. Recall that TOGAGCM shows a very narrow region of high variance and that the area-averaged timeseries covers a broader region. Also apparent in the modelled timeseries are quiescent and active oscillation regimes. For example, in TOGAGCM years 8 to 14 are active while for years 14 to 22 the model is quiescent. HADCM also shows similar behavior but the distinction between active and quiet periods is less clear. This is in contrast to the observations which do not show such long periods of quiescence, at least since 1948.

Having computed these timeseries we then computed power spectra (Fig. 9.7). See Tett et al. (1997) for details of these computations and the procedure followed to give the best-fit to an AR1 model

Apparent in the observed power spectra (Fig. 9.7, a) are two significant but broad peaks; one centered on approximately 48 months and the other on the biennial period. Compare this with the AOGCMs, which have peaks only near the biennial period. In TOGAGCM the strength of this peak is comparable with the maximum peak in the observed data-set. In general the power spectra confirm the picture that we drew earlier when we examined the total power in different period ranges.

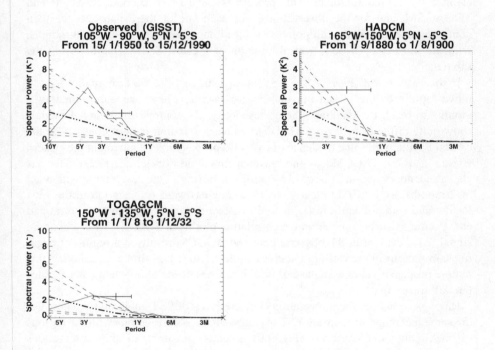

**Fig. 9.7**  Observed and Modelled SST power spectra. Power spectra of SST for the same maximum variance regions used in Fig. 9.6. The thick dot-dashed lines indicate a best-fit red noise power spectrum. The dashed lines show the 95% and 99% confidence limits — lines closer to the red noise spectrum correspond to smaller confidence limits. The bandwidth is shown by the error bar centered near the region of maximum power. The height of each ordinate is the estimated power in each frequency range multiplied by the total number of frequencies.

### 9.4.4 SST and VAT propagation characteristics

So far we have not considered the direction and speed in which SST and VAT anomalies propagate: looking at total power does not enable us to do this. We now examine the propagation properties by computing lagged correlation of SST and VAT across the entire equatorial Pacific with the regions of maximum variability which we considered above. In particular by examining the characteristics of the maximum correlation of VAT with the SST we should be able to diagnose the mean wave speed in the wave guide. Furthermore by doing this we see if there any links between changes in VAT and changes in SST.

We carry out the computations by first taking monthly mean data from the equatorial Pacific Ocean and removing the mean seasonal cycle. Having done this we area average this data between 5°N and 5°S and removed high frequency noise by using a 1-2-1 time smoother. This process is carried out for both HADCM and TOGAGCM, and for the observations. For each one of these data-sets we then compute the correlation at all points and for all lags between -18 and +18 months, with the timeseries of SST from the region of greatest SST variability from the same model.

In the observations(Fig. 9.8), using data for 1962 to 1992, SST anomalies tend to move rapidly to the west from the eastern Pacific. They take approximately 2 months to reach the dateline where they decay. By contrast the coupled models show quite different behaviors. HADCM shows a standing pattern confined to the central Pacific with some evidence of eastward propagation occurring in the eastern Pacific, as does HADCM2 on the seasonal timescale (Tett et al., 1997). There is also a secondary region of large correlations occurring in the east Pacific with a lag of 2 months. TOGAGCM clearly shows a slow eastward propagation in the west Pacific and a significantly faster, but still eastward, propagation in the central and east Pacific. Clearly apparent in the correlation plot from TOGAGCM is a biennial signal. Also evident in the observational and AOGCM results is a region of negative correlations in the western Pacific at slight positive lags. In the two models this feature propagates slowly to the east until it reaches the dateline when it accelerates but still traveling eastward.

Having considered the propagation characteristics of SST anomalies we now use the same technique to examine VAT. Again we look at correlations with the regions of maximum SST variance (Fig. 9.9). Clearly apparent in the observations (Fig. 9.9a) is a eastward traveling signal with maximum correlations at a lag of -1 month to the SST timeseries averaged over the region of maximum variance. Also clearly apparent is a triggering of negative anomalies at 150°E. The models (Fig. 9.9b-c) also show clear evidence of eastward traveling signals. In both HADCM and TOGAGCM simulations negative VAT anomalies[1] are clearly apparent in the west Pacific at the same time that the positive VAT anomalies reach the east Pacific.

---

[1] i.e. opposite sign to the   SST anomalies.

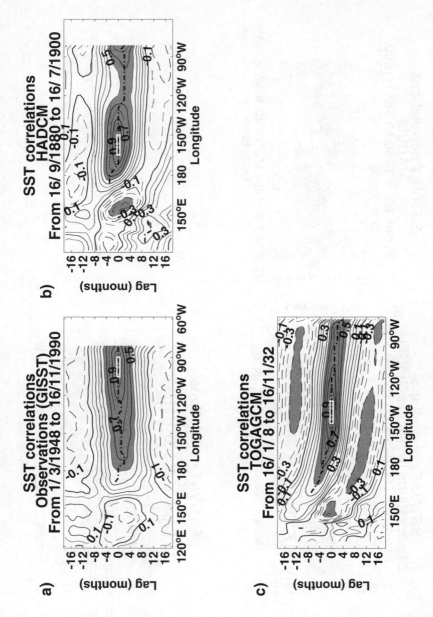

**Fig. 9.8** SST/SST correlations. Lag correlations between SST anomalies (averaged from 5°N to 5°S) across the equatorial Pacific and the region of maximum interannual SST variability (shown by a box in each plot) for (a) observations, (b) HADCM, and (c) TOGAGCM. The contour interval is 0.1 and stippling is used where correlations are less than -0.6 or greater than 0.6. Negative contours are drawn dashed, every odd contour is labeled, and every even contour is bold. The thick dot-dashed line joins the positions of the maximum correlations, where the maximum is greater than 0.4.

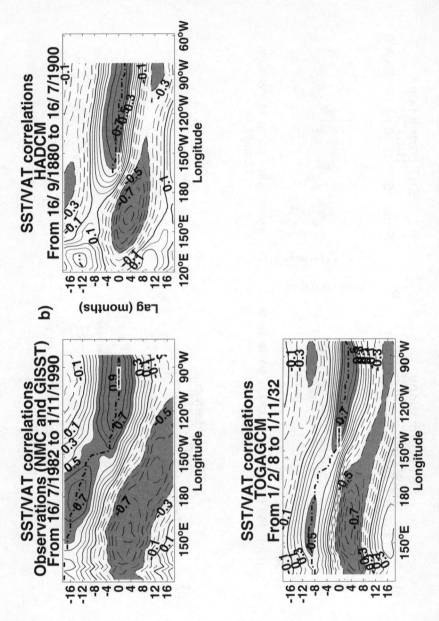

**Fig. 9.9** SST/VAT correlations. Lag correlations between VAT anomalies (averaged from 5°N to 5°S) across the equatorial Pacific and the region of maximum interannual SST variability (shown by a box in each plot) for (a) observations, (b) HADCM, and (c) TOGAGCM. The contour interval is 0.1 and stippling is used where correlations are less than -0.6 or greater than 0.6. Negative contours are drawn dashed, every odd contour is labeled, and every even contour is bold. The thick dot-dashed line joins the positions of the maximum correlations, where the maximum is greater than 0.4.

The speed of propagation changes between models and can be quite different in different places in the same data-set. We estimate the wave speed by examining the slope of maximum correlation and measuring the time lag between maximum correlation at 150°W, where all data-sets show correlations greater than 0.5, and at the coast of South America. This is a distance of $7.78 \times 10^6$m and a free Kelvin wave, traveling at 2.8m/sec would take approximately one month to travel this distance. The estimates that we make are only accurate to the nearest month as they are based on monthly mean data. Examining the observations we estimate that a signal takes approximately 3 months to travel this distance. In TOGAGCM and HADCM the VAT signal also takes about 3 months to travel the same distance.

In the west and Central Pacific analyzed VAT anomalies are moving much slower than in TOGAGCM. Maximum correlations in the west Pacific at 150°E lag those in the far eastern Pacific by 11 months in TOGAGCM and by 17 months in the observations. Some caveats are required when interpreting the NMC analysis data; it is only a 11 year dataset and the results are likely to be dominated by the strong event of 1982/3.

### 9.4.5 Zonal windstress relationships to SST.

In this subsection we examine the relationship between anomalous SST and anomalous zonal windstress.

For all points across the Pacific, between 120°E and 90°W, we form timeseries by meridionally averaging SST and zonal windstress between 2.5°S and 2.5°N. Correlations between the timeseries of anomalous SST at each point and all the timeseries of anomalous $\tau_x$ are computed. Fig. 9.10 shows plots of these correlations for the two models, and observations. Regions of strong local coupling between SST and windstress will be indicated by strong correlations along the diagonal of the diagrams, while strong off-diagonal values indicate substantial non-local interactions. Positive correlations correspond to warm (cool) SST anomalies at the longitude of the SST point associated with westerly (easterly) wind stress anomalies at the longitude of the wind point. Below the diagonals in Fig. 9.10 the SST points lie to the east of the wind points: in this region non-local interactions may be due to the triggering of Kelvin waves by wind stress anomalies that then influence SST to the east, and/or the forcing of wind anomalies to the west of SST anomalies. The latter effect is likely to operate over a limited zonal separation of the wind and SST points.

In Fig. 9.10 the correlations are positive, and largest around and east of the date-line. The highest correlations occur when windstress anomalies are slightly (10 to 15 longitude degrees) to the west of SST anomalies. This suggests that a coupled SST/windstress mode is acting with warm (cool) SST anomalies generating westerly (easterly) windstress anomalies to the west, which in turn can lower (raise) the thermocline and tend to increase (decrease) SST to the east.

These near-local correlations are highest for TOGAGCM (about 0.6), slightly weaker for HADCM (about 0.5), and weaker for the observations (about 0.4) which are less accurately known. For TOGAGCM the region of strong near-local

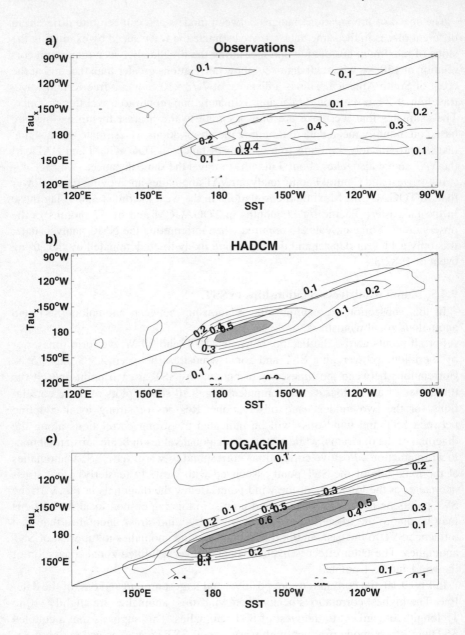

**Fig. 9.10** Correlation coefficients between SST and zonal windstress. A contour interval of 0.1 is used and shading shows where the correlation coefficient is greater than 0.5. The diagonal line shows where the two timeseries have the same coordinates.

interactions extends from the dateline to 120W, while the others are more confined in zonal extent.

For HADCM the correlations are weak away from and below the diagonal in Fig. 9.10, indicating that remote Kelvin-driven connections are weak. By contrast, such connections are substantial for the observations and for TOGAGCM, with values over 0.3 extending well into the east Pacific. For the observations, the pattern of such values indicates that SST in the east Pacific is related to wind stress near the dateline in the region of strong near-local interaction. For TOGAGCM, with its extensive region of strong near-local interaction, SST in the east Pacific is related most to wind anomalies in the eastern portion of the strong interaction zone, with much weaker relation to the strong interactions near the dateline. Although TOGAGCM has stronger remote correlations than HADCM, the remote wind-SST connections are underestimated in both.

For the observations and the GCMs, correlations above the diagonals in Fig. 9.10 are weak, except close to the diagonal where the mechanism is likely to be wind anomalies driven by SST anomalies slightly to the east. In particular, SST anomalies in the west Pacific are not related to wind anomalies in the strong interaction zones.

## 9.5   Decadal variability in ENSO

Observations that are of sufficient length to examine interdecadal variability in ENSO are quite sparse. The best long instrumental records available are surface Pressure measurements: the Southern Oscillation Index (SOI) provides a measure of large scale tropical pressure changes (see Allan et al. (1996) for a review of SOI history and data). Reliable records of SST and winds also extend back to the last century, but early observations are sparse in space and time. (e.g. the Comprehensive Ocean Atmosphere DataSet, Woodruff et al., 1987) Proxy data such as tree rings and coral growth provide evidence on multi-century timescales; see articles in Diaz and Markgraf (1992).

Recent analysis of such observations (Ropelewksi and Jones, 1987; Parker et al, 1994; Mann and Park, 1994; Wang and Ropelewski, 1996; Gu and Philander, 1995) has suggested substantial ENSO variability on decadal scales. There has been a relative maximum in El Niño activity toward the end of the 19th century, a minimum from late '20s to '50s, and relatively large variability to present. There is some evidence suggesting that ENSO variability is higher (lower) when climate mean SST is warmer (colder). Details are sensitive to the choice of SST dataset however.

The mechanism for decadal ENSO variations is not clear. Gu and Philander (1997) have proposed a mechanism involving subduction of water mass anomalies in the sub-tropical Pacific that are advected to the equatorial subsurface region and consequently influence SST variability and associated extratropical connections.

There are several ways of examining interdecadal behavior, using various forms of time series analysis (simple spectra, filters, wavelets, singular spectra, etc.).

When sufficient data is available (as is usually the case in model climate simulations) associated spatial structures can be extracted. Space does not allow a detailed analysis here: we shall simply provide some basic spectral examples.

We first compare the power spectrum of the observed SOI (based on Tahiti and Darwin surface pressure anomalies) with that simulated by the HADCM2 model (Fig. 9.11, (a) and (b)). The model SOI is defined in terms of rectangular regions around Darwin (125°W, 15°S — 130°W, 10°S) and Tahiti (150°E, 20°S — 155°E, 15°S) that represent the centers of action: 1300 years of annual-mean model data are used. Best-fit red noise power spectra are indicated by the dash-dot curves, with dashed lines indicating corresponding 99 and 95% confidence limits.

The most distinct significant peak in the observed spectrum (Fig. 9.11) occurs at about 7 years, with a weak, non-significant spectral peak at approximately 4 years. The HADCM2 SOI spectrum (Fig. 9.11a) has a well defined highly significant peak at approximately 8 years and a broad peak around 3 to 4 years. Qualitatively the observed and model spectra are similar. The absence of peaks at longer periods in the observations and model is striking. For periods from 10 to 200 years the simulated SOI shows no significant departures from white noise. Note that the simulated power is significantly greater than the observed power. HADCM2 overestimates interannual variability in the atmosphere, indicating that the AGCM probably reacts too strongly to SST anomalies.

The power spectrum of HADCM2 surface temperature variability in the central equatorial Pacific (Fig. 9.11c), also has a broad peak at 3-4 years and another peak at approximately 8 years. Generally, the model SST spectrum is quite similar to the SOI spectrum, which indicates strong ocean-atmosphere coupling on interannual scales. The similarity is less pronounced for observed SST and SOI spectra.

According to this limited analysis, HADCM2 (with relatively coarse ocean resolution) is capable of producing low frequency ENSO variability similar to that observed. However, the large amplitude of the model SOI variability indicates that the GCM is too active on these scales.

## 9.6   Summary and Conclusions

Examples of the analysis and comparison of observed and model low frequency variability have been presented. In this chapter we have concentrated mainly on tropical features but the same methods can be applied to look at other features, such as the connections between the tropics and subtropics/extratropics. See Tett et al. (1997) for some more examples.

The interannual variability was first assessed by comparing maps of variance in different period ranges. The AOGCM evidence presented here demonstrates (as found in other AOGCMs) that such models are able to produce significant interannual variability, but they differ in detail from observed behavior and from each other.

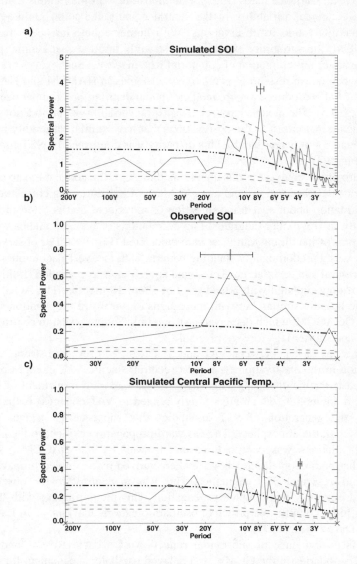

**Fig. 9.11** Power spectra for simulated and observed SOI. Plot (a): SOI from 1300 years of HADCM2 simulation. The model SOI is the pressure difference between "Darwin" (area average for the region 125°W, 15°S — 130°W, 10°S) and "Tahiti" (area average for the region 150°E, 20°S — 155°E, 15°S). The spectrum has been averaged in groups of 10 frequency bins. Plot (b): as (a) except for observed SOI and averaging in groups of 5 frequency bins. Annual mean data are used. Plot (c): simulated SST from HADCM2 in the central Pacific region (180° — 150°W, 5°S — 5°N). The thick dot-dashed lines in each panel indicate a best-fit red noise power spectrum. The dashed lines show the 95% and 99% confidence limits — lines closer to the red noise spectrum correspond to smaller confidence limits. The bandwidth is shown by the error bar centered near the region of maximum power.

For SST, the variance maps (Fig. 9.4) demonstrate that the models tend to concentrate interannual variability in the central equatorial Pacific, unlike the east Pacific maxima found in observations. With higher equatorial ocean resolution (TOGAGCM) the variability extends more strongly into the east Pacific, partially due to the better representation of equatorial Kelvin waves. Such waves are heavily damped when ocean resolution is relatively coarse as in HADCM and HADCM2. TOGAGCM also produced a more realistic spatial distribution of upper ocean variability (Fig. 9.5). The models have relatively little variability at periods longer than 3 years, again in contrast to the observations that have similar variability in the 1+ to 3 year and 3+ to 6 year bands. This disparity was confirmed by SST time series spectra (Fig. 9.7).

A feature of the analysis in this chapter is the use of lag correlations to compare and contrast the observations and models. Maps of equatorial lag correlations provide information about speed and direction of movement that is often difficult to see directly in Hovmoller diagrams. Lag correlations of SST anomalies with SST in the region of maximum variance were calculated (Fig. 9.8). For observed SST, the result was a predominantly standing pattern, with the west Pacific out of phase with the central and east Pacific. The signal at zero lag was preceded (followed) by a signal of opposite sign with a lead (lag) of about 18 months. The AOGCM patterns were broadly similar, but with more signs of westward propagation (particularly in TOGAGCM, which displayed a strong link from the west to central Pacific regions), and shorter lags for sign reversal.

Similar maps of lags between VAT and SST (Fig. 9.9) revealed strong eastward propagation in the observations (i.e. west (central) Pacific VAT changes preceded central (east) Pacific changes). This suggests that eastward movement of thermocline depth anomalies (which are strongly related to VAT changes) is the mechanism for the generation of SST anomalies that subsequently trigger positive feedbacks with the atmosphere. The eastward propagation evident in the observed SST/VAT relations was captured reasonably well by the OAGCMs. We also showed that both coupled models were characterized by a coupled propagation of zonal windstress and SST anomalies. This was in contrast to the observations which did not show this coupled propagation. This picture agrees with that produced by more complicated principal oscillation pattern analysis, as in Latif et al. (1993a); Davey et al. (1994).

There is no evidence of substantial reflection of oceanic Rossby and Kelvin waves at boundaries in the GCMs, so a delayed oscillator mechanism for interannual oscillation that relies on such reflection is not likely to be effective. The model oscillations involve slow propagation of a coupled ocean-atmosphere disturbance, with regeneration in the central-west Pacific.

Interannual behavior is influenced by systematic errors. In TOGAGCM the mean east-west equatorial thermocline slope is weak, zonal wind stress is weak, and the east Pacific SST is too warm. This bias probably allows strong coupled interaction to propagate more easily, and to extend further eastward in TOGAGCM than was evident in the observations or HADCM. A multi-century run of HADCM2 was used to illustrate the analysis of decadal scale behavior. Similar spectral peaks in the observed and simulated SOI spectra were noted (Fig. 9.11). There are significant (relative to best-fit red noise) peaks near a seven year period, but no significant peaks at longer periods. There may be predictability associated with the seven year peak.

The model SST spectrum revealed similar behavior, with interannual ENSO peaks and a significant seven year peak, which suggests that the low frequency behavior involves close ocean-atmosphere coupling.

The model SOI signal is much stronger than observed however, which indicates that the atmosphere is reacting too strongly on these scales. This is a model error that requires further investigation to determine its cause.

The low frequency ENSO variations may be due to slow changes in the background state of the equatorial Pacific. This aspect has not been considered here, but merits attention due to its implications for practical prediction of ENSO and its impacts.

**Acknowledgement**. This research was supported by the EC Environment Research Programme under contract EV5V-CT94-0538

# 10 The Decadal Variability of the Pacific with the MRI Coupled Models

SEIJI YUKIMOTO
*Climate Research Department, Meteorological Research Institute,*
*Tsukuba, Japan*

## 10.1 Introduction

Natural climate variability in the Pacific is one of the most important theme among climate researchers at the Meteorological Research Institute (MRI) in Japan, from a view point of understanding its mechanism and searching predictability for climate around Japan. The institute developed a coupled ocean-atmosphere general circulation model (GCM) (Tokioka et al., 1995), with an emphasis on good representation of natural climate variability especially in the Pacific including El Niño-Southern Oscillation (ENSO). It is shown that ENSO and decadal to interdecadal natural variability in the Pacific are realistically simulated with the coupled GCM (Yukimoto et al., 1996). Analysis of decadal and interdecadal variation of upper ocean and atmosphere revealed in the model may suggest a possible mechanism for such variability. This chapter introduces the spatial and temporal structures of the decadal and interdecadal variability in the Pacific, based on the analysis of MRI model results and observed climate data, and presents some ideas for an inherent mechanism.

## 10.2 Models

The MRI coupled GCM consists of an atmospheric GCM and an oceanic GCM in which an active sea ice model is incorporated. The atmospheric GCM is a grid model with horizontal resolution of 4 degrees in latitude and 5 degrees in longitude. It has 15 vertical levels between the surface and the top (1 hPa) of the model atmosphere. Many ordinary physical parameterization schemes are included; radiation, grid-scale condensation, cumulus convection, planetary boundary layer, hydrological processes over land surface, and so on.

The performance of the atmospheric GCM forced by the observed sea surface temperature (SST) is reported by Kitoh et al. (1995). The model shows not only realistic climatological seasonal variations but also sufficient consistency with the observations in interannual variations, as appearing in the Southern Oscillation Index and anomaly patterns for sea level pressure and 500 hPa geopotential height.

The oceanic component is a world ocean GCM which has a global domain including the Arctic Ocean and the Antarctic Ocean with realistic bottom topography. The model formulation basically follows the ocean GCM of Bryan (1969). Longitudinal grid spacing is set uniformly to 2.5 degrees. Meridional grid spacing is 2 degrees, while it is set to non-uniformly up to 0.5 degrees, within 12 degrees latitudes of the equator in both the hemispheres. Such horizontal grid spacing strategy is chosen to obtain a good representation of the equatorial oceanic waves associated with El Niño. There are 21 vertical layers between the sea surface and the deepest ocean bottom at 5000 m depth with finer resolution near the surface to incorporate the Mellor-Yamada level 2 turbulent closure scheme for simulating the oceanic mixed layer.

The model has an active sea ice model which forecasts thickness and compactness of the sea ice with considerations of thermodynamical processes and dynamical movements. Each sub-component of the model is coupled by exchanging fluxes of heat, freshwater, and momentum, while a climatological "flux adjustment" is employed in order to maintain realistic SST and sea ice distributions. Results from a 150 year integration of this model are introduced in this chapter.

For comparison, results from the atmosphere-slab ocean coupled model are introduced. It has the same the atmospheric GCM and sea-ice model as the coupled GCM, but the ocean model is a slab mixed layer. The slab mixed layer activates the dynamical process only with the thermodynamical process driven by heat flux from the atmosphere. Hereafter, we refer to the former model as AOGCM and the latter one as SGCM for convenience. Comparison of the results from these two models is expected to clarify the dynamical effects of the ocean on the decadal and interdecadal variability of the atmosphere-ocean coupled system.

## 10.3  Spatial structure of the decadal variability

A remarkable decadal shift of the SST in the Pacific Ocean which occurred in the mid-1970's shows a typical basin wide spatial pattern. Over the tropics and subtropics, a wedge shaped warming region is seen with equatorial symmetry with its apex in the western tropical Pacific. In the mid-latitude North Pacific, concurrently, there is a cooling region in the central North Pacific and the Kuroshio Extension, and a warming region along the coast of North America (Fig. 10.1). It is noted that this basinwide pattern of the decadal shift of SST is similar to the spatial empirical orthogonal function (EOF) of the SST variation for the ENSO time scale. Has any decadal change with an ENSO-like basinwide pattern occurred periodically in the past? If not, are they merely coincidental events that happened in the tropical Pacific and the North Pacific independently? Recently, quasi-periodicity of the ENSO-like pattern for the decadal variability has been recognized from a centennial record of the SST observations (Zhang et al. 1997).

First, we take a look at the overall spatial pattern of the SST variations dominant through the time scales from interannual to interdecadal. The leading EOFs of the annual mean SSTs without any filtering (but detrended) are shown in Fig. 10.2. The basinwide pattern of the leading EOF of the observed SST is similar to that for the

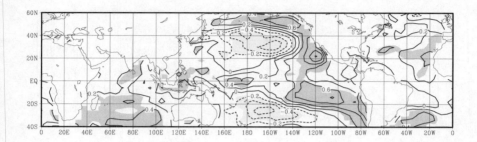

**Fig. 10.1** Differences between the 10-year mean observed SST for 1977-1986 and that for 1967-1976. Contour interval is 0.2 °C and shaded regions denote the 95% level of significance according to the t-test. (Reproduced from GISST (Parker et al., 1994) after Nitta and Yamada,1989)

decadal shift which occurred in the mid-1970's (compare Fig. 10.2a with Fig. 10.1). The following differences are also recognized. There is a larger EOF amplitude in the equatorial eastern Pacific than in the mid-latitude North Pacific, while the decadal SST change in the mid-1970's has a maximum magnitude in the mid-latitude North Pacific, and has a relatively small magnitude in the equatorial region. This feature is due to the fact that the EOF includes the ENSO time scale variability in which El Niño and La Niña generally have their maximum amplitude of the SST in the central to eastern equatorial Pacific.

The basinwide EOF pattern for the AOGCM (Fig. 10.2b) is similar to the observed pattern (Fig. 10.2a). It also has a typical wedge-shaped signal through the tropics and subtropics as observed. The maximum variability in the AOGCM is located around the dateline of the equatorial Pacific. This is considered to be related to a bias in the model ENSO that the SST anomaly is enhanced around the dateline in the mature stage of the El Niño and La Niña (Yukimoto et al. 1996), while the observed El Niño has maximum anomalies in the eastern equatorial Pacific.

Resemblance of the SST variability pattern to the observed pattern is found not only in the AOGCM but also in the SGCM (Fig. 10.2c). The similarity of the pattern in the subtropics and mid-latitudes is notable. For the equatorial SST of the SGCM, however, variability is somewhat different from that of the observations and the AOGCM. It should be remembered that the SGCM does not have any oceanic dynamical mechanism for El Niño.

Spatial patterns of the leading EOF modes for the SST variability are compared between the ENSO time scale (< 12 years) and the decadal time scale (> 12 years, detrended) (Fig. 10.3). The basinwide patterns (wedge-shaped in the tropical-subtropical Pacific and shaped elliptically in the central North Pacific with opposite sign), are seen for both ENSO and decadal time scales. The AOGCM well simu-

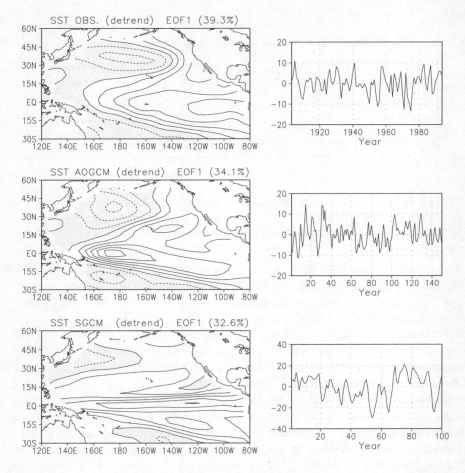

**Fig. 10.2** Spatial patterns (left panels) and temporal coefficients (right panels) for the leading EOFs of annual mean SST (linear trend is removed) for (a) observation for 1903-1994 (GISST, Parker et al., 1994), (b) coupled atmosphere-ocean GCM (AOGCM) for years 1-150, and (c) atmosphere-slab ocean mixed layer GCM (SGCM) for years 1-100.

lates this similarity between the time scales. Differences between the time scales are seen mainly in the relative magnitudes of the variability between the low latitudes and middle latitudes. For the ENSO time scale (Fig. 10.3a upper panel), the variability is smaller in the central North Pacific than in the tropics where the maximum variability is located at the equatorial region. For the decadal time scale (Fig. 10.3a lower panel), on the other hand, a sharp maximum along the equator is not seen, whereas large variability is seen in the central North Pacific. These characteristic inter time scale differences are also simulated in the AOGCM (Fig. 10.3b).

a                                                    b

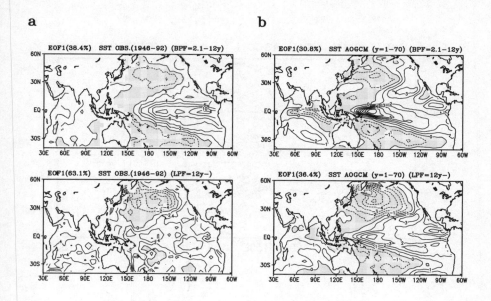

**Fig. 10.3** Comparison of eigenvectors for the leading EOFs of the SSTs between the ENSO time scale (< 12 years) (upper panels) and the decadal time scale (> 12 years) (lower panels) for (a) observation, and (b) AOGCM, respectively. Values are in proportion to amplitude of variance, since EOFs are calculated for variance matrix.

It is well known that winds at the sea surface change SST directly through heat fluxes and indirectly through advection, such as the Ekman current and the Ekman pumping. To investigate the spatial patterns of the variability of the sea surface wind is, therefore, important in clarifying the spatial structure of the SST change associated with the atmosphere-ocean coupled variability. The spatial patterns of the leading EOF modes for the surface wind stress variability in the AOGCM are shown in Fig. 10.4, for the ENSO time scale (Fig. 10.4a) and the decadal time scale (Fig. 10.4b), respectively. Each temporal coefficient of the EOF is highly correlated (with coefficients of 0.82 and 0.65 for ENSO time scale and decadal time scale, respectively) with that of SST for the corresponding time scale. The overall spatial pattern for the decadal time scale is similar to that for the ENSO time scale. Westerly wind anomalies are seen in the west-central equatorial Pacific during the warm period of equatorial SST for both time scales. In the mid-latitudes, both patterns have a dominant westerly wind anomaly and a cyclonic wind anomaly on its northern side, suggesting the intensification of the Aleutian Low in winter. In the subtropical eastern region of the North Pacific, a cyclonic wind anomaly around Hawaii is also consistently seen for both time scales, though that for the ENSO time scale is not clearly separated from the northern cyclonic anomaly.

a

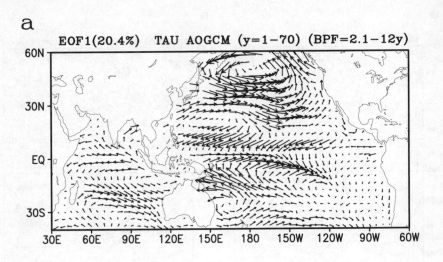

b

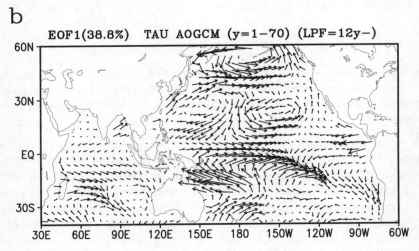

**Fig. 10.4** Eigenvectors of the leading EOFs of the surface wind stress of the AOGCM for (a) ENSO time scale, and (b) decadal time scale. Arrow size is in proportion to amplitude of variance, since EOFs are calculated for the variance matrix.

The most prominent variability of the SST in the Pacific Ocean has wedge-shaped spatial patterns that are similar between the ENSO and the decadal time scales. It is shown that the AOGCM well simulates these spatial patterns. Furthermore, the SGCM (which does not include any oceanic dynamical processes) also simulates the pattern with certain resemblance. It is found from the AOGCM result that the SST variability is correlated with the surface wind variability with spatial patterns that are also similar between the ENSO and the decadal time scales. For the ENSO time scale, correlation between anomalies of SST tendency and surface heat flux is larger than 0.7 in the mid-latitude North Pacific in the AOGCM (Yuki-moto et al., 1996), which implies that the surface heat flux accounts for a large part of the SST variation. Furthermore, the spatial pattern of the surface heat flux varia-tion for the decadal time scale has qualitative resemblance to that for the ENSO time scale. These results indicate that the overall spatial structure of the decadal variability in the Pacific Ocean is maintained or enhanced by the atmosphere-ocean mixed layer coupled system, in which thermodynamical processes with a short time scale (probably less than a year) are dominant. However, this hypothesis is not sufficient to explain the differences between the ENSO and the decadal time scales in the contrast of the SST variability in the tropics/subtropics versus the mid-lati-tudes.

## 10.4  Temporal structure of the decadal variability

To cause a regular oscillation or a quasi-periodical variation of SST, there should be a negative feedback mechanism which reverses a state with one sign to another state with opposite sign. This negative feedback mechanism is considered a pace-maker or a regulator of the variability, if the positive feedback mechanism would play the role of an engine to sustain or enhance the variation. Investigating the tem-poral structure of the variability will be a key to elucidation of these mechanisms.

Variability of the SST with the basinwide spatial pattern as shown in Fig. 10.2, appears to be dominant over the time scales from ENSO to multi-decades (see also EOF time series in Fig. 10.2). In order to examine more detailed temporal structure of this SST variability, power spectra of the leading EOFs are calculated (Fig. 10.5). For the observed SST (solid line), the spectrum has several peaks between the periods of 2 and 7 years, corresponding to the ENSO time scale, and a bulge between the periods of 10 and 20 years. The result from the AOGCM (dashed line) shows good agreement with that from the observations in the overall spectral property (except in the periods less than 3 years), though their individual peaks are not necessarily identical. For the result from the SGCM (dotted line), on the other hand, the overall spectral property is quite different. The variability in the decadal to interdecadal time scales is much larger than those of the observed and the AOGCM, while that of the ENSO time scale is as large as that of the AOGCM.

How about the spectral structures in which all the modes are included ? Fig. 10.6 shows the meridional variation of power spectra zonally averaged over the Pacific sector for SSTs. For both the observations and the AOGCM, in low latitudes, they show relatively flat spectral density for the time scales from ENSO to decadal. In the mid-latitudes, on the other hand, they show a larger power in the interdecadal time scale. This feature is remarkable in the latitudes of the Kuroshio Extension found in the observations and the AOGCM. The SGCM shows quite different spec-tral structure, particularly in the decadal to interdecadal time scale.

It is suggested in Fig. 10.6 that the AOGCM is capable of simulating the tempo-ral structure of the observed SST variability as well as the spatial structure. The

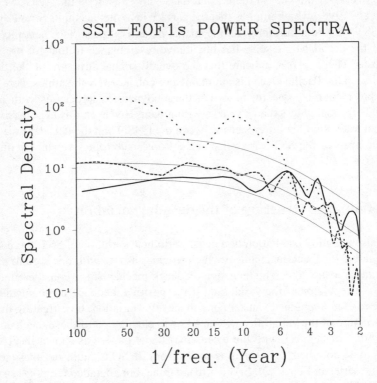

**Fig. 10.5** Power spectra of the leading EOFs of SSTs for the observations (solid line), the AOGCM (dashed line), and the SGCM (dotted line). Unit is relative. Thin lines show the red spectrum best-fitted to the spectrum for AOGCM and its 95% confidence limits.

SGCM (which has the same atmospheric component, but has no dynamical ocean), however, is not capable of simulating the temporal structure of the variability. This result indicates that the temporal structure of the variability is associated with dynamical processes in ocean layers beneath the mixed layer.

## 10.5  Two distinct decadal modes in the AOGCM

An oscillatory SST variation requires a positive feedback mechanism to maintain or enhance SST variation against dissipation and a negative feedback mechanism to terminate and reverse the anomaly within a characteristic time scale. The temporal structure of the variation is determined by the combination of the strength of coupling and the delay time in the feedback mechanisms. This is typically seen in the theory of the delayed oscillator in the ENSO cycle (Suarez and Schopf, 1988). From the results described in section 10.4, it is suggested that the temporal structure of the variability is related to dynamical processes in the sub-surface ocean.

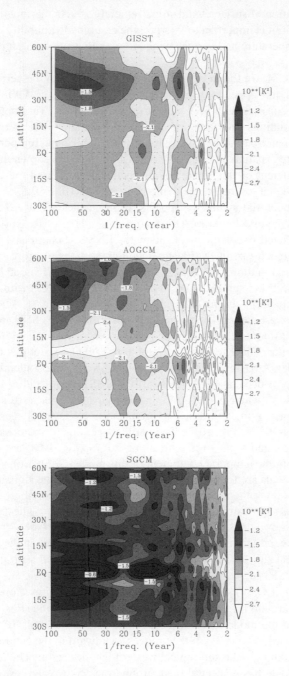

**Fig. 10.6** Meridional variations of power spectra zonally averaged over the Pacific sector (150°E - 120°W) of the SSTs for (a) observation, (b) AOGCM, and (c) SGCM. Values are common logarithm of the power of temperature (unit is K²).

Since the variation of surface wind stress remotely affects thermocline depth by its curl component, it is important to interpret the combined variability of SST, subsurface ocean temperature and surface wind stress with spatial structure and temporal evolution.

In the AOGCM, we found two distinct decadal modes in the vertically averaged temperature of the upper 600 m (hereinafter referred to as VAT) of the Pacific Ocean, which covaries with SST and surface wind stress (Yukimoto et al., 1998). To examine variations with temporal evolution, a complex EOF (CEOF) analysis is applied to the band-pass (15 to 32 years) filtered VAT data. Temporal evolution of the reconstructed spatial pattern of the first CEOF mode is shown in Fig. 10.7. The temporal coefficient of this mode has a spectral peak around the 21-year period, coincident with that for the first EOF mode of the SST (Fig. 10.5). A wedge-shaped pattern with equatorial symmetry appears. It propagates westward in the northern subtropics and the mid-latitudes. This feature reminds us of Rossby wave propagation in the delayed oscillator mechanism for ENSO (Schopf and Suarez, 1988). However, its velocity is much slower, since the latitudes of propagation are much higher, compared to those for ENSO. This interdecadal VAT mode has a good correlation with SST and surface wind stress variation (Fig. 10.7) with consistent basinwide spatial patterns. In the tropical Pacific, a westerly (easterly) wind anomaly appears in the west-central Pacific and a positive (negative) SST anomaly appears in the central-eastern Pacific, when the positive (negative) VAT anomaly appears in the eastern Pacific. These features are similar to ENSO, except that their latitudinal extent is broader. A basinwide spatial pattern of SST is almost identical to the leading EOF mode of decadal SST (see Fig. 10.3). SST varies rather locally keeping the spatial pattern, whereas, the VAT signals propagate westward across the basin with nearly constant magnitudes. The propagating VAT signal seems dominant at latitudes $20°$ N-$30°$ N (Fig. 10.7). The vertical sections of the ocean temperature, surface heat flux and wind stress curl along $20°$ N (Fig. 10.8) are examined with regressions from the temporal coefficient of this mode (CEOF1 of VAT). Westward propagating signals are found in the sub-surface layer. As the signals migrate, their maxima are found at the western end of the surface wind forcing (a negative wind stress curl leads to a positive VAT) detached from the surface. The signal of the surface wind stress curl does not migrate along with the subsurface signal, but it varies as an standing oscillation with maximum amplitude centered in the central Pacific. The signals near the surface, on the other hand, correspond with the surface heat flux where the SST signal has a large amplitude.

The second CEOF mode of the interdecadal VAT variation shows signals rotating clockwise around the subtropical gyre in the North Pacific (Fig. 10.9). A signal appears around the model Kuroshio Extension region (around $40°$ N) and slowly migrates eastward ($\phi=0°$-$60°$). It then moves southward in the central North Pacific ($\phi=120°$), extending to the subtropical western Pacific region. The temporal coefficient of this mode has a spectral peak around the 15-16 year period, the same as that for the second EOF mode of SST. The spatial pattern of SST, which appears in the phase lag (see $\phi=0°$), is almost identical to the second EOF mode of SST. This

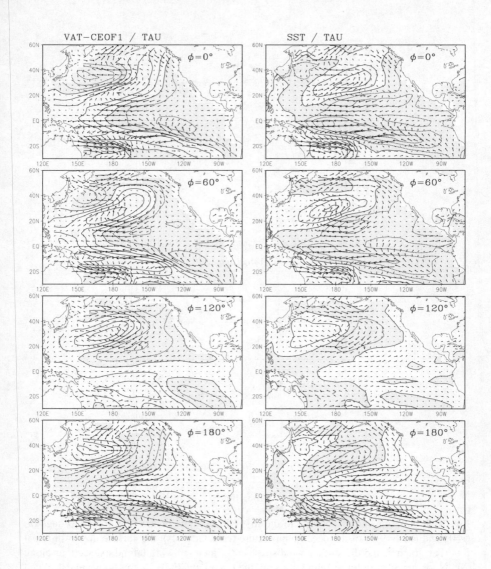

**Fig. 10.7** Reconstructed temporal evolution of spatial patterns of the first complex-EOF for the band-pass (periods from 15 years to 32 years) filtered VAT (600 m) variation (left panels), and corresponding variations of SST at each phase lag obtained from linear regression on the temporal CEOF (right panels). Arrows (common between left and right panels) denote variations of the surface wind stress the same as for the SST. Regressions are calculated against band-pass (periods from 15 to 32 years) filtered data.

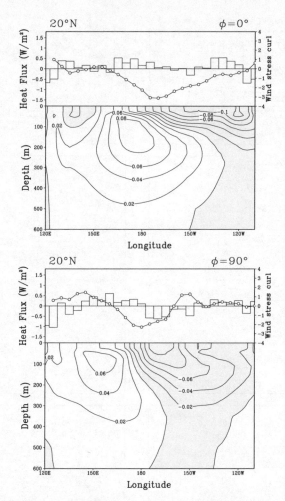

**Fig. 10.8** Variations corresponding to the first CEOF mode of the VAT for phase lags (a) $\phi=0°$ and (b) $\phi=90°$, of the ocean temperature (contour plots), the surface downward heat flux (bar graphs), and the surface wind stress curl (line graphs) in latitudinal sections along 20° N, obtained from linear regressions on the temporal CEOF. Regressions are calculated against band-pass (periods from 15 to 32 years) filtered data.

decadal VAT mode is correlated with variations of SST and surface wind stress, but specifically around the model Kuroshio Extension region. It is recognized that a strong positive (negative) SST signal around the Kuroshio Extension region is accompanied by westerly surface wind weaker (stronger) than normal (Fig. 10.9, $\phi=0°$). Fig. 10.10 shows the meridional sections of temperature in the central North Pacific (180°E) regressed on this mode. A positive signal at the surface in the mid-latitudes ($\phi=60°$) moves down and southward into the subtropical subsurface

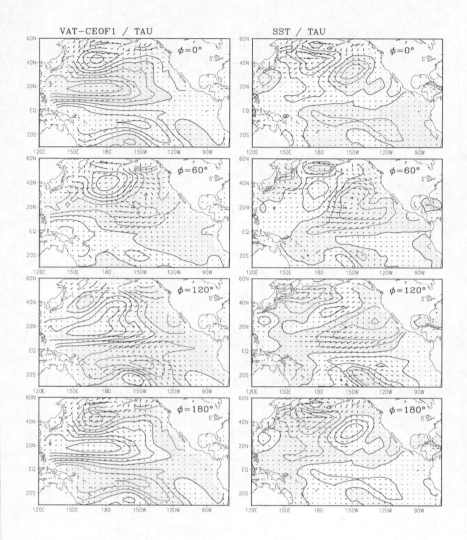

**Fig. 10.9** Same as Fig. 10.7, but for the second CEOF.

deeper than 200m ($\phi$=120°) along the isopycnal surfaces around 1025 kgm$^{-3}$ (25$\sigma$). This vertical structure with temporal evolution suggests subduction of the mid-latitude surface anomaly into the subtropical subsurface along the ventilated isopycnal surface. This feature is also supported observationally by the analysis of Deser et al. (1996). The temperature anomaly injected into the subsurface layer at the central subtropical Pacific extends westward and then appears to move into the Kuroshio region.

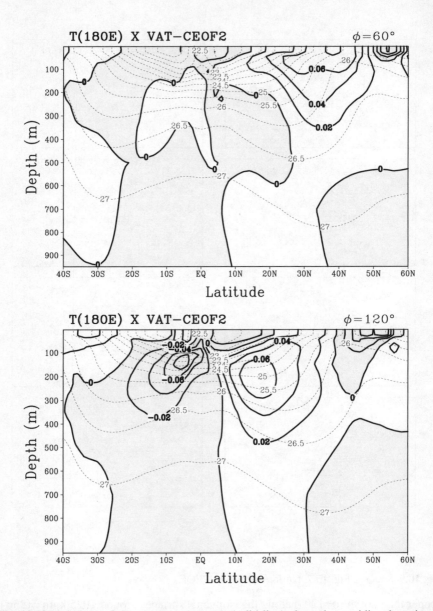

**Fig. 10.10** Variations of the ocean temperature (solid line) along the meridional section at 180°E, corresponding to the first CEOF mode of the VAT, obtained from linear regressions by the temporal CEOF against band-pass (periods from 15 to 32 years) filtered data. Dashed line plot denotes climatological annual mean potential density in the same section in the AOGCM.

## 10.6 Discussion about the inherent mechanism

We found two atmosphere-ocean coupled modes for the decadal variability in the Pacific Ocean. The first one is ENSO-like decadal variability. The positive feedback which maintains this mode is associated with the variations of tropical SST and the atmospheric circulation. Relationship between SST and the atmospheric circulation is similar to the SST-wind feedback for ENSO. The tropical SST anomaly accompanies atmospheric convective activity in the tropical western Pacific, which induces the mid-latitude atmospheric circulation anomaly with teleconnection patterns, and consequently forces the mid-latitude ocean. These processes are consistent with the claims that the decadal variability of the SST in the mid-latitude North Pacific is primarily attributed to the tropical SST variation, as argued by Graham et al. (1994), Lau and Nath (1994), and others.

The negative feedback part of the first mode is associated with propagating trans-Pacific signals with decadal travel time. Wedge-shaped signals formed by the wind forcing in the eastern Pacific propagate westward as subsurface signals in the subtropics and the mid-latitudes. The time scale of the negative feedback is dependent on the travel time of the signals to transverse the basin. The simulated velocity of the signal (~1.5 cm/s at 30° N) is smaller than that for the free first baroclinic Rossby wave (~ 3 cm/s at 30° N) inferred from the model ocean stratification. This discrepancy may be related to the indication that the subsurface signals are not free waves but are forced waves modulated by slow changes in the forcing which is producing the waves.

The second mode has SST variability distinct in the Kuroshio Extension region. The positive feedback of this mode is associated with the local air-sea interaction of SST and the atmospheric circulation in the mid-latitudes. If the negative SST anomaly around the Kuroshio Extension region, for example, leads to intensified westerly wind, it enhances surface cooling and southward advection of cold water due to the Ekman current. Consequently, the negative SST anomaly is further reinforced. Regression maps show this evidence (Fig. 10.9).

The selection of time scale of this mode is associated with signals rotating clockwise around the subtropical gyre in the North Pacific. This point was first claimed by Latif and Barnett (1994) that the interdecadal variability in the North Pacific is primarily attributed to atmosphere-ocean coupled instability in the mid-latitudes. The signals reinforced at the surface around the Kuroshio Extension region are subducted into the subsurface and advected along the subtropical gyre circulation in the North Pacific. The time scale of the feedback is primarily determined by the advection speed along the subtropical gyre circulation. However, the slow wave-like property of the variation of the subsurface ocean and lagged spin up of the subtropical gyre should be also considered, since the variability of this mode also accompanies significant surface wind curl variations in the subtropical Pacific (see Fig. 10.9) and associated change of the subtropical gyre circulation. At the present study, it is difficult to ascertain what extent can the two modes be regarded as independent. There may be a possible linkage between the two modes, since both modes accompany notable subsurface signals which enter the subtropical Western Pacific.

The above hypothesis for the mechanism contains a number of processes which should be proved or verified. These subjects will be pursued in future studies. Of course, the model does not perfectly simulate real climate variability. It is certain, however that the coupled models introduced in this chapter are powerful tools to pursue such study.

# 11 Dynamics of Interdecadal Variability in Coupled Ocean-Atmosphere Models

MOJIB LATIF

*Max-Planck-Institut fur Meteorologie,*
*Bundesstrasse. 55, D-20146 Hamburg, Germany*

## 11.1 Introduction

The causes of interdecadal climate variability can be quite different, including external and internal forcing mechanisms. On the one hand, external forcing mechanisms have been discussed for a long time. Variations in the incoming solar radiation, for instance, were proposed as one of the major sources of interdecadal variability (e. g. Labitzke (1987), Lean et al. (1995)). Cubasch et al. (1996) show indeed that some climate impact of the sun might exist on time scales of many decades and longer, but it is fairly controversial at present how strong the fluctuations in the solar insulation actually are. The forcing of interdecadal climate variability by volcanos is well established and therefore less controversial, and major volcanic eruptions can be easily seen in regional and globally averaged temperature records (e. g. Robock and Mao (1995)).

On the other hand, interdecadal variability arises from interactions within and between the different climate sub-systems. The two most important climate sub-systems are the ocean and the atmosphere. Non-linear interactions between different space and time scales can produce interdecadal variability in both the ocean (e. g. Jiang et al. (1995), Spall (1996)) and the atmosphere (e. g. James and James (1989)) as shown by many modeling studies. More important, however, seem to be the interactions between the two systems. The stochastic climate model scenario proposed by Hasselmann (1976) is a "one-way" interaction: The atmospheric "noise" (the high-frequency weather fluctuations) drives low-frequency changes in the ocean, leading to a red spectrum in the ocean's sea surface temperature (SST), for instance, in analogy to Brownian motion. It has been shown that the interannual variability in the midlatitude upper oceans is consistent with Hasselmann's (1976) stochastic climate model (e. g. Frankignoul and Hasselmann (1977)). This concept has been generalized recently by Frankignoul et al. (1996) who incorporated the wind-driven ocean gyres into the stochastic climate model concept, which extends the applicability of the stochastic climate model to interdecadal time scales. The atmospheric noise can excite also damped eigenmodes of the ocean circulation on interdecadal to centennial time scales, as shown, for instance, by the model studies of Mikolajewicz and Maier Reimer (1990), Weisse et al. (1994), and Griffies and Tziperman (1996).

"Two-way" interactions between the ocean and the atmosphere in the form of unstable ocean-atmosphere interactions were also proposed to cause interdecadal climate variability (e. g. Latif and Barnett (1994), Gu and Philander (1996), Chang et al. (1996)). Similar to the El Nino/Southern Oscillation (ENSO) phenomenon (e. g. Philander (1990), Neelin et al. (1994)), ocean and atmosphere reinforce each other, so that perturbations can grow to climatological importance. The memory of the coupled system (which resides generally in the ocean) provides delayed negative feedbacks which enable continuous, but probably damped, oscillations that are forced by the internal noise within the coupled ocean-atmosphere system.

The intention behind this review paper is to summarize the mechanisms that lead to interdecadal variability in coupled ocean-atmosphere models. The problem of interdecadal variability is approached by using a hierarchy of coupled models. Some mechanisms were investigated with relatively complex coupled models in which at least one component is represented by a general circulation model (GCM) (e. g. Delworth et al. (1993), Latif and Barnett (1994), Groetzner et al. (1996), Saravanan and McWilliams (1996), Chen and Ghil (1996), Xu et al. (1996)), while other mechanisms were studied using relatively simple or conceptual models (e. g. Gu and Philander (1996), Chang et al. (1996)). It should be noted that this chapter does not attempt to provide a complete overview of all mechanisms that can lead to interdecadal variability. Emphasis is given to those modes which were either identified in coupled ocean-atmosphere models, or for which ocean-atmosphere interactions are crucial. Several "ocean-only" modes (e. g. Weaver and Sarachik (1991), Weaver et al. (1993), Weisse et al. (1994), Jacobs et al. (1994)), for instance, identified in uncoupled ocean model simulations are not described here. Sensitivity studies with coupled models (e. g. Meehl (1996)) are also not described in this review article.

Furthermore, this chapter does not aim to provide a complete overview of the observational work documenting interdecadal variability. Observations are presented only to motivate some of the modeling studies and to verify some of the modeling results. The reader is referred to the observational papers of e. g. Folland et al. (1986), Dickson et al. (1988), Mysak et al. (1990), Deser and Blackmon (1993), Kushnir (1994), Trenberth and Hurrell (1994), Mann and Park (1994), Levitus and Antonov (1995), Zhang et al. (1996), Mantua et al. (1997) and references therein for further information on the observational aspects of interdecadal variability. A fairly comprehensive overview of the different aspects of interdecadal variability (including theoretical, modeling and observational studies) can be found in the recent book published by Anderson and Willebrand (1996).

The paper is organized as follows. The interdecadal variability that originates in the tropics is described in section 11.2. Interactions between the tropics and midlatitudes that may lead to interdecadal variability are discussed in section 11.3. The interdecadal variability which is associated with the mid-latitudinal ocean gyres is described in section 11.4, while the interdecadal variability which arises from variations in the thermohaline circulation is summarized in section 11.5. A brief overview of studies dealing with the predictability at decadal time scales is given in section 11.6. The paper is concluded with a discussion in section 11.7.

## 11.2 Interdecadal variability originating in the tropics

### 11.2.1 The Tropical Pacific

The tropical Pacific Ocean has a prominent role in forcing global climate anomalies. This is due to the existence of the El Nino/Southern Oscillation (ENSO) phenomenon, the strongest interannual climate fluctuation (see, for instance, Philander (1990)). It is likely that interdecadal fluctuations in tropical Pacific SST will also have a significant impact on regional and global climate. Interdecadal northeastern Australian rainfall anomalies, for instance, were shown to be highly correlated with interdecadal fluctuations in tropical Pacific SST (Latif et al. (1997a)).

ENSO is a classical example of an inherently coupled air-sea mode which is characterized by a dominant period of about 4 years. The spectrum of tropical Pacific SST, however, is relatively broad due to the presence of noise and/or nonlinear interactions between the annual cycle and the dominant ENSO mode. Here, results are shown from the study of Eckert and Latif (1996) who investigated the role of the stochastic forcing on the low-frequency variability in the tropical Pacific.

A series of experiments were performed with a hybrid coupled model (HCM), consisting of an oceanic general circulation model (OGCM) and a statistical atmosphere model. The latter is an equilibrium atmosphere (see, for instance, Barnett et al. (1993)) which does not simulate any internal variability. The HCM simulates self-sustained oscillations with a period of about 5 years at sufficiently strong coupling (Fig. 11.1a). The spectrum of the simulated SST anomalies, however, is unrealistic when compared to that derived from the SST observations, with one relatively strong and narrow peak at the dominant ENSO frequency (Fig. 11.1d). The inclusion of stochastic forcing leads to a much more realistic simulation (Fig. 11.1b), and the spectrum of the simulated SST anomalies is now in better agreement with the spectrum of the observed SST anomalies. In particular, the stochastically forced HCM simulates considerable interdecadal variability. As can be inferred from the experiment with the uncoupled ocean model (zero coupling strength) forced by noise only (Fig. 11.1c), the level of the simulated interdecadal variability is much lower relative to the coupled case. Thus, the coupling between ocean and atmosphere is important in generating not only ENSO itself but also some part of the interdecadal variability in the tropical Pacific by amplifying the noise in the system. However, the level of the interdecadal variability simulated in the coupled integration with noise is lower than that observed, indicating that other processes such as the non-linear interactions between different time scales or interactions with phenomena outside the tropical Pacific may be also important in generating interdecadal variability in the tropical Pacific (see also section 3).

Similar results are obtained from the Lamont intermediate coupled model (ICM), described in detail by Zebiak and Cane (1987), which is also a regional model of the tropical Pacific. The Lamont model simulates not only considerable interdecadal but also centennial variability (Fig. 11.2). Since the atmospheric component of the Lamont model generates internal variability, it is likely that the occurrence of

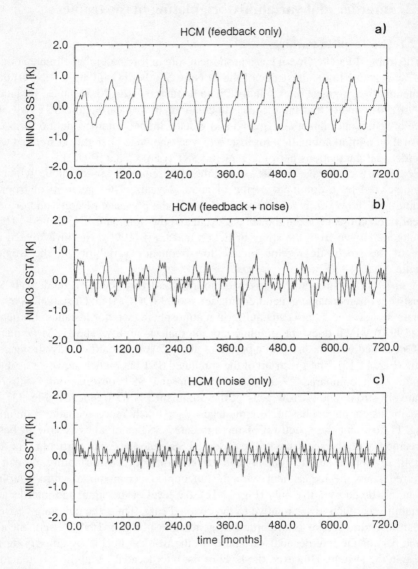

**Fig. 11.1** Time series and spectra of SST anomalies (C) averaged over the Nino-3 region (150 W-90 W, 5 N-5 S) as simulated by the Eckert and Latif (1996) hybrid coupled model (HCM). a) Nino-3 SST anomaly time series simulated in the unperturbed coupled control integration, b) Nino-3 SST anomaly time series simulated in the coupled integration with noise, c) Nino-3 SST anomaly time series simulated in the integration without atmospheric feedback (uncoupled ocean model) but with noise included. d) Spectra of the time series shown in a)-c). The spectrum of the observed Nino-3 SST anomaly as derived from the GISST data set of the British Meteorological Office is shown for comparison.

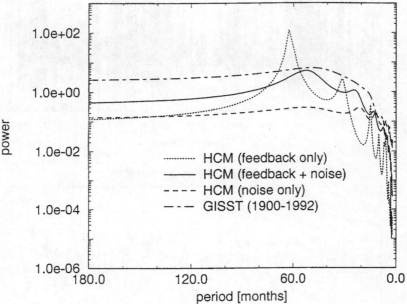

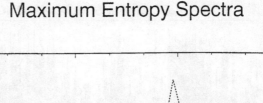

**Figure 11.1 (continued)**

the low-frequency variability is also related, at least partly, to the stochastic forcing, as described above. Non-linear effects, however, might be also important (e. g. Münnich et al. (1991), Jin et al. (1994)). In summary, coupled processes within the tropical Pacific climate system can produce considerable interdecadal variability which will have global climate impacts through atmospheric teleconnections (e. g. Horel and Wallace (1981), Glantz et al. (1991)).

## 11.2.2 The Tropical Atlantic

Interdecadal variability in the tropical Atlantic evolves rather differently than that in the tropical Pacific, which is probably due to the different basin geometries. While the zonal wind stress/SST feedback is the dominant growth mechanism in the equatorial Pacific and the surface heat flux acts as a damping, the surface heat flux has a much more active role in the tropical Atlantic (see Chang et al. (1996) for a good summary and references therein). A weak ENSO-like mode, however, exists also in the equatorial Atlantic (e. g. Zebiak (1993), Carton and Huang (1994)), with a time scale of about 2.5 years (Latif et al. (1996a)).

The existence of a decadal dipole mode with opposite SST anomalies north and south of the Intertropical Convergence Zone (ITCZ) has been postulated by several

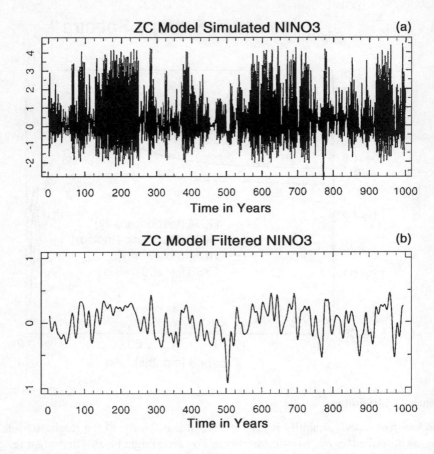

**Fig. 11.2** Nino-3 SST anomalies (C) as simulated by the Zebiak and Cane (1987) intermediate coupled model (ICM) in a 1000-year control integration. a) Raw time series based on monthly values, b) low-pass filtered Nino-3 SST anomaly time series in which variability with time scales longer than approximately 10 years is retained

authors (e. g. Mehta and Delworth (1995), Chang et al. (1996)). The SST dipole influences the rainfall over parts of South America, as shown by Moura and Shukla (1981), and there is some evidence that Sahelian rainfall also is affected by the dipole (see, for instance, Lamb and Peppler (1991) for a good overview). However, it is controversial as to whether there exists a distinct time scale, and even the exist-ence of the dipole structure itself is questioned in some studies (e. g. Houghton and Tourre (1992), Enfield and Mayer (1996), Mehta (1997)). Carton et al. (1996), for instance, accept the existence of the spatial dipole structure, but they argue that its dynamics can be described to first order by the stochastic climate model of Hassel-mann (1976). Chang et al. (1996) on the other hand regard the SST dipole as a "true" oscillatory mode with a period of approximately 12-13 years.

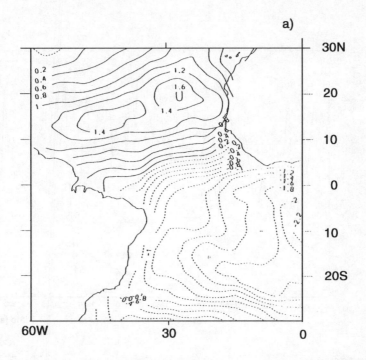

**Fig. 11.3** Second most energetic Empirical Orthogonal Function (EOF) of observed (GISST) tropical Atlantic SST anomalies for the period 1904-1994. a) EOF pattern (dimensionless), b) corresponding time series (°C), c) spectrum [(°C) Wa] of the principal component shown in b). The EOF mode shown accounts for about 23% of the variance, and the EOF analysis was performed on annual mean values. The linear trend was removed prior to the EOF analysis

In order to get more insight into the nature of the interdecadal variability in the tropical Atlantic, observed (GISST) annual mean SSTs were analyzed for the period 1904-1994. An EOF (Empirical Orthogonal Function) analysis was conducted, and the second most energetic EOF mode accounting for about 23% of the variance is associated with the north-south dipole structure (Fig. 11.3a) discussed above. A spectral analysis of the corresponding principal component (Fig. 11.3b) shows indeed enhanced variability at a period of approximately 13 years (Fig. 11.3c), which supports Chang et al.'s (1996) hypothesis. It should be noted, however, that the EOF mode accounts for a relatively large fraction of the variance in the Southern Hemisphere only. While the local explained variances amount to typically 40% in the southern center of action, they amount to about 20% only in the northern center of action. Furthermore, as will be shown below (section 11.4.2), a mode in the North Atlantic with a period of about 12 years was identified by Deser and Blackmon (1993). It is unclear to which extent the tropical dipole and

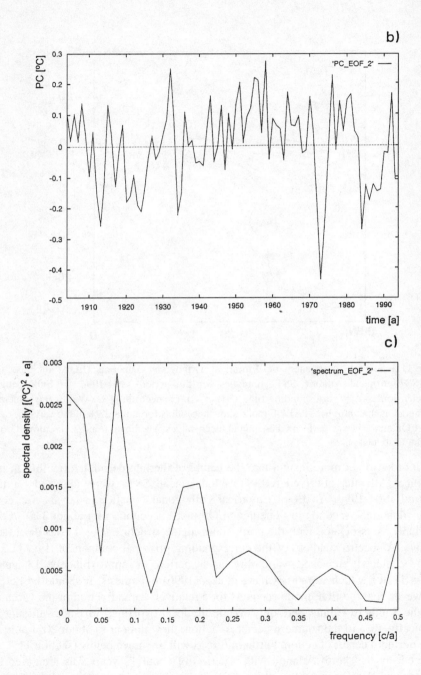

**Figure 11.3 (continued)**

the midlatitude mode reflect the same phenomenon, or whether and how they interact with each other.

Chang et al. (1996) derived a plausible theory which would explain physically
the "dipole-oscillation" and its spatial structure by conducting a series of experiments with an intermediate coupled model (ICM) and a hybrid coupled model.
Here, results are shown from the intermediate coupled model. The ICM simulates
self-sustained interdecadal oscillations for certain parameters, and the simulated
spatial structure resembles that observed closely (Fig. 11.4).The mechanism for the
oscillation can be summarized as follows. Suppose the system is in a state similar
to that shown in Fig. 11.4a, with warm SST anomalies north and cold SST anomalies south of the ITCZ.The associated wind response will be such that the trade
winds are weakened over the warm SST  anomaly in the north and strengthened

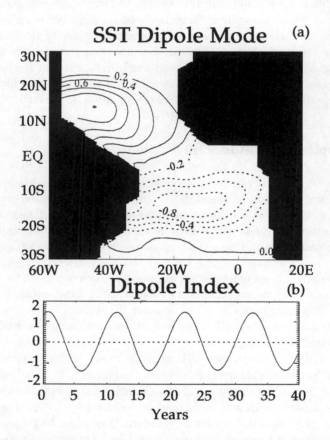

**Fig. 11.4** The tropical Atlantic dipole mode as simulated by the intermediate coupled model
(ICM) of Chang et al. (1996). Shown is a) the characteristic SST anomaly pattern as
obtained from a regression analysis of the SST anomalies upon a dipole index which is
shown in b). The dipole index is defined as the difference of the SST anomalies in the two
centers of action. See Chang et al. (1996) for details.

over the cold SST anomaly in the south. This will lead to surface heat flux anomalies that reinforce the initial SST anomalies in both poles of the dipole. Thus, the air-sea interactions over the tropical Atlantic are unstable. The horizontal advection of heat by the steady ocean currents will counteract the growth in SST by transporting anomalously cold (warm) water to the north (south). As pointed out by Chang et al. (1996), the system gives rise to self-sustained oscillations, when the positive and negative feedbacks balance properly. In this scenario, the Atlantic dipole is an inherently coupled air-sea mode which might be predictable several years ahead. Preliminary results from decadal forecast experiments with a simplified (hybrid) coupled ocean-atmosphere model support this picture (P. Chang, pers. comm., 1996).

In summary, unstable air-sea interactions can produce interdecadal variability in both the tropical Pacific and Atlantic Oceans. The characters of the interdecadal variabilities, however, are quite different in the two oceans. While the interdecadal variability in the Pacific is dominated by zonal asymmetries, the interdecadal variability in the Atlantic is dominated by meridional asymmetries. Furthermore, while the dynamical feedback between the zonal wind stress and SST is crucial in the Pacific, it is the thermodynamic feedback between the surface heat flux and SST which is important to the growth of anomalous conditions in the Atlantic.

## 11.3 Tropics-Midlatitudes Interactions

A hypothesis for the generation of interdecadal variability based on tropics-midlatitudes interactions was proposed by Gu and Philander (1996). Their theory is based on the existence of a shallow meridional circulation which links the tropics to the extra-tropics. Certain pathways were identified in the ocean through which information can flow from the extra-tropics towards the tropics (e. g. Liu et al. (1994), Liu and Philander (1995), Lu and McCreary (1995)). These pathways which are mainly located in the central and eastern Pacific are given by surfaces of constant densities (isopycnals). Temperature anomalies at the surface in the extra-tropics can affect the tropics by traveling along the isopycnals into the equatorial thermocline, where the anomalies are upwelled to the surface. Unstable air-sea interactions will further amplify the signal.

Suppose a warm interdecadal SST anomaly exists in the equatorial Pacific (Fig. 11.5a). The local and remote atmospheric responses to this SST anomaly will be similar to those observed during El Nino situations, i. e. westerly wind anomalies over the western equatorial Pacific and strengthened westerlies in midlatitudes (Fig. 11.5b). The intensified westerlies will force a negative SST anomaly in the North Pacific, mostly through enhanced heat loss from the ocean to the atmosphere. The cold anomaly will be subducted and flow equatorward along isopycnals. Once it has reached the equatorial region, it will be upwelled to the surface, and the SST tendency will be reversed. Gu and Philander (1996) show by means of a simple box model that their subduction-mechanism can lead to continuous inter-

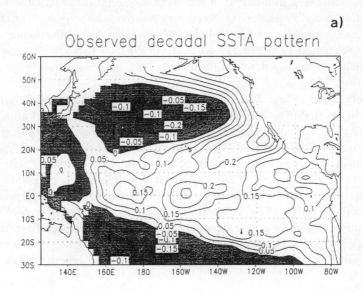

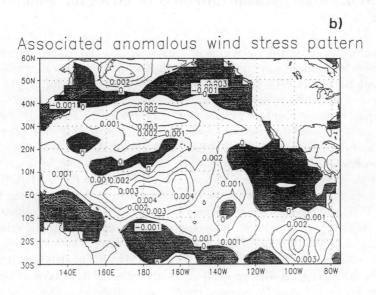

**Fig. 11.5** a) Spatial pattern of the leading EOF mode of low-pass filtered (retaining variability with time scales longer than 5 years) observed (GISST) SST anomalies in the Pacific for the period 1949-1990. The EOF mode accounts for 44% of the variance. b) Associated regression pattern of observed (da Silva et al. (1994)) zonal wind stress anomalies (N/m) to the principal component of the leading EOF mode. Shaded areas denote negative values in both panels.

decadal oscillations with a period of the order of several decades. Observational evidence for Gu and Philander's (1996) theory is provided by the work of Deser et al. (1996) who investigated the evolution of the subsurface temperatures in the Pacific for the period 1970-1991. Deser et al. (1996) show that a negative tempera-ture anomaly propagated slowly equatorward along isopycnals during the period analyzed (Fig. 11.6).

The theory of Gu and Philander might explain the interdecadal variability observed in the ENSO activity and predictability (e. g. Balmaseda et al. (1995)). The ENSO cycle was not well developed, for instance, during the 1990s (e. g. God-dard and Graham (1997), Ji et al. (1997), Latif et al. (1997a)), and most ENSO pre-diction models failed to forecast correctly the evolution of tropical Pacific SSTs during this period (e. g. Latif et al. (1997b)). It is well known from ENSO theory (Zebiak and Cane (1987), Neelin et al. (1994)) that the instability characteristics depend on the position of the equatorial thermocline. A deeper thermocline, for instance, might inhibit the coupled system to oscillate, and this might have been the case during the 1990s. The theory of Gu and Philander (1996), however, needs still to be proven by means of more complex coupled ocean-atmosphere models.

## 11.4 Midlatitude Decadal Variability Involving the Wind-driven Gyres

A mechanism for interdecadal climate variability that can lead to interdecadal climate cycles in the North Pacific and North Atlantic Oceans was proposed by Latif and Barnett (1994), Latif and Barnett (1996), Latif et al. (1996b), and Groetzner et al. (1996) by analyzing the results of a multi-decadal integration with the coupled ocean atmosphere general circulation model "ECHO-1" described by Latif et al. (1994). Similar results were obtained by Robertson (1996) for the North Pacific and Zorita and Frankignoul (1996) for the North Atlantic who investigated the interdecadal variability simulated in the ECHAM1/LSG CGCM described by von Storch (1994). The interdecadal modes derived from the ECHAM1/LSG cou-pled model, however, are much more damped relative to those simulated by the ECHO-1 coupled model. The same seems to be true for the modes identified in the GFDL (Geophysical Fluid Dynamics Laboratory) coupled model described by Manabe and Stouffer (1996). There is gyre-related interdecadal variability in the GFDL coupled model, but it differs (like the ECHAM1/LSG model) from that sim-ulated by the ECHO-1 model, probably due to the differing degrees of air-sea cou-pling, as discussed below. Pronounced interdecadal variability is also simulated by the Hadley Centre CGCM in the North Pacific and North Atlantic Oceans (Tett et al. (1996)), but the mechanisms that lead to the interdecadal variability have not been investigated yet.

The gyre modes are generated by large-scale ocean-atmosphere interactions in midlatitudes and must be regarded as inherently coupled modes, as originally sug-gested by Bjerknes (1964). The memory of the coupled system, however, resides in

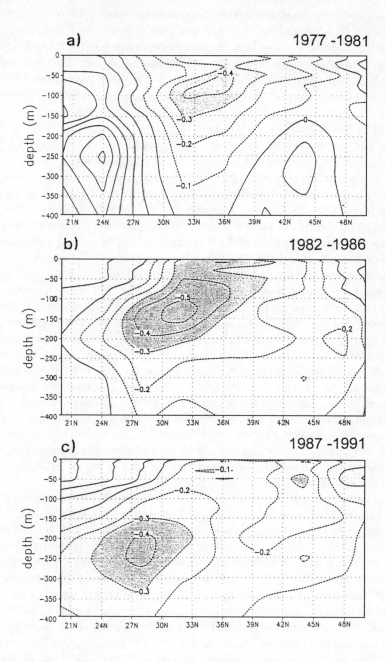

**Fig. 11.6** Anomalies of zonally averaged (170 W-145 W) temperature (C) as function of depth and latitude in the Pacific for different time periods. a) 1977-1981, b) 1982-1986, c) 1987-1991. Anomalies of less than -0.3 C are shaded. As shown by Deser et al. (1996), the negative temperature anomalies spread along isopycnals. See Deser et al. (1996) for details.

the ocean and is associated with slow changes in the subtropical ocean gyres. The existence of such interdecadal cycles provides the basis of long-range climate forecasting at decadal time scales in midlatitudes (see section 6). A detailed description of the gyre-mechanism can be found in Latif et al. (1996b) who describe a series of coupled and uncoupled numerical experiments and in Weng and Neelin (1997) who derive analytical prototypes for ocean-atmosphere interactions in midlatitudes.

### 11.4.1 The North Pacific Gyre Mode

Latif and Barnett (1994) introduced the hypothesis of the interdecadal North Pacific coupled mode. When, for instance, the subtropical ocean gyre is anomalously strong, more warm tropical waters are transported poleward by the western boundary current and its extension, leading to a positive SST anomaly in the North Pacific (Fig. 11.7). The atmospheric response to this SST anomaly is the PNA (Pacific North American) pattern which is associated with changes at the air-sea interface that reinforce the initial SST anomaly, so that ocean and atmosphere act as a positive feedback system. The atmospheric response, however, consists also of a wind stress curl anomaly which spins down the subtropical ocean gyre, thereby reducing the poleward heat transport and the initial SST anomaly. The ocean adjusts with some time lag to the change in the wind stress curl, and it is this transient ocean response that allows continuous oscillations. The corresponding evolution in upper ocean heat content in the North Pacific, as derived from the Latif and Barnett (1994) simulation is shown in Fig. 11.8. The upper ocean heat content anomalies as reconstructed from the leading Complex EOF (CEOF) mode show a clockwise rotation around the subtropical gyre, which can be partly attributed to baroclinic Rossby wave adjustment and advection of temperature anomalies by the mean gyral circulation. Xu et al. (1996) was able to reproduce the results of Latif and Barnett (1994) with a hybrid coupled model of the North Pacific, confirming that the tropics play a minor role. As will be shown below, this characteristic evolution of temperature anomalies at subsurface levels can be exploited for the prediction of decadal climate changes (Fig. 11.22).

Observations support the existence of the interdecadal cycle in the North Pacific/ North American climate system. Latif and Barnett (1996), for instance, demonstrate the existence of a 20-year periodicity by analyzing station data of surface temperature and rainfall over North America since 1860. This is consistent with the study of Haston and Michaelson (1994) who reconstructed Central Californian coastal rainfall for the last 600 years. Robertson (1996) found some evidence of a 30-year periodicity in North Pacific SST. It should be kept in mind, however, that time scale estimates are subject to large uncertainties due to the relatively short observational records available. Finally, a similar rotation in upper ocean heat content to that described by Latif and Barnett (1994) was identified by Zhang and Levitus (1997) in the observations (Figs. 9, 10, and 11). They investigated the anomalous three-dimensional temperatures in the North Pacific during the period 1960-1990 by means of EOF analysis. The two leading EOF modes of the combined temperature variability show a consistent phase relationship to each other,

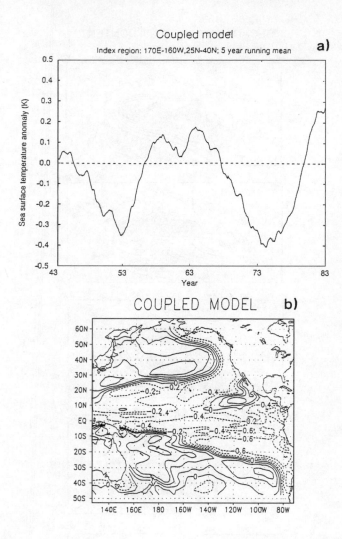

**Fig. 11.7** a) Low-pass filtered (retaining variability with time scales longer than 5 years) anomalous sea surface temperature (C) averaged over the North Pacific (170 E-160 W, 25 N-40 N) as simulated by the ECHO-1 CGCM. b) Spatial distribution of correlation coefficients of the index time series shown in (a) and low-pass filtered SST anomalies in the Pacific. See Latif et al. (1996b) for details

with EOF mode 2 leading EOF mode 1 by several years (Fig. 11.9). Thus, the temperature variability reconstructed from the two leading EOFs can be described as a cyclic sequence of patterns (Fig. 11.10 and Fig. 11.11). In particular, the observations show a clockwise rotation of the subsurface temperature anomalies, similar to that in the ECHO-1 CGCM (Fig. 11.8).

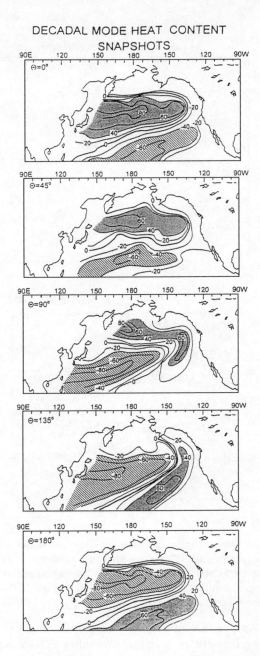

**Fig. 11.8** Evolution of anomalous heat content in the ECHO-1 CGCM. The individual panels show the heat content anomalies at different stages of the decadal cycle, approximately 2.5 years apart from each other. The heat content anomalies shown are reconstructions from the leading CEOF mode of low-pass filtered (retaining variability with time scales longer than 5 years) heat content anomalies. See Latif and Barnett (1994) for details.

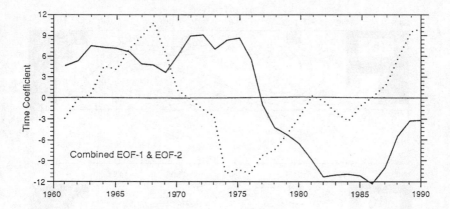

**Fig. 11.9** Time coefficients of the two leading EOFs of the observed low-pass filtered (retaining variability with time scales longer than three years) three-dimensional temperature anomalies in the North Pacific. The first EOF accounting for 27% of the variance is denoted by the solid line, while the second EOF accounting for 17% of the variance is denoted by the dashed line. See Zhang and Levitus (1997) for details

### 11.4.2 The North Atlantic Gyre Mode

Similar modes are simulated by some coupled models in the North Atlantic. Zorita and Frankignoul (1996) describe such a mode simulated by the ECHAM1/ LSG model, while Groetzner et al. (1996) identified it in the ECHO-1 model. Delworth (pers. comm., 1996) found a weak gyre mode in the GFDL (coarse-resolution) coupled model which was identified by performing a CEOF analysis of anomalous dynamic height. According to the theory of Latif and Barnett (1994), the period of the North Atlantic gyre mode should be shorter relative to that of the North Pacific gyre mode, and observations support this (e. g. Deser and Blackmon (1993)). The characteristic time evolution, SST and 500 hPa patterns associated with the North Atlantic interdecadal mode as simulated by the Latif et al. (1994) coupled model (ECHO-1) is shown in Fig. 11.12. The main anomalies are a positive SST anomaly centered near 35 N and a negative SST anomaly to the northwest near 50 N (Fig. 11.12b). The atmospheric response pattern is reminiscent of the NAO (North Atlantic Oscillation), with centers of action near Iceland and the Azores (Fig. 11.12c). The evolution in the upper ocean heat content is very similar to that simulated in the North Pacific (not shown), with a clockwise rotation of the anomalies around the subtropical gyre (Groetzner et al. (1996)). The period of the simulated oscillation amounts to about 17 years in the ECHO-1 model (Fig. 11.12a), which is longer than observed, while it amounts to about 10 years in the ECHAM1/LSG model (Zorita and Frankignoul (1996)).

Deser and Blackmon (1993) identified an interdecadal mode in the North Atlantic by analyzing surface quantities observed in this century (Fig. 11.13), and this

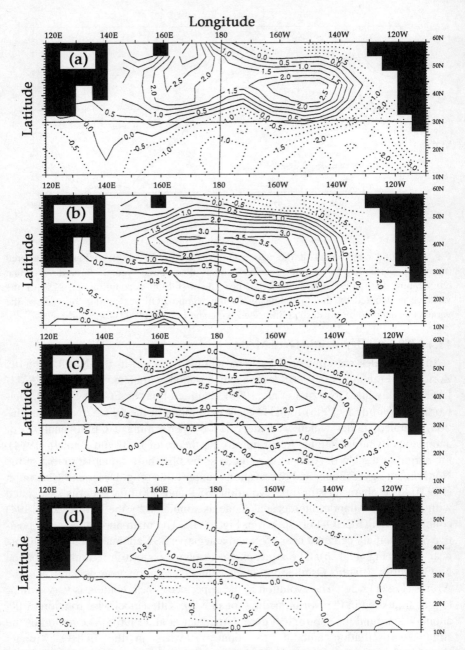

**Fig. 11.10** Spatial patterns of the leading EOF mode of the observed three-dimensional temperature variability. The individual panels show the loadings at a) sea surface, b) 125m, c) 250m, and d) 400m depths, respectively. The corresponding EOF time series is shown in Fig. 11.2 to Fig. 11.9. See Zhang and Levitus (1997) for details

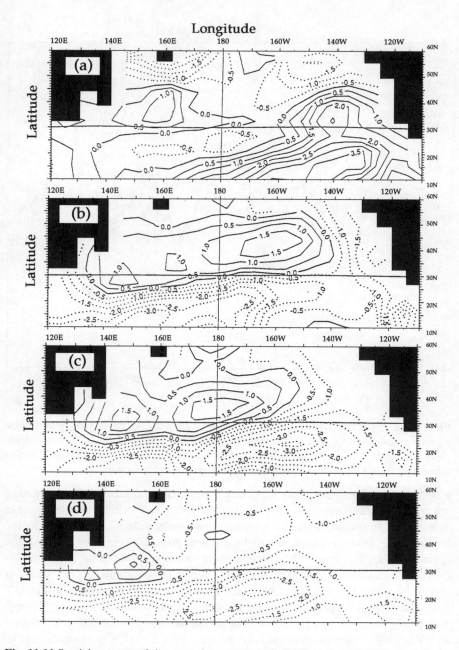

**Fig. 11.11** Spatial patterns of the second most energetic EOF mode of the observed three-dimensional temperature variability. The individual panels show the loadings at a) sea surface, b) 125m, c) 250m, and d) 400m depths, respectively. The corresponding EOF time series is shown in Fig. 11.9. See Zhang and Levitus (1997) for details.

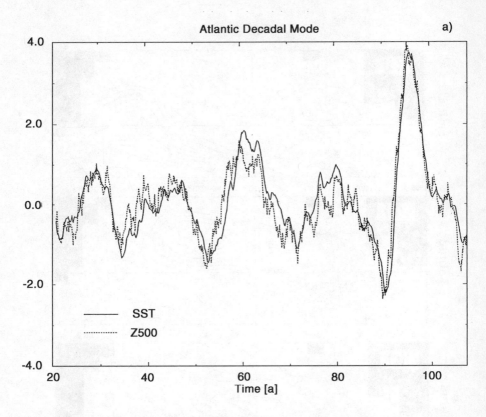

**Fig. 11.12** Canonical Correlation Analysis (CCA) of low-pass filtered (retaining variability with time scales longer than 5 years) North Atlantic SST and 500 hPa height anomalies simulated by the ECHO-1 CGCM. a) Canonical time series, b) canonical SST (C), and c) canonical 500 hpa (gpm) height pattern. See Latif et al. (1996b) for details.

interdecadal mode shares many aspects of the interdecadal variability simulated in the coupled GCMs investigated by Zorita and Frankignoul (1996) and Groetzner et al. (1996). The spatial surface air temperature anomaly pattern, as expressed by the leading EOF mode, is characterized by a dipole structure (Fig. 11.13a), and the spectrum (Fig. 11.13c) of the corresponding principal component (Fig. 11.13b) shows a statistically significant peak at a period of approximately 12 years. It should be noted, however, that although a corresponding EOF analysis of the SST anomalies revealed a peak at the same frequency, it was not statistically significant. The atmospheric pressure pattern accompanying the interdecadal surface air fluctuations is very similar to that simulated by the coupled models and reminiscent of the NAO (not shown). Likewise, subsurface temperature anomalies at ocean weather ship "C" in the North Atlantic at 125 m depth (Levitus et al. (1994)) show some remarkable oscillatory behavior during the last few decades, with a period of

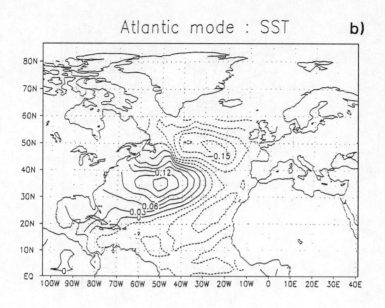

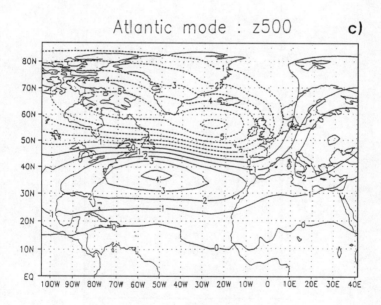

**Figure 11.12 (continued)**

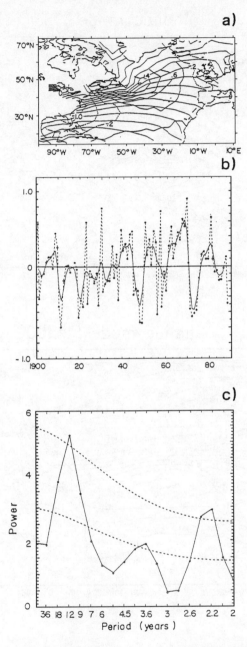

**Fig. 11.13** Leading EOF mode of observed North Atlantic (winter) surface air temperature anomalies. a) EOF pattern (dimensionless), b) corresponding principal component (C), c) spectrum of the principal component shown in (b). The EOF mode accounts for 21% of the variance. See Deser and Blackmon (1993) for details

the order of about 15 years (Fig. 11.14). As pointed out by Latif et al. (1996b) and Groetzner et al. (1997), these variations might be forced by the atmosphere in response to variations in the subtropical gyre circulation.

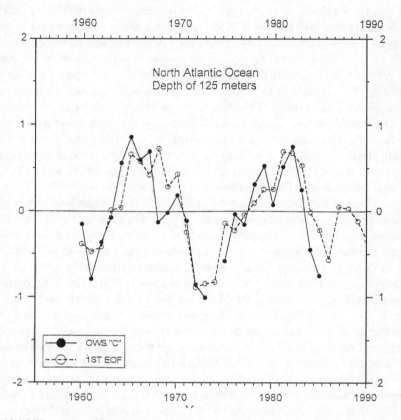

**Fig. 11.14** Time series of subsurface temperature anomalies at 125m (C) measured at ocean weather ship "C" (35.5 W, 52.5 N) and time series of the first EOF (multiplied by -1) of North Atlantic temperature anomalies at the same depth. See Levitus et al. (1994) for details.

### 11.4.3 Relative Roles of Stochastic and Deterministic Processes

A competing hypothesis for the generation of interdecadal variability was offered by Frankignoul et al. (1996). They extended the stochastic climate model theory of Hasselmann (1976) to include the variations in the wind-driven ocean gyres. While the coupled interactions proposed by Latif and Barnett (1994) will be associated with a preferred period, the scenario of Frankignoul et al. (1996) predicts a red spectrum with a high-frequency w decay that levels off at low frequency. Different coupled models show quite different qualitative behaviors. The GFDL and the ECHAM1/LSG coupled models, for instance, do not exhibit statistically significant peaks in the spectra of the simulated SSTs in the regions of the subtropical gyres in the North Pacific and North Atlantic, while the MPI (Max-Planck-Institut fur Meteorologie) coupled model ECHO-1 does exhibit significant decadal peaks in both the North Pacific and North Atlantic.

These differences can be understood partly in terms of the different air-sea coupling strengths (Münnich et al. (1997), Neelin and Weng (1997)). In order to explore this further, Münnich et al. (1997) coupled a shallow-water ocean model to a simple atmospheric feedback. Münnich et al. (1997) investigated three cases. The simple coupled model simulates a damped interdecadal oscillation in the control case. When noise is added to the coupled model, (Fig. 11.15, upper) a well-defined spectral peak is retained in the simulation which is superimposed on a red background (Fig. 11.15, lower). This situation corresponds probably to the MPI-(ECHO-1) model. When the noise-forced integration is repeated, but with the coupling between ocean and atmosphere turned off, the spectrum of the simulated thermocline depth anomalies is basically red and very close to that predicted by Frankignoul et al.'s (1996) theory (Fig. 11.15, middle). The GFDL and ECHAM1/LSG coupled models reside probably in this latter parameter regime.

Thus, a critical factor determining the character of the interdecadal variability in midlatitudes is the sensitivity of the atmosphere to midlatitude SST anomalies. This issue is still a matter of intense scientific discussion. There are, however, some indications, that the resolution of the atmosphere model needs to be sufficiently high, in order to represent the atmospheric response adequately which involves changes in the transient activity (e. g. Palmer and Sun (1985)). This is supported by a new integration performed at GFDL. When the GFDL coupled model is run at higher resolution (R30), the nature of the interdecadal variability simulated in the North Pacific seems to change relative to the old run (which used a R15 atmosphere model), and the simulated spatial patterns and time scales are in qualitative agreement with that simulated by the MPI-(ECHO-1) model (Delworth, pers. comm., 1996) which employed an atmosphere model with comparable resolution (T42). In summary, fluctuations in the wind-driven ocean gyres are likely to cause interdecadal climate fluctuations in both the North Pacific and North Atlantic Oceans. The importance of the air-sea coupling and relative roles of stochastic and deterministic processes, however, need to be addressed in more detail.

## 11.5  Midlatitude Interdecadal Variability Involving the Thermohaline Circulation

One can separate conceptually the oceanic wind-driven from the oceanic thermohaline circulation. While the wind-driven circulation is forced by wind stress variations, the thermohaline circulation is forced by density anomalies. In reality, however, both circulations are coupled to each other, especially in the North Atlantic. Modes were described in the last section which are to first order "pure" gyre modes. In this section, modes are described which involve mainly the thermohaline circulation as an active component. Delworth et al. (1993) analyzed the multi-cen-

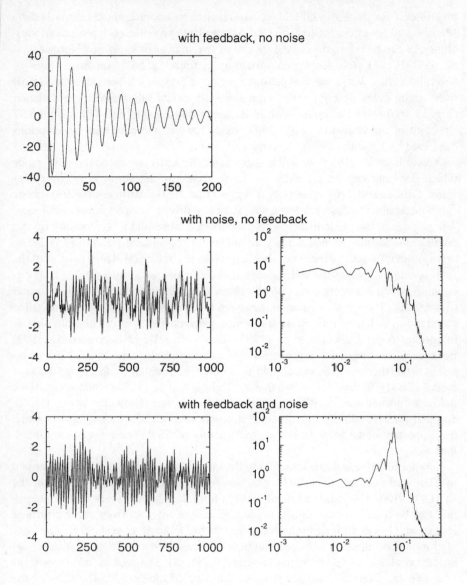

**Fig. 11.15** Interdecadal variability experiments with a simple coupled model. The upper panel shows the thermocline perturbation simulated in a 200-year control integration which was perturbed initially. The middle panels show the time series of the thermocline perturbation and the corresponding spectrum obtained in a 1000-year run without atmospheric feedback but with the inclusion of stochastic forcing. The lower panels show the thermocline perturbation and the corresponding spectrum when both the atmospheric feedback and stochastic forcing are included. See Münnich et al. (1997) for details.

tury integration with the GFDL (coarse-resolution) coupled model described by
Manabe and Stouffer (1996). The CGCM simulates pronounced interdecadal variability, as can be inferred from the spectra of two indices of simulated North Atlantic SST (Fig. 11.16a). Enhanced variability is found at 56 N on time scales of
several decades, with a spectral peak at a period of 50 years. Likewise, a meridional
overturning index describing the strength of the model's thermohaline circulation
(Fig. 11.16b) shows a similar peak in its spectrum (Fig. 11.16c). The spatial SST
structure of the Delworth et al. (1993) mode agrees favorably with observations
described by Kushnir (1994), as shown in Fig. 11.17.

Delworth et al. (1993) describe their mode basically as an ocean-only mode
which does not depend critically on the feedback by the atmosphere. In a later
study, Griffies and Tziperman (1995) suggest that the Delworth et al. (1993) mode
can be described to first order as a stochastically driven damped linear oscillation.
The physics of the mode involves both the thermohaline and a horizontally (gyral)
circulation. Salinity-related density anomalies in the sinking region (52 N-72 N)
drive anomalies in the strength of the thermohaline circulation. The transport of the
salinity anomalies into the sinking region is controlled by a horizontal (gyral) circulation which is mainly controlled by upper ocean temperature anomalies further
to the south. The gyral circulation, however, is not in phase with the thermohaline
circulation, and it is this phase difference between the two circulations that is
important to the Delworth et al. (1993) oscillation. The Delworth et al. (1993)
mechanism was basically reproduced by Lohmann (1996) who studied the variations of the thermohaline circulation in an OGCM coupled to an energy balance
model (EBM). It should be noted that the Delworth et al. (1993) mode seems to be
linked to pronounced oscillations in the Greenland Sea (Delworth et al. (1997)),
and that these oscillations involve the atmosphere as an active component, which
has important consequences to the predictability of SST anomalies in the Greenland Sea.

Another interdecadal mode involving the thermohaline circulation was described
by Timmermann et al. (1997) who analyzed a multi-century run with the
ECHAM3/LSG CGCM. Preliminary results from the first 320 years of this integration can be found in Latif et al. (1997c). As can be inferred from the overturning
index (which was defined in the same way as in Delworth et al. (1993)), the coupled model simulates pronounced interdecadal variability, with a statistically significant peak at a period of about 35 years Fig. 11.18). The peak is fairly robust, as
was revealed from the full 800 years of the coupled integration. The level of the
interdecadal variability simulated is similar to that in the GFDL (coarse-resolution)
CGCM. However, in contrast to the Delworth et al. (1993) mode described above,
the Timmermann et al. (1997) mode appears to be an inherently coupled ocean-
atmosphere mode.

Let us consider the situation ten years before the maximum overturning occurs.
A well developed negative SST anomaly covers almost the entire North Atlantic
(Fig. 11.19b) at this time. The atmospheric response as expressed by the anoma-
lous sea level pressure (SLP) is also well developed (Fig. 11.19a) and is associated

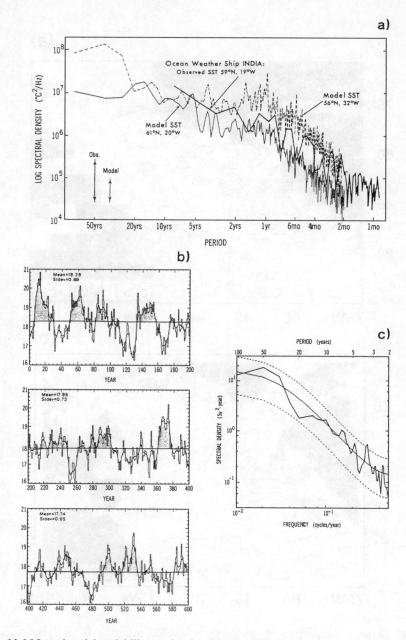

**Fig. 11.16** Interdecadal variability as simulated by the GFDL (coarse-resolution) CGCM in the North Atlantic. a) Spectra of two model SST indices and one observed SST index (the locations are given in the figure). b) Time series of an overturning index (Sv) describing the strength of the model's thermohaline circulation. c) Spectrum of the first 200 years of the overturning index shown in (b). See Delworth et al. (1993) for details

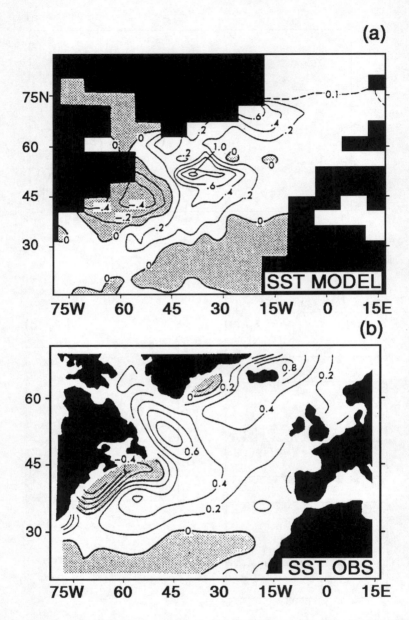

**Fig. 11.17** Comparison of interdecadal North Atlantic SST anomalies (C) as a) simulated by Delworth et al. (1993) and b) derived from observations by subtracting the periods 1950-1964 (warm) and 1970-1984 (cold) by Kushnir (1994). See Delworth et al. (1993) for details.

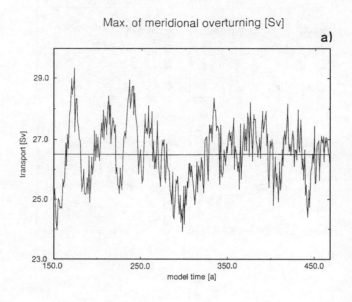

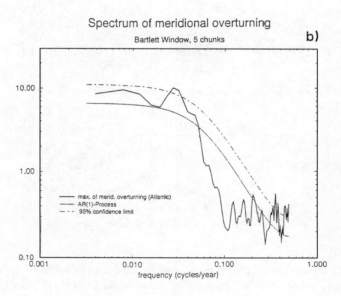

**Fig. 11.18** a) Time series of an overturning index (Sv) describing the strength of the thermohaline circulation in the experiment with the ECHAM3/LSG CGCM. The overturning index is defined in the same way as in Delworth et al. (1993). b) Spectrum of the overturning index (Sv Wa) shown in (a). See Timmermann et al. (1997) for details

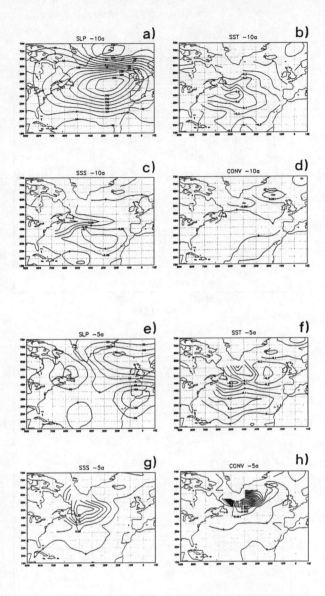

**Fig. 11.19** . Composites describing the anomalous conditions in different quantities at two different stages of the interdecadal mode simulated by the ECHAM3/LSG CGCM. Panels (a)-(d) show the anomalies 10 years prior to the maximum overturning, while panels (e)-(h) show the anomalies 5 years prior to the maximum overturning. The quantities shown are anomalous SLP (Pa), SST (C), SSS (psu), and potential energy loss by convection (W/m). See Timmermann et al. (1997) for details.

with fresh water flux anomalies (not shown) which force positive salinity anomalies off Newfoundland and in the Greenland Sea (Fig. 11.19c). The convection is near normal at this time (Fig. 11.19d). The salinity anomalies amplify and propagate into the region of strongest convection which is located slightly south of Greenland (Fig. 11.19g), where they strengthen the convection five years later (Fig. 11.19h). This will eventually lead to an anomalous strong thermohaline circulation, enhanced poleward heat transport in the upper ocean, and the generation of a large-scale positive SST anomaly, which completes the phase reversal. The atmospheric response to this positive SST anomaly will be associated with fresh water flux anomalies which in turn force negative salinity anomalies off Newfoundland and in the Greenland Sea, opposite to those shown in Fig. 11.19c, which will eventually initiate the next cold phase. Some elements of this feedback loop were also identified in observations by Wohlleben and Weaver (1995).

The role of the air-sea coupling in generating interdecadal variability associated with the thermohaline circulation is, as in the case of the gyre modes, rather different in the different models. While the Delworth et al. (1993) mode seems to be an ocean-only mode, the Timmermann et al. (1997) mode appears to be an inherently coupled air-sea mode. The differences between the two model simulations may be well explained by the different sensitivities of the respective atmospheric component models to midlatitudinal SST anomalies. At this stage of understanding, however, we are not in the position to decide which scenario is the more realistic one.

It is interesting to note that salinity anomalies play an important role in both the Delworth et al. (1993) and Timmermann et al. (1997) interdecadal modes. Some simpler coupled models give quite different results. Although similar interdecadal modes are simulated by these simplified models, the temperature effects dominate the salinity effects. The hybrid coupled model of Chen and Ghil (1996), for instance, does not even include salinity effects, while the coupled model of Saravanan and McWilliams (1996) is based on an idealized geometry. These are severe limitations in the formulations of these two particular simplified coupled models. The Coupled GCMs on the other hand employ relatively large flux corrections, especially in the heat and fresh water fluxes. The effects of all these model limitations on the simulation of interdecadal variability are relatively unknown, and much more work is needed in this research area.

## 11.6  Predictability at Decadal Time Scales

The problem of the predictability at decadal time scales has two aspects. The first aspect involves the interdecadal variations in the predictability of interannual phenomena, such as ENSO. As shown in a number of studies (e. g. Balmaseda et al. (1995), Chen et al. (1996), Latif et al. (1997b), Ji et al. (1997)), there is considerable variability in the ENSO prediction skill from decade to decade. The 1970s and the 1990s, for instance, were less predictable than the 1980s, as can be inferred from the ENSO skill scores presented by Chen et al. (1996) (Fig. 11.20). Some of the interdecadal modes described above may well influence the nature of the ENSO variability and hence the ENSO prediction skill by slowly modulating the background conditions in the tropical Pacific. This issue, however, has not been addressed in great detail so far.

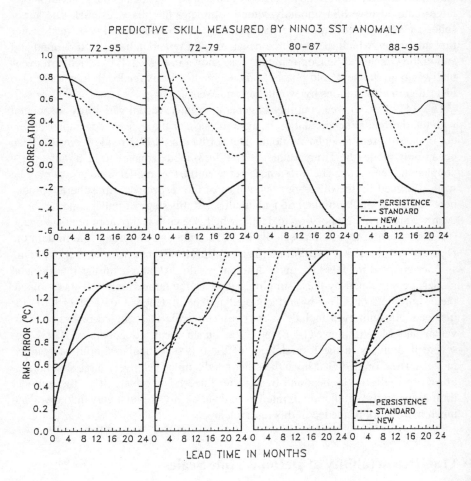

**Fig. 11.20** Correlation skills and RMS (root mean square) errors (C) obtained in predicting eastern equatorial Pacific SST anomalies in the Nino-3 region with two versions of the coupled model of Zebiak and Cane (1987). Shown are the correlation skills and RMS errors for different time periods. The top panels show the correlation skills, while the bottom panels show the RMS errors. Please note the marked variations in skill from decade to decade. See Chen et al. (1996) for details.

The other aspect of the predictability at decadal time scales involves the prediction of the interdecadal phenomena themselves. Venzke and Latif (1997) investigated the predictability of the North Pacific gyre mode. They forced the ECHAM3 (T42) AGCM by observed SSTs for the period 1949-1990 and used the simulated wind stresses and heat fluxes to drive an oceanic GCM. The simulated upper ocean heat content in the western North Pacific, as expressed by the simulated sea level, shows a remarkable lead-lag relationship to the observed SST anomalies in the central North Pacific (Fig. 11.21), with the anomalous heat content leading the anomalous SST by several years. This suggests that decadal North Pacific SST anomalies are predictable several years ahead. In particular, the strong decadal SST change in the mid-seventies was preceded by a corresponding decadal change in upper ocean heat content about five years earlier. This feature was also noticed by Zhang and Levitus (1997) who investigated observed surface and subsurface temperature observations (see Fig. 11.9 and Fig. 11.10).

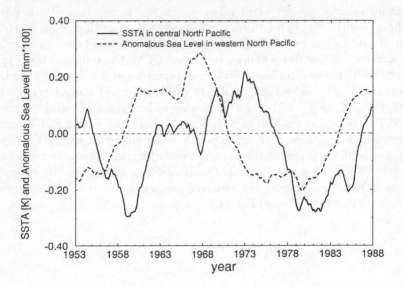

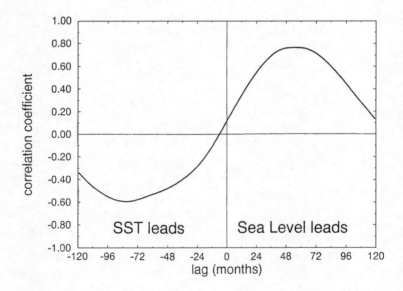

**Fig. 11.21** a) Time series of anomalous sea level in the western North Pacific as simulated by an oceanic GCM and of observed anomalous SST in the central North Pacific. b) Lagged cross correlations between the anomalous western Pacific sea level and the anomalous central Pacific SST. Both time series were smoothed by applying a five-year running mean filter. See Venzke and Latif (1997) for details.

Griffies and Bryan (1996) investigated the predictability of the Delworth et al. (1993) mode described above by conducting ensembles of classical predictability experiments with the GFDL (coarse-resolution) CGCM. Those were performed in such a way that the coupled model was restarted from initial states obtained from a control integration, but with atmospheric perturbations superimposed. The divergences of the trajectories within the forecast ensembles provide informations on the predictability of the coupled system. Griffies and Bryan (1996) derived basically an upper limit of predictability, since they assumed perfect oceanic initial conditions.

Their results indicate that the state of the interdecadal mode described by Delworth et al. (1993) as measured by subsurface quantities is predictable up to lead times of one to two decades (Fig. 11.22). The SST fluctuations associated with the decadal mode, however, appear to be predictable up to lead times of a few years only. The Delworth et al. (1993) mode seems to be connected also to the Arctic climate system, and air-sea interactions may be important in this region (Delworth et al. (1997)). This has important consequences for the predictability of SST anomalies in the Greenland Sea which appear to be predictable about 5-10 years in advance (Delworth, pers. comm., 1996).

Similar decadal predictability experiments are currently conducted with the ECHAM3/LSG CGCM at MPI (Groetzner, pers. comm., 1997), in order to study the predictability of the interdecadal mode described by Timmermann et al. (1997). The Delworth et al. (1993) and Timmermann et al. (1997) interdecadal modes appear to have quite different characteristics. While the former can be described to first order as an ocean-only mode, the latter depends crucially on the air-sea coupling. It will be interesting to investigate, how this difference manifests itself in the predictability of the two modes.

Ensembles of uncoupled AGCM experiments prescribing observed SSTs and sea ice distributions for several decades are currently performed at different institutions worldwide, in order to study the predictability of the atmosphere at decadal time scales. An example from the Hadley Centre AGCM is described here (Rowell and Zwiers (1997)). A 6-member ensemble of integrations was conducted for the period October 1948 to December 1993, each experiment starting from different initial conditions. Rowell and Zwiers (1997) found that the tropics are generally more predictable than the extra-tropics. The extra-tropics appear predictable during particular seasons and in particular regions only. Climate anomalies over North America and Central Europe, for instance, appear to be predictable in summer but not in winter. Some of these results, however, will be model-dependent, and many more such experiments are needed, in order to gain further insight into the nature of the atmospheric response to interdecadal variations in the boundary conditions.

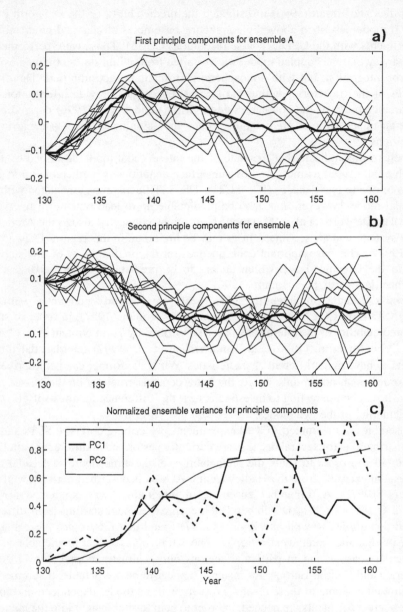

**Fig. 11.22** .Results of the predictability experiments conducted by Griffies and Bryan (1996). The top two panels (a) and (b) show the evolutions of the first two leading EOFs of dynamic topography which describe basically the Delworth et al. (1993) mode. The thick lines denote the ensemble means, while the thin lines denote the individual realizations. The normalized ensemble variances are given in panel (c). See Griffies and Bryan (1996) for details

## 11.7 Discussion

Coupled ocean-atmosphere models simulate a wide range of interdecadal variability, and many aspects of the simulations show an encouraging similarity with the interdecadal variability observed. Different competing hypotheses, however, were put forward to explain the interdecadal variability in the simulations. Some models, for instance, are largely consistent with Hasselmann's (1976) stochastic climate model scenario, exhibiting relatively featureless red spectra in some oceanographic key quantities. Other models simulate interdecadal oscillations, with statistically significant spectral peaks superimposed on the red background.

One major problem in verifying the different mechanisms proposed is certainly the lack of an adequate observational database, which inhibits us from verifying rigorously the different hypotheses for the generation of interdecadal variability. Reliable surface observations over the oceans exist for approximately the last 100 years, while reliable land observations may exist for the last 150 years. High-quality subsurface measurements exist for the last few decades only. Thus, there is a strong need to make more use of paleoclimatic data, in order to enhance our understanding of the interdecadal variability. However, it should be possible within the CLIVAR period to make good progress in the understanding of the short-term decadal variability that occurs on time scales of about 10 years by an integrated analysis of all available data sets (including paleo-observations, direct observations and model results) and the collection of new observations.

The role of large-scale air-sea interactions in causing interdecadal variability must be addressed in more detail. One key factor determining the nature of the simulated interdecadal variability, for instance, appears to be the atmospheric response to midlatitudinal SST anomalies. Some atmosphere models seem to be relatively insensitive, while other atmosphere models exhibit a detectable response to extratropical SST anomalies. We are at a relatively early stage in the understanding of the dynamics of this atmospheric response, and further numerical experimentation with state-of-the-art AGCMs will help to advance our understanding. Similarly, the oceanic adjustment in response to changes in the fluxes at the air-sea interface, especially the fresh water flux, is not well understood yet. In particular, it is not clear how well the coarse-resolution OGCMs (which are commonly used in coupled ocean-atmosphere models) simulate the "real" ocean.

There are, however, many other problems that need to be addressed in the modeling community, such as the role of flux corrections or the relative importance of salinity and temperature anomalies in the generation of interdecadal variability. Likewise, the relative roles of the tropics and the extratropics, of stochastic and deterministic processes need to be investigated in more detail. Furthermore, the dependence of the results on model parametrizations and parameters has not yet been adequately explored. The problem of decadal predictability has been addressed so far in very few studies only, and a co-ordinated international decadal prediction program under the auspices of CLIVAR (Climate Variability and Predictability) would be desirable.

It is likely that the interdecadal variability has some impact on the interannual variability on the one hand, but that the interdecadal variability itself is influenced by centennial variability and global warming on the other hand. The nature of these interactions are largely unexplored, and much more work is needed during the CLI-VAR program to reach a better understanding of these interactions. The anomalous 1990s in the tropical Pacific, for instance, which were characterized by persistent warm conditions and a weak ENSO activity may be a good example of how the different time scales interact with each other. Some studies argue that the 1990s are already an expression of global warming (e. g. Trenberth and Hoar (1996)), while other studies attribute the anomalous 1990s to interdecadal variability (e. g. Latif et al. (1997a)) which affected the statistics of the ENSO cycle (Gu and Philander (1996)). The bottom line is that models that are applied to the greenhouse warming problem need to simulate realistically both the interannual and interdecadal variability. This, however, is generally not the case. Thus, coupled model development is still a major issue.

**Acknowledgments.** The author would like to thank Dr. T. Delworth for his valuable comments on an earlier version of the manuscript. Drs. A. Grotzner, M. Münnich, D. Rowell, S. Zebiak and Messrs. C. Eckert, A. Timmermann, S. Venzke provided important input to this review paper. This work was supported by the EC Environment Research Programme under contract EV5V-CT94-0538 and the German government under grant no. 07VK01/1.

# 12 Simulations of the Variability of the Tropical Atlantic

M.J.ORTIZ BEVIA, W. CABOS NARVAEZ
*Departamento de Fisica, Universidad de Alcala de Henares,*
*Apdo 20, 28880 Alcala de Henares, Madrid, Spain.*

J. M. OBERHUBER
*DKRZ, Bundestrasse 55, D-20146 Hamburg.*

## 12.1 Introduction

In a number of observational studies, the seasonal as well as the interannual variability in the tropical Atlantic has been analysed (Arnault, 1987; Houghton, 1993; Weingartner and Weisberger, 1987, 1991; Servain, 1993). The seasonal cycle of the tropical Atlantic ocean has been successfully simulated by simple and more complete general circulation ocean models (see the review by Duchene and Frankignoul (1990)). Details of this circulation specially difficult to simulate are the subject of recent studies with high resolution models (Johns et al. (1990), Schott and Boening (1991)).

The most important features of the interannual variability of the tropical Atlantic are the episodic warmings and coolings of the temperatures in the Gulf of Guinea. These are characterized by the Gulf of Guinea index (hereinafter GG index), shown in Fig. 12.1. This index was produced by averaging the SST anomalies, previously smoothed with a three points running mean, to the region (20W - 10E, 3N - 12S) (Servain (1993)). In Fig. 12.1 we also show the Niño3 index, built also form SST anomalies, that were similarly smoothed and averaged to the region (150W - 90W, 5S - 5N). The observed Atlantic warming events take place every years, and never reach more than $1.8°$ K.

With respect to the Pacific case, the intensity is reduced and the maxima occur more towards the center of the basin. The relationships with ENSO events have been investigated in early observational studies (Wright, 1987). The correlation between both indices is significant, with a maximum value of 0.65, obtained with the GG index lagging the Niño3 index by three months. But there were Atlantic warmings not preceded by any ENSO signal. Precipitations in North-East Brazil are heavily correlated with the GG index, while only weakly with El Niño, although the region is nearer to the coast of Peru than to the Gulf of Guinea.

At seasonal time scales, the circulation in the tropical Atlantic has some traits in common with those of the other tropical oceans, as for instance, the equatorial system of currents or the eastern basin upwelling in summer and fall of the northern hemisphere. Other effects, as the mainly meridional displacement of the Intertropical Convergence Zone (ITCZ) are particular of this basin. While many of the features of the seasonal Atlantic variability are well documented and explained in the literature (see for instance Philander and Pacanowski, 1986), the mechanism that determines the interannual variability and its relationships with ENSO are currently being discussed.

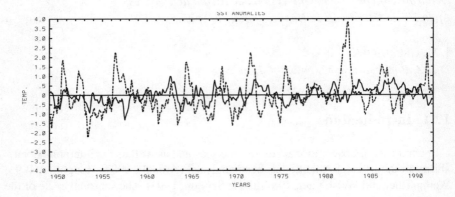

**Fig. 12.1** The GG Index (dashed line), from 1950 to 1992 against the Niño3 index (solid line), for the same years.

In the last years there has been an increased interest in the interannual variability of the Atlantic basin and the tropical Atlantic has received renewed attention. Because of the poor coverage, specially of subsurface data, the knowledge of this variability through observations is necessarily limited. Model simulations could be very valuable to understand the physical interactions that determine it. In intermediate coupled models of the equatorial Atlantic (Zebiak, 1993), the spatial and temporal characteristics of the equatorial warmings are explained by a coupled mode. In these simulations, there is another coupled mode, with a decadal time scale of variability, and apparently not related to the Gulf of Guinea warmings, known as the dipole mode (Ping Chang et al., 1997).

Some of the modeling studies focus on the relationships between ENSO and the Atlantic events. For instance, Delecluse et al. (1994) use an oceanic general circulation model (OGCM) forced with the response of an atmospheric general circulation model to observed SST, and Latif and Barnett (1995), a fully coupled general circulation model. Links between the Atlantic and Pacific warmings have also been the aim of recent observational studies (Curtis and Hastenrath (1995), Wagner (1996)). In both of these studies, there is a new stress on the importance of the wind anomalies in the north tropical Atlantic in controlling the northward SST gradient. This gradient in turn, has a significant part in the GG index variability (Cabos et al., 1998). The mechanisms of interaction between ENSO and the tropical Atlantic are traced in Endfield and Mayer (1997) to a weakening of the northern trades and the following sea surface temperature warming, persistent although weak. North of this region, the enhancement of the subtropical gyre detected is unrelated to any SST local anomaly.

Only modeling studies with coupled general circulation models give a simulation of the ocean-atmosphere interactions that is exhaustive and free from observational errors, but simulations of particular events, obtained by forcing an OGCM with atmospheric observations, can also be valuable. In the case of the tropical Atlantic, there are only two datasets of observations covering an extended period: the ECMWF analysis and the subjectively analyzed observations by Servain et Legler (1996). In a series of papers, Huang and others (Carton and Huang, 1994; Huang et al., 1995; Huang and Shukla, 1997) have recently used both of them as forcing of an OGCM to simulate the equatorial Atlantic variability for the years 1980-1989.

In the present work we also simulate the interannual equatorial Atlantic variability by forcing an OGCM with atmospheric observations. In order to understand the generation of the events, we focus on the features that are common to several of them. The events compared are those simulated in the decade 1980-1989, where three warmings (1981, 1984 and 1988) took place. In the first of our simulations, hereafter referred to as E1, where the forcing data came from the ECMWF analysis, the 1984 and 1988 events were successfully reproduced, while the 1981 warming was missed. To investigate the reasons of this failure, we have carried out our second simulation (run S1), where the Servain et al. data were used for the anomalous atmospheric forcing. Two other simulations (E2 and S2) were intended to test the influence of the mid-latitude anomalies in the generation of the events.

Coupled model studies have shown that the "heat content", that is, the thermal energy stored in the ocean upper layers, is the optimal variable to monitor the generation of the anomalous warmings of the tropical oceans (Chao and Philander (1993), Latif and Graham (1992)). Nevertheless, the onset stage of the warmings has been also related to anomalies in other surface variables, like mixed layer velocity convergence (Latif and Graham, 1992; Latif et al., 1996). Can the anomalies of the surface variables be related to the heat content anomalies in a simple way? To check this point, we reduce the evolution of the heat content anomalous field to those of a few modes, and, through simple statistical modeling, we try to connect them with other surface variables, like mixed layer velocities, or wind stresses.

The structure of the paper is as follows. The simulation of the seasonal variability is described in section 12.2, where details of the model layout and the forcing used for this simulation are also given. In section 12.3, different simulations for the period 1980-1989 are described, and its interannual variability, represented by the GG index, is compared. In this section, we also include a description of a coupled run (OPYC-T42), whose interannual variability in the tropical Atlantic will be used to test the generality of the analysis of the forced simulations. In section 12.4, we study the onset and evolution of the events in different simulations, through the analysis of the anomalous heat content field. In section 12.5, we include details of the systematic statistical analysis followed in Cabos et al. (1998) to relate the evolution in time of the anomalous heat content to those of the anomalies of other oceanic variables or of the forcing fields. The results presented through this paper are discussed in section 12.6.

## 12.2 Simulation of the seasonal variability

Our simulation of the seasonal cycle is satisfactory (Cabos et al. (1998)), and similar to the one reported in Carton Huang (1994), and in Huang et al. (1995). The model used in this work is an updated version of the model developed by Oberhuber (1993a). An updated summary can be found in the Appendix of Cabos et al. (1998). It consists in a number of isopycnal ocean layers fully coupled to a surface bulk mixed layer model, that in turn is coupled to a sea-ice model. Isopycnal models use Lagrangian coordinates in the vertical, in contrast to level models, where the grid points are at fixed depth levels. The model solves the full primitive equations for mass, mass flux, temperature and salinity in spherical geometry with a realistic equation of state. The layers' interchange is due to diapycnal mixing and convection. The interior ocean is coupled to the mixed layer through the entrainment/detrainment processes. A potential vorticity and enstrophy conserving scheme is implemented. In the horizontal, the equations are discretized on an Arakawa B-grid. A time integration scheme consisting of a semi-implicit scheme combined with a predictor-corrector technique is used in order to achieve large time steps. Table 12.1 includes the values of some of the relevant parameters adopted in the present simulation.

| Parameter | Typical Value | Description |
|:---:|:---:|:---:|
| $m_0$ | 0.5 | Coefficient for wind stirring |
| $\delta t$ | 12 hr | Simulation Time Step |
| $h_B$ | 20m | Solar radiation Penetration depth |
| $\mu$ | 1.67 | Ekman layer constant |
| $c_r$ | 0.0012 | Surface Drag Coefficient |

**Table 12.1** Model parameters, typical value and description

The model domain covers 56°S to 65°N, and 80°W to 20°E. The grid varies zonally from 2.0° to 1.5° and meridionally from 2.6° to 0.5°. The center of the zonal focus is in the GG, while the meridional one is on the equator. The model has eleven layers, the time step is half a day and the model uses non-slip boundary conditions along the coasts. The domain is open to the north and south.

The main difference between the present version of the model and the one used in others simulations, as for instance by Miller et al. (1994), consists in the way the boundary conditions in the open ocean are imposed. For the present simulation we have used a new version of the model that incorporates an open boundary formulation, due to Kauker and Oberhuber (1996). The main point of this formulation is that on the boundary the pressure is imposed but near it, the flow is fully computed and not prescribed at all. Boundary conditions, namely the layers thicknesses, potential temperatures and salinities are taken from the output of a global OGCM of higher resolution (a global version of OPYC at horizontal resolution corresponding to grid of the T106, forced with climatological winds and fluxes). In this way, the boundary values are interpolated only horizontally, avoiding the errors introduced by interpolating in the vertical. The barotropic part of the sea level at the boundaries was estimated from the barotropic transport of this global run using the geostrophic approximation. Nevertheless, because of the inclusion of two open boundaries, one is left with the problem of determining the differences between mean sea levels at these. In this case, the value of this difference in the global model can be taken only as a first indication. Because of the disparities in the global and regional model resolution are conveyed into their physics, this difference in mean sea level must be finally tuned.

The data sets required to force the model are air temperature, relative humidity, cloudiness, wind stress, the time averaged absolute wind speed and its standard deviation, and surface salinity. The wind stress data are basically the climatology of the ECMWF analysis from 1980 to 1989, corrected at the equator with the Hellermann-Rosenstein climatology. The other atmospheric forcing data required for the forcing come basically from COADS, although other global data sets were also used. The model was forced with heat fluxes derived from observations according to Oberhuber (1988) plus a relaxation to the climatology of the observed SST.

$$Q_s = Q_{obs} - \left(\frac{dQ}{dT}\right)_{obs} (T_{obs} - T_S) \qquad (12.1)$$

where $T_{obs}$ and $T_S$ are the observed and simulated climatological SST. The relaxation coefficients $(dQ/dT)_{obs}$ are also computed following Oberhuber (1988). During the last year of the spinup, the daily values of $Q_s$ were saved, and used afterwards as the forcing in the anomalous run.

We can characterize the Carton and Huang (1994), and Huang et al. (1995) simulation in terms of its vertical discretization of the model (fixed levels), the physical parameterization used, the model domain and layout and the parameterization of the heat fluxes. Only the winds stresses are common to our simulation of the tropical Atlantic. In the case of the wind forcing, the differences can (they include a high frequency component) be considered small. And nevertheless both models are successful at simulating the tropical Atlantic variability at two different (annual and interannual) time scales.

Observations show that the ITCZ is nearer the Gulf of Mexico in September, in March-April it is nearest to the equator and at its most eastern position. Sea surface temperatures (hereinafter SST) in the east reach then its maximum value. All the surface currents except the North Brazil current are weak and the zonal slope of the thermocline is almost horizontal. In May, as the ITCZ moves north, the intensification of the south trades is followed by a strengthening of the South Equatorial Current. Due to intense upwelling, the thermocline shoals in the east, a strong North

Equatorial Countercurrent is observed, and measurements show that an undercurrent also exits. SST reach its minimum value in the western part of the basin in July and a month later in the Gulf of Guinea. The SST errors of our simulation of the seasonal cycle, computed by subtracting the climatological monthly mean from the AMIP (Atmospheric Models Intercomparison Project) from the simulated SST are below 1° C even in the critical months (June-July-August). The seasonal evolution of the simulated currents is also captured. The error in the simulated magnitude of the currents is roughly lower than 20% in most of the domain (Richardson and Reverdin, 1987). For instance, for the North Brazil current, the estimated value is below 86 cm/s while observations show maximum values of 110 cm/s at some months. Other magnitudes like mixed layer depth or the heat content have reasonable spatial structure and values. A more detailed account can be found in Cabos et al.(1998).

## 12.3  The simulation E1 of the interannual variability.

Using the last year of the spinup as initial state, the model was forced with wind stresses from the ECMWF analysis for the period 1980-1989. The heat fluxes were obtained from the last year of the spinup plus a relaxation to the simulated climatology given by the expression

$$Q_s = Q_{obs} - \left(\frac{\overline{dQ}}{dt}\right)(T_{obs} - T_S) \qquad (12.2)$$

where $T_a$ is the SST obtained in this experiment and $T_S$ the simulated SST climatology from the last year of the spinup and $\overline{(dQ/dT)} = 0.3(dQ/dT)_{obs}$. The 0.3 is a tuning factor to obtain anomalies of the order of magnitude than the observed ones.

In Fig. 12.2a we show values of the Atlantic index simulated (dashed line) with our ocean model and the observed one (thin line). We see that the model captures quite well the warming events of 1984 and 1988 but the 1981 event is simulated as one of cooling. In our Fig. 12.3 the peak phase of the simulated warmings (middle row) is compared with observations (top). The equatorial character of the simulated warming is a trait shared with Huang et al. (1995), as well as the higher values of the anomalies with respect to observations. From a comparison month by month of the observed and simulated 1988 event (not shown) it can be seen that the starting of the episodes is well modeled. Later, when the anomalies extend to the tropics along the African coast the model does not perform well, although for the right season (winter 1989) it reproduces the cooling that closes the event.

The evolution of the cooling events (observed and simulated) compares well with the reverse of the warming events (observed and simulated respectively): the modeled cooling also anticipates the observed one. An interesting feature, common to the Huang et al. (1995) simulation and ours, is the misrepresentation of the 1981 event. This common trait supports Huang et al. (1995) conclusion that a trend in the

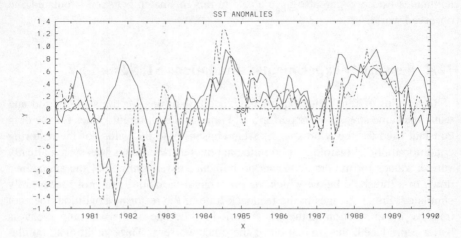

**Fig. 12.2** (a) The thin continuous line represent the observed GG Index. The dashed line, the Index simulated with the E climatology and the E anomalous stresses for all the domain (run E1). The thick continuous line, the Index simulated with the E anomalous stresses only in the tropical band, superimposed on the S climatology (run E2).

## (b)

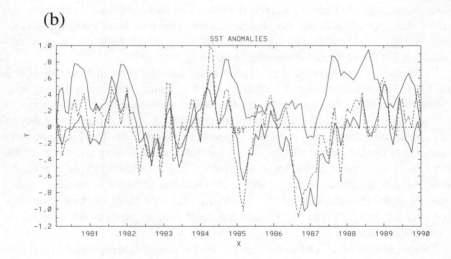

**Figure 12.2 (continued)** (b) The thin continuous line represent the observed GG Index. The dashed line, the Index simulated with the S climatology, the S anomalous stresses in the tropical band and the E anomalous stresses for the rest of the domain (run S1). The thick continuous line, the Index simulated with the S anomalous stresses only in the tropical band, superimposed on the S climatology (run S2).

ECMWF winds is responsible for this failure. The magnitude of the simulated SST anomalies (too big) can also have a role in this mismatch between simulated and observed world.

## 12.4  Sensitivity experiments: simulations S1, S2 and E2.

In a series of sensitivity experiments, additional runs for the same period and using the same stresses were performed. For instance, different values for the drag constant used for the wind stresses were chosen. Also the values of the restoring condition to the climatology used in the anomalous heat fluxes was systematically varied. Values for the drag constant do influence the magnitude of the anomalies, down to a threshold below which the interannual variability was not successfully simulated. Small changes in the feedback-time of the restoring condition does not improve the simulation of the whole period, while one of the warming events is better reproduced, the simulation of the other worsens. Three additional simulations of the interannual variability for this region were devised to assess the influence of possible errors of the ECMWF analyses in the failure of the 1981 warming event, as pointed out in Carton and Huang (1994). The wind stresses described in Servain et al. (1987) were the only other set of observations available as an alternative to the ECMWF wind stresses forcing. These wind stresses, produced from observations from 1964 to 1994 through subjective analysis, have an acceptable spatial coverage. Preliminary statistical analyses on them, shown the existence of a trend, that appears as the first component in an analysis of variance of the anomalies. But this trend practically disappears after 1980, which made the data set suitable as wind stress forcing during the last decade.

Unfortunately, the Servain et al. stresses cover only the tropical band, from 20°S to 30°N. The practical issue of including or not an anomalous forcing for the mid-latitudes, could not be separated from theoretical considerations. The importance of the forcing at mid-latitudes to simulate correctly the interannual variability of the equatorial Atlantic, was a point hard to decide in advance. Therefore, we have carried out two additional simulations of the ocean model, using anomalies of the Servain et al. stresses as forcing at the tropical band. In one of them, hereinafter referred to as S1, these anomalies are blended with the ECMWF anomalies, used in the rest of the domain. In the other, referred as S2, anomalous stresses cover only the tropical (from 20°S to 30°N) band and there is no anomalous wind forcing in the mid-latitudes.

The simulation of the GG index (dashed line) in run S1 is represented against the observations (thin line) in Fig. 12.2b. We can see an initial warming during 1980, that has no correspondence in the observations. The simulation of the 1981 warming is improved with respect to the run E1. The 1982 cooling is also simulated, while the 1984 warming occurs half a year earlier than in the observations. This warming event decays quite quickly, and afterwards an unrealistic cooling event

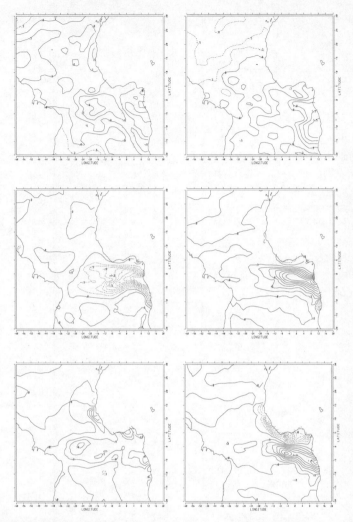

**Fig. 12.3** Peak phase of the warmings of 1981 (right) and 1984 (left) from Servain data (top row), from the simulation E1 (middle row) and from the E2 (bottom row)

starts. Fig. 12.4 shows the SST simulation of the 1981 warming event (bottom left) and of the same stage of the 1984 event (bottom, right), in the run S1.

The GG index of run S2 represented in Fig. 12.2b with thick line, is not very different from the one of run S1. Values for the GG index during 1980 are closer to the observations than those of the S1 simulation. The magnitude of the 1981 warming is also underestimated. There on, the GG index of both simulations are almost indistinguishable. For the fourth simulation, referred as E2, the forcings were built by combining Servain climatology with the anomalous ECMWF stresses only in

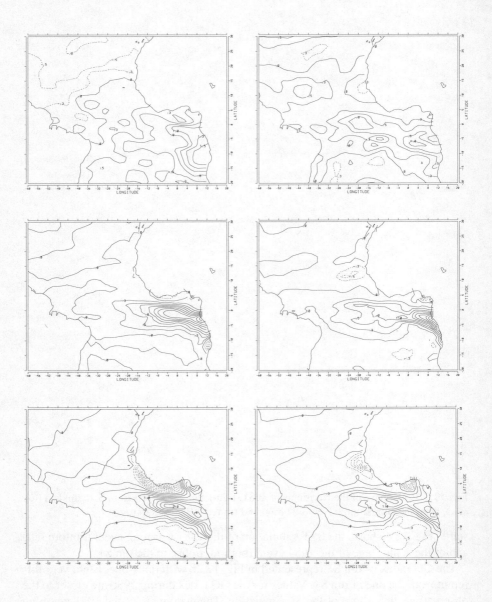

**Fig. 12.4** Peak phase of the warmings of 1984 (right) and 1988 (left) from Servain data (top row), from the simulation E1 (middle row) and from the E2 (bottom row)

the tropical band. The GG index of run E2 is represented in Fig. 12.2a with a thick line and is very similar to the index of the run E1.

## 12.5   The tropical Atlantic variability in a coupled general circulation model.

To test the generality of some of the results obtained in the next section, we were allowed to analyze the interannual variability at the tropical Atlantic region that appear in 40 years of an extended global run with a coupled ocean-atmosphere general circulation model. The atmospheric component of this coupled run is the ECHAM4 general circulation model developed at Max Planck Institut fur Meteorologie (Hamburg). It is the current generation of ECHAM models which evolved under extensive reconstruction in parameterization, to adapt them to the needs of climate experiments. The model is run in a grid of approximately 2.8°. The oceanic component is the global version of the same oceanic model used here, the OPYC (Oberhuber, 1993). The scalar points of the oceanic model coincide with those of the atmospheric model except equatorward of 36, where the meridional grid is gradually decreased to reach 0.5° equatorward of 10°. The models are coupled quasi-synchronously, exchanging averaged quantities once a day. Details of this run can be found in Roeckner et al. (1996) or Bacher et al. (1997). For the purpose of comparison of the present paper we have analyzed a time slice of 40 years, from year 236 to year 275 of the present day climate simulation. The (filtered) GG index for this run is represented in Fig. 12.11 (middle row) with solid line. We see that the index presents traits of interannual variability that are common to the observed ones, albeit with reduced magnitude of the anomalies. The same can be said of the simulated Niño3 index (also filtered), represented in Fig. 12.11 (top row), also with solid line.

## 12.6   Statistical Comparison of the simulated interannual variability.

An useful variable to consider the start of the anomalous events is the heat content $hc$ stored in the upper ocean at each horizontal grid point $(x, y)$ and at each instant of time $t$ as:

$$hc(t, x, y) = \rho c_p \int_{h_0}^{0} (T(t, x, y, z) - T_0(t, x, y)) dz \qquad (12.1)$$

where $T_0(t, x, y)$ is the temperature at a reference depth $h_0 = 360m$ and the same horizontal grid point and time $t$. It estimates the relative thermal energy stored in the upper ocean active layer. t

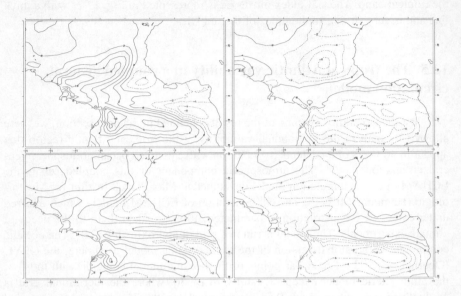

**Fig. 12.5** Antecedent conditions (top) and initial stage (bottom) of the 1984 (left) and the 1988 warmings in the simulation E1.

In the ENSO case, changes of this variable anticipate the appearance of the anomalous SST (Chao and Philander, 1993). This feature is also present in our Atlantic simulation, as shown by Fig. 12.5 where the antecedent conditions to the 1984 and 1988 warmings are presented. In the case of the 1984 event (Fig. 12.5, left), antecedent conditions like the ones found for the ENSO case, are clearly identifiable in the previous autumn (above): there is a heat content accumulation in the western part of the basin, with centers north and south of the equator and a strong gradient along it. In the next season (winter 1984, below) the heat accumulation in the eastern equator referred by Huang et al (1995) as the initial stage of the events appears in a pattern hereinafter referred as "onset pattern". There is no equivalent to the "antecedent pattern" in the seasons previous to the 1988 event (Fig. 12.5, right). Antecedent conditions of this event can be traced back as far as the spring of the preceding year, where warm water seems to accumulate at the eastern basin, while the heat content in the tropics is anomalously low. Just at the equatorial region, between $8°$ S and $8°$ N, this pattern has similarities with the "onset pattern" and therefore we will call it "preonset pattern". Both conditions, "antecedent" and "preonset", are traceable to anomalies of the currents, both outside and at the equator, that in turn are related with anomalies in the wind forcing. For instance in the autumn of the 83 there is a strengthening of the North Equatorial Current (NEC) and negative anomalies in the southern gyre while in the spring of 87 the situation is reverse: The anomalies of the currents in those two regions are positive. In both cases there are also important disturbances along the equator.

The characteristics of the simulated interannual variability can be summarized through a statistical analysis that captures the most relevant features of this variability in both, space and time. To this aim we use a version of the technique known as POP (Principal Oscillation Pattern) analysis (Storch et al., 1990). Propagating features can be better noticed in the POP analysis of the anomalous heat content. Anomalies of this field were built by subtracting the seasonal mean (the mean for each month) from the simulated value at the 15th of each month. The anomalous field was then smoothed, using a three points running mean average. Previous to the POP analysis, we proceed to the customary reduction of the degrees of freedom of the field by an expansion in terms of the Empirical Orthogonal Functions (EOF) of the field, and further truncation.

### 12.6.1 The simulations E1 and E2.

For the run E1, the four first EOFs retained explain 90% of the variance of the field. POP 2 shown in Fig. 12.6 (top left) is similar to the antecedent pattern we

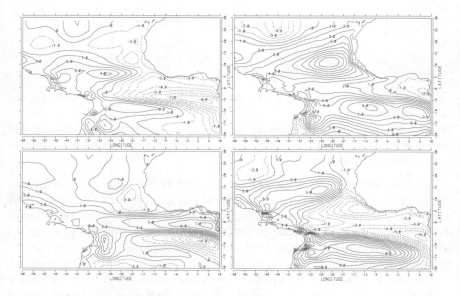

**Fig. 12.6** Pair of POP 2/1 (left) and 3/4 (right) from the simulated heat content anomalies in run E1.

have discussed above while the POP 1, represented in the same figure, bottom left, correspond to the onset stage. Their time coefficients, shown in the top row of Fig. 12.7, compare well to the POP evolution scheme: P2 → P1 → −P2 → −P1. Maximum positive values in POP 2 precede the 1984 event by one year, and its maximum negative values yield the peak-mature stage of the episode. The POP 1 has maximum value in spring of 1984. The differences between the two warm

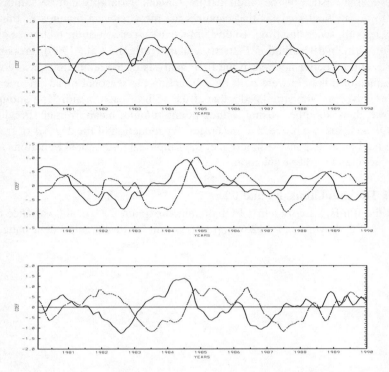

**Fig. 12.7** Empirical time coefficients of the pair 1/2 of the POP of the simulated heat content anomalies in the run E1, E2 and S1. (top to bottom). (Time coefficient 1, solid line, time coefficient 2, dashed line)

events, 1984 and 1988, lie in the contribution of POP 3/4: while their coefficients (see top row of Fig. 12.8) have a different sign in the reconstruction of the 1984 event, they contribute with the same (negative) sign to the 1988 warming. It seems important at this point to remember that POPs belonging to different pairs are not orthogonal and therefore that the total variance explained by the two pairs cannot be obtained simply by adding the variance explained by each pair. Therefore, the contribution of each pair to the evolution of the anomalous heat content field has to be estimated in a less straightforward way. For this, three different anomalous heat content indices are built by averaging anomalies of the heat content through the same box used for the GG index. First, the anomalous heat content field is obtained from a reconstruction of the field with only the first four EOFs (10% of the variance of the field is then lost). This reconstructed field is identical to the one obtained with all four POPs. Second, the field is reconstructed with the pair of POPs 1/2, and last, the field is reconstructed with only the pair of POPs 3/4. Those are the indices represented in Fig. 12.10 (bottom row) with the short-dashed line, solid line, and long-dashed line, respectively. We can see from them how the evolution of the heat content in the region of interest is basically accounted for by the

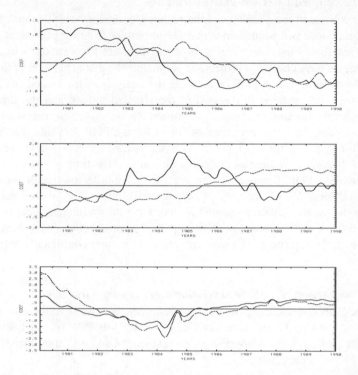

**Fig. 12.8** Empirical time coefficients of the pair 3/4 of the POP of the simulated heat content anomalies in the run E1, E2 and S1. (top to bottom).

first pair of POPs. The second pair correct this picture, introducing the differences between the 1984 and 1988 events.

The POP time coefficients of the heat content anomalies of the run E2, shown in Fig. 12.10 and Fig. 12.11, middle rows, show similarity with those of run E1, and there is also a strong resemblance among the spatial patterns. But in this case, the contribution of the POP 1 is important also before the 1988 event and how the pair 3/4 contributes also to the reconstruction of the 1984 event.

## 12.6.2 Simulations S1 and S2.

In the case of the S1 run, the output is represented by the four first POP of the anomalous ocean heat content. The first pair of POP are represented in Fig. 12.9, and their evolution in time is plotted in Fig. 12.7 and Fig. 12.8 (bottom rows). Time coefficients of this conform well with the POP propagating scheme and the spatial patterns are similar to the first pair of patterns detected in the analysis of the E simulations.

**12.6.3 The coupled ECHAM4/OPYC run.**

The time series of the coupled simulation have, compared with the ones from the forced simulation, two peculiarities that are important from the statistical point of view: its length and the time sampling interval. Because we want to focus on common traits, we have chosen to project the anomalous heat content field of this simulation into the first pair of POP obtained in the analysis of the forced run E1. The patterns were interpolated to the grid of the coupled model, and the time series were previously filtered, to remove the variability associated to time scales lower than 18 months. The time evolution obtained for the POP 2 in this way is represented the Fig. 12.11 (top row) with dashed line against the Niño3 index of the coupled run (filtered), represented with solid line. The time coefficient of the projection onto POP 1 appear in the Fig. 12.11 (middle row) with dashed line against the GG index of the coupled run (filtered), represented with solid line. In this same figure we have represented an index for the anomalous heat content in GG (solid line), and the same index for a partial reconstruction of the heat content anomalous field in terms of the two patterns and the time coefficients obtained by projection.

**12.6.4 Comparison of statistical relationships among variables**

To model this relationship in the simplest way, we use a multivariate lagged regression. Let $s_j(t)$ be the time coefficient of the $j$-th POP of the heat content anomalies, and $r_k(t)$, the $k$-th principal component of some of the other simulated fields or of the forcing fields.

$$s_j(t) = \sum_{k=1}^{p} a_{jk}^l r_k(t-l) \qquad (12.2)$$

where $l$ is a fixed time lag. The value of $l$ is allowed to vary in a range from -6 to +6 months. For a certain $j$, to take into account a $r_m(t)$ and a lag $l$, two statistical tests have to be passed. Only the $a_{jm}$ that are significantly different from 0 will be considered and in any case, the variance explained by the signal has to be greater than the residuals. For those regressions that passed both tests, we proceed to visual inspection. i

Next the anomalous fields of SST, mixed layer depth and mixed layer zonal and meridional velocities were expanded in terms of their EOFs. The first four principal components of each field, explaining an amount of variance from 60% (the mixed layer depth) to a 80% (SST anomalies), were combined (multivariate regression) to model the time coefficients of the four first POPs of the heat content. None of the selected fields was able to provide a satisfactory simulation of all the four time coefficients. The field that performed best was the meridional mixed layer velocity: the appearance of the warmings in the GG is due, in part, to an anomalous convergence of the meridional velocities in the mixed layer. An explanation for such relative failure can be the difference in the time scales of variability that we are trying to relate through the multivariate regression.

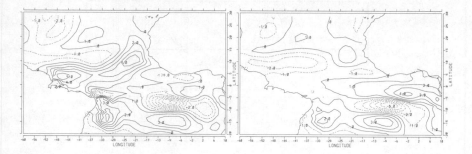

**Fig. 12.9** Pair of POP 2/1 (left) from the simulated heat content anomalies in run S1

This lack of similarity is more evident between the forcing wind field and the output field (heat content). The forcing anomalies cannot even be reduced to a few modes. Therefore, we try to define for each grid point a variable that will measure its persistence rather than the forcing. Such variable is defined as

$$\theta_w = \int_{t-t_d}^t wc(\vec{r}_H, t')dt' \qquad (12.3)$$

where $wc$ stands for the forcing variable (that is zonal or meridional wind or equivalently the wind stress curl and its divergence), $\vec{r}_H$ locates the grid point and $t_d$ is the characteristic decorrelation time at $\vec{r}_H$. For each of these variables, seasonal anomalies were computed as usual. The anomalous values of the integrated forcing variables were then computed using expression (7.3).

We then proceed to a reduction of the number of degrees of freedom of the field through an expansion in terms of EOF. The variability of the forcing field used in the E1 run is well represented in terms of its six first EOFs, explaining from roughly 70% to 60% variance of the anomalies of the integrated wind fields.

When the regression of the $r_m(t-l)$ at $l$ lag gives a satisfactory simulation of the $s_j(t)$, the regression coefficients allow for the identification of a spatial structure of the forcing whose temporal evolution gives an estimation of the heat contents with a certain lag. These patterns are known as associated patterns and are given by

$$\vec{v}_j = \sum_{k=1}^p a_{jk}^l \vec{e}_k^l \qquad (12.4)$$

where $\vec{v}_j$ is the atmospheric pattern associated to the $j$-th POP, $\vec{e}_k^l$ the k-th EOF, and $a_{jk}^l$, the coefficients determined in regression (7.3) at lag $l$.

The analysis was originally carried out with the wind forcing resolved in terms of wind stress divergence and its curl (Cabos et al., 1998). An example of the best fit of each of the POP time-coefficients, when the forcings are described in terms of the integrated divergence and curl, are presented in our Fig. 12.12, while the corre-

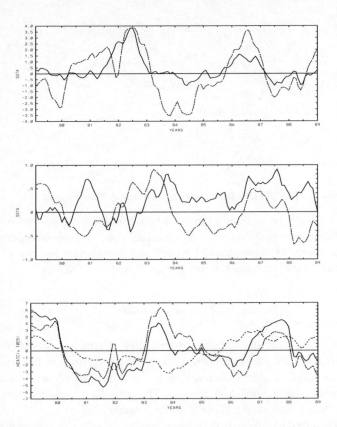

**Fig. 12.10** Top panel, the observed Niño3 Index, filtered as to eliminate the variability associated to time scales lower than 18 months (solid line). Time coefficient the POP 2 of the first pair identified in the analysis of run E1 (dashed line). Middle panel, the observed GG Index filtered as above (solid line). Time coefficient of the POP 1 of the first pair identified in the analysis of run E1 (dashed line). Bottom panel, reconstruction of the GG Index of the anomalous heat content using the first pair of POP patterns and its time coefficients

sponding associated patterns are shown in Fig. 12.13. According to these last, the POP 1, 2 and 3 are preceded (lags -6 to -1) by extraequatorial signals in the Northern Hemisphere (north of 18° N) that maintain themselves for different lags.

In general, the multivariate lagged regression gives slightly better fit in the case of the integrated convergence. But there are cases, where both, the fit and the characteristics of the associated pattern of the integrated curl, are better (for instance, for the time coefficient of POP 3 in the E1 run). On the other hand, while the main signals in the first EOFs of the integrated convergence lie in the equatorial band, the first EOF of the integrated curl have important signals in the above mentioned region of the Northern Hemisphere. Therefore, a good fit in terms of the integrated curl indicates that the time evolution of this mode of the heat content is clearly determined by the extratropical signals in the wind forcing, and not merely the

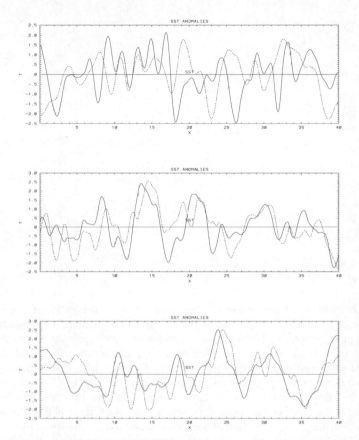

**Fig. 12.11** Top panel, the simulated Niño3 Index of the coupled run, filtered as to eliminate the variability associated to time scales lower than 18 months (solid line). Time coefficient of the projection of the (filtered) anomalous heat content of the coupled run on the POP 2 of the first pair identified in the analysis of run E1 (dashed line). Middle panel, the simulated GG Index of the coupled run, filtered as above (solid line). Time coefficient of the projection of the (filtered) anomalous heat content of the coupled run on the POP 1 of the first pair identified in the analysis of run E1 (dashed line). Bottom panel, reconstruction of the anomalous heat content GG Index using the projection time coefficients and the two POP patterns.

result of the integration of the atmospheric noise in this region by the ocean model. In cases where both fields fit equally well the heat content time coefficient (as in POP 1/2), they allow for a complementary interpretation.

The extraequatorial signal that appear in the associated pattern 1 (the pattern associated to the evolution given by the time coefficient 1), centered in the Sargasso Sea, indicates sustained convergence. The associated signal in the integrated wind curl confirm the existence of a low pressure in this region. The extraequatorial signal in the associated pattern 2 is less clear and located in the middle Atlantic. The

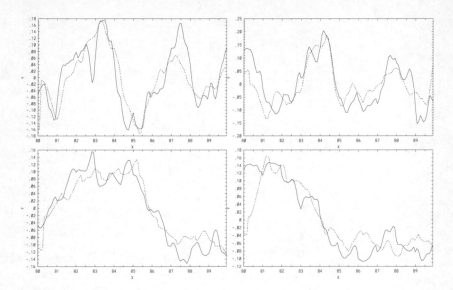

**Fig. 12.12** Evolution of the empirical time coefficients of the POPs of the simulated anomalies of heat content (solid line) of the E1 run and fit obtained through multivariate lagged regression from the PC of the integrated forcing fields. Top left: the modeled time coefficient correspond to the POP 2 of the anomalous heat content, the forcing field is the windstress convergence, the optimal lag -2 month. Top right: the modeled time coefficient correspond to the POP 1 of the anomalous heat content, the forcing field is the windstress convergence, the optimal lag -4 months. Bottom right: the modeled time coefficient correspond to the POP 3 of the anomalous heat content, the forcing field is the windstress curl, the optimal lag -3 months. Bottom left: the modeled time coefficient is of the POP 4 of the anomalous heat content, the forcing field is the windstress divergence, the optimal lag -1 months.

signal in the associated pattern 3 is near the coast of Africa, weaker and noisier. The pair 1/2 of the associated patterns of the forcing fields in the S runs show similar signals in the subtropics and north of Bahamas (see Fig. 12.14).

Although all the associated patterns on both fields (convergence and curl) have important variability at the equator, the associated pattern obtained by fitting with the integrated divergence and curl, do not show any clear variability at the equator. Evolution from associated pattern 2 to 1 seems to be related to a persistent strengthening of the zonal winds at latitudes that coincide roughly with the maxima of the positive anomalies of heat content in POP 2 (see Fig. 12.6)

## 12.7 Discussion.

In the presented work we present several simulations of the interannual variability of the Tropical Atlantic for the decade 1980-1989, where three warming events

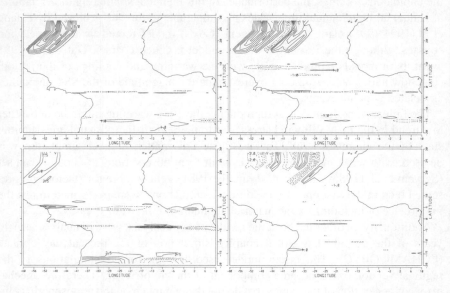

**Fig. 12.13** Patterns of the integrated forcing field associated to the pair of POP 1(left top)/ 2 (right top) and to the pair of POP 3 (right bottom)/ 4(left bottom) of the anomalous heat content in the run E1.

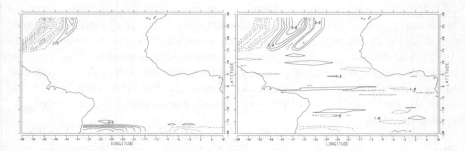

**Fig. 12.14** Patterns of the integrated forcing field associated to the pair of POP 1(left)/ 2 (right) of the anomalous heat content in the run S1.

(1981, 1984 and 1988) where observed. The simulations could provide an exhaustive information (surface and subsurface) on the mechanism of generation of the events. Through their analysis, we want to understand the part played by the different oceanic mechanism and also by some characteristics of the forcings.

The simulations presented here were produced by forcing the same regional version of an ocean general circulation model with data sets built from observations, among these, the interannual anomalies of wind stresses. To simulate an acceptable annual cycle, most of the Atlantic domain had to be included. In the first of these simulations, where the anomalies of the ECMWF wind reanalysis were used for the anomalous forcings, the performance of our model at simulating the variability at annual and interannual time scales is comparable to the one achieved by Huang et al.(1994,1995) using a very different (Cox's) model forced by the same wind stresses, although the resolution of our model is much coarser. Out of the 1981 event, both models reproduce quite well the warming and coolings of the period, somehow anticipated. And both overestimate the magnitude of the SST anomalies associated with the events.

The failure of both models at reproducing the 1981 warming might be due either to a trend in the ECMWF wind stresses or to the unrealistic (climatological) initial state of the anomalous run. To investigate this matter, another simulation was devised, with wind anomalous stresses built from observations of another data set (Servain et al. (1986)). In this new run, the 1981 event is more satisfactorily represented than in the first one. Two additional simulations, were performed to test the importance of including extraequatorial anomalies in the wind stresses.

Additionally, in order to check the results of our analysis, we will analyse also 40 years of the output of a global coupled simulation of the present day climate (ECHAM4/OPYC). The ocean model used for the forced simulations is the regional version of the oceanic component of the coupled run. Therefore, mechanisms of generation of the events that do not depend much on the atmospheric component are expected to be similar.

The generation of the events is best understood with the help of a variable that captures the thermal structure of the ocean, like the heat content. A statistical analysis of this field allows to reduce the number of variables to just only four patterns and their corresponding time coefficients. The first pair of modes shows propagating features, and together they give a representation of the events similar to the one found for the Pacific one: a redistribution of warm waters across the basin (Wyrtki, 1985). The second pair do not show propagating characteristics. The two first modes detected in our analysis give a reconstruction of the Gulf of Guinea index of the heat content that is more satisfactory in the case of the 1984 than in the 1988 event. To the reconstruction of this last, the contribution of the second pair of patterns is important.

On the other hand, the description of the events in terms of only four variables allow us to find easily a connection with ENSO. The pattern 2 of the first pair obtained from our statistical analysis of the two different forced runs, is the one that shows an accumulation of warm waters to both sides of the equator and negative anomalies at the gulf of Guinea. The time coefficient of this pattern in run E1, is represented in our Fig. 12.8 (top row). Its similarities with the observed Niño3 SST index, represented in the same figure, are evident. The warm SST anomalies in the equatorial Pacific, then, induce some changes that lead to increasing the heat content of the upper layers of the north tropical Atlantic, north of the equator. Because there is not a direct interaction between the two oceans, this feature suggest an atmospheric link. The time coefficient of POP 1, represented in Fig. 12.8 (middle row), is obviously related to the observed (SST) Gulf of Guinea index, also represented in the same figure. Nevertheless, out of the equator, the correction made by the second pair of modes to the evolution given by the first one must be substantial. Snapshots of the heat content at stages previous to the appearance of the surface warmings, show that in the case of the 1988 event, the heat content is drained from the western part of the equatorial Atlantic, without been previously accumulated there. Our analysis of the anomalies of the heat content in the other forced run (where the anomalies of wind stress from Servain et al.(1986) were used to build the wind forcing) detect a first pair of pattern similar to the one found in the ECMWF run.

We have used the anomalous heat content field for the same domain in the coupled ECHAM4/OPYC3 run, to check the generality of the two first modes detected by our analysis of the forced run, and the connection of the second of these modes with Niño3 SST (Fig. 12.10). Propagating characteristics can be yet observed, and although the similarities between the second time coefficient and the simulated Niño3 index are here less evident, there are periods, as happened in the second decade, where both curves remain particularly close. It can be therefore concluded that the first pair of pattern detected by our analysis of the heat content field represent a mechanism of generation of the events that is quite frequent, although not the only one.

To explain the generation of the events, we try to relate, in a simple (linear) way, the anomalies of heat content to those of the surface variables and to the forcings. Only two of the four POP time coefficients that give the time evolution of the heat content anomalies can be expressed in terms of the principal components of surface variables (SST, mixed layer depth, u and v mixed layer velocity). Those fits point to the convergence of the meridional mixed layer velocity as one of the important mechanism that precedes the appearance of warm waters in the GG.

Due to the different time scales of variability, it is not possible to relate the anomalies of the heat content to those of the forcings using the simple multivariate lagged regression model. We define then some related to them that measure the persistence of the anomalies at each point of the grid. The evolution in time of these new fields is related to the heat contents through the linear model. The fitting of the POP time coefficients of heat content by the new atmospheric variables are quite good. Through regression coefficients, we determine four patterns of the forcing variables, whose time evolution match these of the POP of the anomalous heat content and that are associated to the time evolution of the POPs of the anomalous heat content.

In our analysis, the wind data were represented by its divergence and curl. The first pair of associated patterns show wavelike atmospheric activity at two narrow bands, north and south of the equator, at roughly the latitudes where the maxima of the heat content anomalies in the POP 2 are located. Persistent atmospheric anomalies in the northwest part of the basin are of relevance in the first three associated pattern. The evolution of the fourth POP is connected to a signal north of 18° N, near the coast of Africa. The extratropical signal is clearer in the pattern associated to the POP 2: the divergence and the curl pattern indicate a strengthening of the gyre near the northern boundary of the domain, with convergence south of it. In the pattern associated to POP 1, the signal in the convergence field, weakened and noisier, has traveled eastward, toward the center of the basin and near the tropic.

Our analysis points, therefore, to two different ways in which warming events in the Atlantic can be generated. The first one is evident in the generation on the 1984 event, and is represented by the first pair of propagating patterns detected in the heat content field. The second POP, with an accumulation of heat content (warm waters) in the western part of the basin, north and south of the equator, and negative anomalies at the Gulf of Guinea, appears first. Its time evolution (very similar to the observed Niño3 index in the forced run and to the simulated Niño3 index for some time slices of the coupled run) suggest that this mode is forced by the Pacific warmings. Our statistical analysis detects extratropical atmospheric signals in the northern hemisphere that provide an interbasin link and can explain the anomalous heat content north of the equator. These atmospheric signals are centered at 20° N, 50° W, a region of downwelling, which strength is controlled by the wind curl, feed the subsurface tropical waters and influence through oceanic mechanisms the heat content at the equator. The first POP, representing the onset of the events, appear later and its evolution in time is well modelled by a convergence of mixed layer velocities at the GG that only in part is related to the atmospheric convergence. This first POP is connected also to an extratropical signal, located eastward of the one associated with the second POP.

By the second mechanism, warm waters accumulate slowly in the Gulf of Guinea and north of the equator, near the coast of Africa. In this last region, the accumulation is related to an anomalous atmospheric convergence. The transport by the countercurrent of these anomalies, slowly reinforces the Gulf of Guinea incipient warming and originates the warm event.

**Acknowledgments.** Thanks are due to J. Servain, that made his datasets available to us, and to the Laboratoire d' Oceanographie Dinamique et Meteorologie (LODYC) in Paris, for the FOQUAL-SEQUAL current measurements. The simulations reported in this paper were performed under contract EV5V-0124 and EV5V-CT94-0538C of the EC Environment Research Programme. Part of the computer time used was allotted at DKRZ(Hamburg) by the European Climate Computer Network (ECCN). When this time was exhausted, the Spanish Meteorological Office made additional CRAY hours available to us. The experiments with the S winds were supported by CRAY additional hours by the Spanish CICYT.

Finally, we are grateful to the Deutsches Klimarechenzentrum GmbH (DKRZ) for providing the output of the coupled experiments using the ECHAM4/OPYC3 model.

# 13 North-Atlantic Decadal Climate Variability in a Coupled Atmosphere/Ocean/Sea-Ice Model of Moderate Complexity

R. J. HAARSMA, F. M. SELTEN, J. D. OPSTEEGH, Q. LIU AND A. KATTENBERG

*KNMI, De Bilt, The Netherlands*

## 13.1 Introduction

Decadal variability in SST has the largest signals in the extra-tropical regions of deep and intermediate water formation of the North-Atlantic ocean: the Greenland Sea, the Labrador Sea and the area close to Bermuda (Deser and Blackmon 1993, Kushnir 1994). Unstable air/sea interactions might well occur in this area, potentially affecting climate variability on various time-scales.

In his analysis of Atlantic air/sea interactions Bjerknes (1964) suggested two mechanisms of air/sea interactions. The first mechanism is that ocean SST responds to changing wind patterns due to changes in the sensible and latent heat fluxes and the entrainment of cool water from below. This mechanism has been shown to operate on the interannual time scale in a number of studies with coupled GCM's and has been confirmed in observational data (Palmer and Sun 1985, Wallace et al 1992, Delworth 1996, Verbeek 1997).

On the decadal time-scale Bjerknes suggests that SST anomalies are not directly forced by wind stress anomalies but are the visible images of changing ocean currents induced by variability in deep water formation that result in variations in oceanic heat transport. The atmosphere should somehow respond by compensating for the changes in oceanic heat transport in order to keep the total heat transport in the climate system approximately constant. How this is actually accomplished is not yet well understood and cause and effect are still unclear. Changes in the atmospheric circulation during the cool phase of the great salinity anomaly in the sixties have been reported (Knox et al 1988, Shabbar et al 1990) but have not been related to changes in atmospheric heat transport.

With extended observational datasets the existence of a correlation between time-series of Atlantic SST anomalies and atmospheric variables on decadal time scales has been confirmed and the associated Atlantic SST anomaly pattern has been reconstructed (Kushnir 1994, Folland et al 1986, Deser and Blackmon 1993). Grötzner et al (1996) performed a canonical correlation analysis (CCA) for observed SST and sea level pressure (SLP) data over the North Atlantic. The leading CCA mode showed clear variations on a decadal time scale. The corresponding

SLP pattern closely resembles the NAO pattern. The SST pattern is similar to one of the dominant EOF's of SST variability.

Recent simulations with coupled GCM's (Delworth et al 1993, Von Storch 1994, Latif and Barnet 1994, Grötzner et al 1996, Robertson 1996) clearly show the existence of a relation between decadal variability in atmosphere and ocean. The associated ocean variability can not entirely be explained by local mechanisms, such as for instance the integration of white noise atmospheric forcing by the ocean leading to red noise SST variability (Hasselmann 1976). Nonlegal effects are important and point to an active role for the ocean in the air/sea interaction (Bryan and Stouffer, 1991). These conclusions are not yet firmly founded on a number of consistent model simulations. Different models usually give substantially different results. Nevertheless the vague contours of a picture of North-Atlantic air/sea interaction on decadal time-scales is starting to emerge both from the analysis of observational and model data. The explanation of decadal variability in terms of cause and effect relationships is however far from complete.

One of the obstacles in modeling atmosphere ocean interaction on decadal timescales is that experiments with GCM's are still fairly expensive to run for periods of order thousand years. The KNMI-DICE project focused on studying the nature of air/sea interaction in the North-Atlantic area on timescales of 10 to 50 years, which we will refer to as decadal timescales. We have analysed the air-sea interaction processes associated with this variability. This has been done within the context of the KNMI climate model (ECBILT), which is a model of moderate complexity. The relative simplicity of ECBILT makes it feasible to perform many long term simulations under simplified and controlled conditions. This approach makes it possible to pinpoint crucial cause and effect relationships involved in the interaction between atmosphere and ocean.

The atmospheric part of ECBILT is a spectral T21 global three level quasi-geostrophic model with simple parametrizations for the adiabatic processes (Haarsma et al 1997). The model is realistic in the sense that it contains the minimum amount of physics that is necessary to simulate the mid-latitude planetary and synoptic-scale circulations in the atmosphere as well as its variability on various time-scales. So we believe that it can be used to study the fundamental nature of air/sea interaction in the extratropics. As an extension to the quasi-geostrophic equations, an estimate of the neglected ageostrophic terms in the vorticity and thermodynamic equations is included as a time and spatially varying forcing. This forcing is computed from the diagnostically derived vertical motion field. With the inclusion of the ageostrophic terms the model is able to qualitatively simulate the Hadley circulation correct. This results in a drastic improvement of the strength and position of the jet stream and the transient eddy activity. Despite the inclusion of these additional terms the model is, because of the quasi-geostrophic approximation, two orders of magnitude faster than AGCMs. It has been coupled to a GFDL type ocean model and a thermodynamic sea-ice model. The model runs on present generation workstations, taking 0.2 hr cpu time for the simulation of 1 yr (Power Indigo of Silicon Graphics). A similar approach of using a climate model of intermediate complexity has been adopted by Saravanan and McWilliams (1997), although their model is more simple. They use a 2-layer primitive equation model for the atmosphere, which is coupled to a zonally averaged sector ocean model.

After a brief description of the climate of the separate subsystems and of the coupled system we will focus on the following two questions:

**1)** Can we detect coupled modes of North-Atlantic decadal variability in simulations with the coupled system? If yes what is the structure and the spectrum of these coupled modes and how do they compare with the observations and with the modes generated by 'state-of-the-art' AOGCM's?

**2)** To what extent are these modes truly coupled modes or is the coupling one-way in the sense that the atmospheric part of the variability pattern is generating the oceanic part without a substantial feedback from the ocean to the atmosphere?

As stated earlier this last question is an important topic in the current debate on decadal variability. For instance Latif and Barnett (1994) and Grötzner et al (1996) hypothesize that this feedback is essential for their modes of variability, whereas Saravanan and Mc Williams (1997) demonstrate that for the modes generated in their model the coupling is one-way and that the atmospheric forcing can be considered as basically stochastic in nature. To address this question for the modes generated by ECBILT we performed additional experiments in which we made changes to the coupling.

This report summarizes the three components of the climate model, i.e. the atmosphere, the ocean and the sea-ice model in section 13.2 and describes briefly the climate of the atmosphere and ocean models when being run separately using prescribed climatological boundary conditions in section 13.3 and 13.4 respectively. Section 13.5 focuses on the North-Atlantic climatology and variability of the coupled model. A decadal mode of variability is described and analysed in section 13.6 followed by some conclusions in section 13.7.

## 13.2 Description of the model

The climate model contains three separate subsystems (atmosphere, ocean and sea-ice), which exchange heat, moisture and momentum. The dynamical component of the atmospheric model was developed by Molteni (Marshall and Molteni 1993). The physical parametrizations are similar as in Held and Suarez (1978). For a detailed description of the atmosphere-, ocean- and sea-ice model the reader is referred to Lenderink and Haarsma (1994 and 1996) and Haarsma et al (1997).

### 13.2.1 Atmospheric model

**Equations.** The dynamical behavior of the atmospheric model is governed by the quasi-geostrophic potential vorticity equation. The equation in isobaric coordinates is:

$$\frac{\partial q}{\partial t} + \mathbf{V}_\psi \cdot \nabla q + k_d \nabla^8 (q - f) + k_R \frac{\partial}{\partial p}\left(\frac{f_o^2}{\sigma}\frac{\partial \psi}{\partial p}\right) = \qquad (13.1)$$

$$-\frac{f_o R}{c_p}\frac{\partial}{\partial p}\left(\frac{Q}{\sigma p}\right) - F_\zeta - \frac{\partial}{\partial p}\left(\frac{f_o Q}{\sigma}\right)$$

where $q$ is defined as:

$$q = \zeta + f + f_o^2 \frac{\partial}{\partial p}\left(\frac{1}{\sigma}\frac{\partial \psi}{\partial p}\right) \qquad (13.2)$$

The variable $\zeta = \nabla^2 \psi$ is the vertical component of the vorticity vector, $\psi$ is the streamfunction, $\mathbf{V}_\psi$ is the rotational component of the horizontal velocity, $f$ is the Coriolis parameter, $f_0$ is $f$ at 45° North and South, $k_d \nabla^8(q - f)$ is a highly scale selective diffusion, $k_R$ is a Rayleigh damping coefficient, $R$ is the gas constant, $p$ is pressure, $c_p$ is the specific heat for constant pressure, $\alpha = -(\alpha/\theta)(\partial\theta/\partial p)$ is the static stability, $\alpha$ is the specific volume, $\theta$ is the potential temperature, $Q$ is the diabatic heating, $F_\zeta$ contains the ageostrophic terms in the vorticity equation and $F_T$ is the advection of temperature by the ageostrophic wind.

The ageostrophic forcing $F_\zeta$ reads:

$$F_\zeta = \mathbf{V}_\chi \cdot \nabla(\zeta + f) + \zeta D + \omega \frac{\partial \zeta}{\partial p} + \mathbf{k} \cdot \nabla \omega \times \frac{\partial \mathbf{V}_\chi}{\partial p} \qquad (13.3)$$

where $\mathbf{V}_\chi$ is the divergent component of the horizontal wind, $D$ is the divergence of the horizontal wind, $\omega = dp/dt$ is the vertical velocity in isobaric coordinates, and $F_T$ reads:

$$F_T = \mathbf{V}_\chi \cdot \nabla\left(\frac{\partial \psi}{\partial p}\right) \qquad (13.4)$$

Estimates of $F_\zeta$ and $F_T$ will be used to represent the effect of the ageostrophic circulation on the tendency of the geostrophic vorticity. This will be explained in the next subsection.

**Ageostrophic forcing.** We have included the ageostrophic terms $F_\zeta$ and $F_T$ in the potential vorticity equation (13.1) in order to improve the simulation of the Hadley circulation. This appears to have large consequences for the strength and position of the jetstream and for the extra-tropical transient eddy activity. We include these terms by computing the vertical velocity diagnostically from the thermodynamic equation after the temperature tendencies have been computed. Next the horizontal divergence is derived from the continuity equation, using the boundary conditions for $\omega$ at the top and bottom of the model atmosphere and the divergent wind $\mathbf{V}_\chi$ is derived from the divergence fields. From $\mathbf{V}_\chi$ and $\omega$ the ageostrophic terms $F_\zeta$ and $F_T$ are computed. In the computation of the tendency $\partial\psi/\partial t$ from the potential vorticity equation (13.1) we use as a first guess the values of $F_\zeta$ and $F_T$ as determined in the previous timestep. The tendency $\partial\psi/\partial t$ is computed and the corresponding divergent circulation diagnosed. Computing $F_\zeta$ and $F_T$ from this divergent circulation a new estimate for the part of $\partial\psi/\partial t$ forced by the ageostrophic terms is computed. This starts an iteration process. We may iterate until $\partial\psi/\partial t$ has converged. In practice the adjustments in $\partial\psi/\partial t$ after the first iteration are substantially smaller than 10%. After 2 iterations the adjustments are smaller than 1%. For reasons of efficiency we run the model without iterating. So we use the ageostrophic terms from the previous timestep as an additional forcing in ECBILT. A similar idea has been put forward by Bengtsson(1974).

**Model discretisation.**The model is spectral with horizontal truncation T21. It has three vertical levels, at 800 hPa, 500 hPa and 200 hPa respectively, on which the potential vorticity equation (13.1) is applied. At the top of the atmosphere ($p = 0$ hPa) the rigid lid condition $\omega = 0$ is applied. The lower boundary condition for $\omega$ is:

$$\omega = -\rho_s g\left(\frac{C_d}{f_0}\zeta_s - \mathbf{V}_\psi^s \cdot \nabla h\right) \qquad (13.5)$$

where $\omega_s = dp/dt$ is the vertical motion field and $\zeta_s$ the vorticity at the top of the boundary layer, for which we take $\zeta$ at 800 hPa, $C_d$ is the surface drag coefficient, $\rho_s$ is the density at the surface and $\mathbf{V}_\psi^s$ the rotational velocity at the surface, $h$ is the orographic height.

**Damping timescales.** For the Ekman dissipation we assume

$$k_E = \frac{\rho_s g C_D}{f_0} = \frac{1}{\tau_E}(1 + \alpha_1 LS(\lambda, \phi) + \alpha_2 FH(h)) \qquad (13.6)$$

where $\tau_E$ is the Ekman damping time scale, which is assumed to be 6 days. The expression between brackets parameterizes the effect of surface roughness and the orographic height. The fraction of land within a grid box is given by $LS$ and $FH = 1 - exp(-h/1000)$. The $\alpha_i$ are constants: $\alpha_1 = \alpha_2 = 0.5$.

For the scale selective diffusion coefficient we take $k_d = (1/\tau_d)a^8(21 \cdot 22)^{-4}$, where $a$ is the earth radius. We have chosen $\tau_d$ equal to 2 days.

The Rayleigh damping represents the effect of temperature relaxation. For the damping coefficient we have taken $k_R = 1/\tau_R$, with $\tau_R = 30$ days.

**Moisture.** The diabatic heating in the atmosphere is caused by the radiative heating, the release of latent heat and the exchange of sensible heat with the earth. The parametrizations of these processes is described in Haarsma et al (1997).

The release of latent heat is strongly connected with the transport of moisture and convection. Changes in the specific humidity are described by a single equation for the total precipitable water content between the surface and 500 hPa. Above 500 hPa the atmosphere is assumed to be completely dry. The equation reads:

$$\frac{\partial q_a}{\partial t} = -\nabla_3 \cdot (\mathbf{V}_a q_a) + E - P \qquad (13.7)$$

where $q_a$ is the total precipitable water content between the surface and 500 hPa, $\mathbf{V}_a$ the transport velocity for $q_a$, $E$ evaporation and $P$ precipitation. For the horizontal component of $\mathbf{V}_a$ 60% of the sum of the geostrophic and ageostrophic velocity at 800 hPa is taken. The vertical component of $\mathbf{V}_a$ is $\omega_{500}$. Moisture which is advected through the 500 hPa plane by the vertical velocity $\omega_{500}$ is removed by precipitation which falls through the underlying layer.

Below 500 hPa precipitation occurs when $q_a > 0.8 q_{max}$ where $q_{max}$ is the vertically integrated saturation specific humidity below 500 hPa. The relative humidity $r$ is assumed to be constant throughout the lower atmospheric layer.

Precipitation and convective adjustment were parameterized according to Held and Suarez (1978). The hydrological cycle is closed by using a bucket model for soil moisture. Each land gridpoint is connected to an ocean gridpoint to define the river runoff (see Fig. 13.1). Accumulation of snow occurs in case of precipitation when the surface temperature is below zero.

### 13.2.2 Ocean model

The model equations are given by the heat and salt budgets, continuity equation, a quadratic equation of state, the hydrostatic approximation, and the horizontal momentum equation. For a detailed description of the ocean model we refer to Lenderink and Haarsma (1994,1996) and Haarsma et al (1997). In this paper we will only describe the simplifications we have applied to the momentum equations.

**Splitting in barotropic and baroclinic modes.** The momentum equations are split into a barotropic (depth averaged) and a baroclinic part (deviation from barotropic velocity). For the wind driven barotropic flow we adopted the Stommel model:

$$\kappa_s \nabla^2 \psi + \frac{1}{a^2 \cos \phi} \frac{\partial f}{\partial \phi} \frac{\partial \psi}{\partial \lambda} = \frac{\mathbf{k} \cdot \nabla \times \tau}{\rho_0} \qquad (13.8)$$

where $\psi$ is the barotropic streamfunction, $\kappa_s$ is a linear friction coefficient (set to $8 \times 10^{-6} s^{-1}$) and $\tau$ is the windstress. The baroclinic zonal velocity is defined by $\hat{u} = u - \bar{u}$ where $u$ is the vertically averaged zonal velocity. The equation for $\hat{u}$ reads:

$$\frac{\partial \hat{u}}{\partial t} - f \hat{v} = -\frac{1}{\rho_0 a \cos \phi} \left( \frac{\partial p^*}{\partial \lambda} - \frac{1}{H} \int_{-H}^{0} \frac{\partial p^*}{\partial \lambda} dz \right) + \nu \nabla^2 \hat{u} + \frac{\partial}{\partial z} \left( \kappa \frac{\partial \hat{u}}{\partial z} \right) \qquad (13.9)$$

where $\nu$ and $\kappa$ are the horizontal and vertical eddy viscosities respectively (set to $8 \times 10^5 m^2 s^{-1}$ and $50 \times 10^{-4} m^2 s^{-1}$ respectively) and

$$p^* = \int_z^0 \rho(\lambda, \phi, z) g dz . \qquad (13.10)$$

**Discretization.** The model is discretized on an Arakawa B-grid. It has a uniform depth of 4000 m divided in 12 vertical layers. The vertical velocity $w$ is computed at the intersection between two layers. The atmospheric forcing is added to the two top layers of 30 and 50 m respectively.

For the Arctic ocean and a few shelf seas like the Mediterranean sea and the sea of Okhotsk, the momentum equations are omitted and in the tracer equations only the diffusive terms and atmospheric forcing are retained. This is done because of the coarse resolution of the ocean model which inhibits a faithful simulation of the ocean currents in these small seas. The depth of these diffusive "lakes" is set at 1000 m. They exchange heat and salt with the ocean through diffusivity. The diffusion coefficients for the diffusion in the "lakes" and for the exchange with the ocean can be chosen independently of the diffusion coefficient of the ocean model. The horizontal diffusion coefficient was set to $1 \times 10^3 m^2 s^{-1}$, the vertical to $3 \times 10^{-5} m^2 s^{-1}$. The lakes are depicted in Fig. 13.1 as areas of light shading.

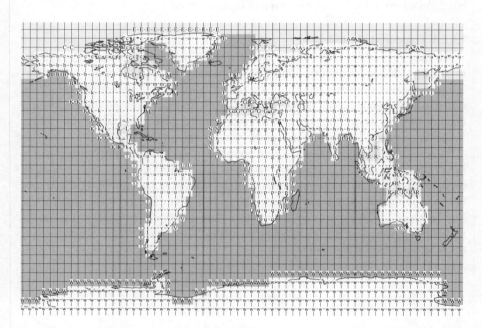

**Fig. 13.1** Land-sea mask of ECBILT. Dark shading corresponds to ocean grid boxes, light shading to lake grid boxes. River runoff is accumulated in land areas indicated by the same lowercase character and is dumped into the sea in the area indicated by the corresponding uppercase character.

**Boundaries.** The no-slip condition and the no-normal flow condition are employed, i.e. the velocities $u$ and $v$ are set to zero on the land-sea boundary. There is no flux of heat and salt across the boundary. For the horizontal advection this is achieved by the no normal flow condition. For horizontal diffusion the normal derivative of the tracer is set to zero.

Due to the absence of bottom topography the Antarctic circumpolar current will be strongly underestimated. In order to correct for this deficiency the boundary value of the streamfunction $\psi$ at Antarctica is set to generate a barotropic mass transport of 100 Sv through the Drake passage in the absence of wind forcing.

### 13.2.3 Sea-ice model

Sea-ice is formed if the ocean temperature drops below -2 °C. Sea-ice affects the heat budget of the ocean and the atmosphere by strongly reducing the heat exchange between those two subsystems. The salinity budget of the ocean is affected by sea-ice by means of brine rejection and melt water. The effect of sea-ice on the momentum transfer between atmosphere and ocean is neglected. The sea-ice model is based on the zero layer model of Semtner(1976). This model computes the thickness of sea-ice from the thermodynamic balances at the top and the bottom of the ice. The ice model consists of a single layer of sea ice with a variable thickness. On top of the ice there is a variable layer of snow.

### 13.2.4 The coupled model

The time step of the atmospheric model is 4 hrs, whereas the ocean model has a time step of 1 day, implying 6 atmosphere time steps for one ocean time step. During these 6 time steps the ocean surface is kept constant. The fluxes of heat, moisture and momentum across the ocean surface are integrated during this period and used during the next time step of the ocean model. The time step of the sea-ice model is chosen to be equal to that of the atmosphere model.

The horizontal discretization of grid points in zonal direction (5.625 degrees) is the same for the scalar gridpoints (*T, q, S*) of the atmosphere and ocean model. In meridional direction the discretization is slightly different because of the difference between a regular lat-lon and a Gaussian grid. In this case the values have to be interpolated. The winds of the atmosphere have to be interpolated to the vector points of the Arakawa-B grid of the ocean model. The sea-ice model is computed at the scalar points of the ocean model.

## 13.3  Climate of the atmosphere model

We have been running the model atmosphere for a period of 500 years with seasonal cycle included both in the solar forcing and in the SST's. For the solar forcing we used daily averaged values. The climatological SST's are from the COADS dataset (Woodruff et al 1987). We have compared the results of the simulation with the NCEP-NCAR data which are available on CD (Kalnay et al 1996).

Fig. 13.2 displays the simulated zonally averaged zonal wind for winter (a) and summer (c) and their observed counterparts (b and d). The simulated jets agree well with the observations. The most significant difference is that the location of the jet maximum in the winter hemisphere lies approximately ten degrees too far poleward. The strength of the summer jet maximum in the northern hemisphere is substantially too weak and also the easterlies are slightly too weak. The winter jets are accompanied by a Hadley circulation (not shown) which is too weak by a factor of

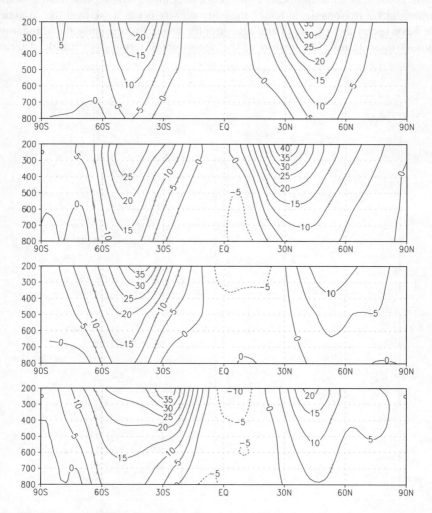

**Fig. 13.2** Simulated zonally averaged zonal wind [m/s] for winter (a1) and summer (a2) and their observed counterparts (b1 and b2).

two and not confined enough in latitudinal extent. Especially the downward branch in the winter hemisphere extends too far into the extra-tropics. This may be associated with the errors in the latitudinal location of the jet maxima in the winter hemispheres. We may have to go to higher resolution to partly cure this problem.

Fig. 13.3 displays the NH winter mean geopotential height at 500 hPa and the standard deviation of daily geopotential height values, both for the model (left) and for the NCEP-NCAR data (right). The simulation of the stationary planetary wave pattern is much too weak and dominated by a wavenumber one pattern. The Icelandic low is almost absent. The maxima in the stationary waves are too far to the north, consistent with the location of the jet maximum. The representation of the stormtracks is reasonable, although the eddy activity is too weak over the Atlantic. We have tested the impact of the ageostrophic terms by repeating the simulation without these terms. The neglect of the ageostrophic terms significantly deterio-

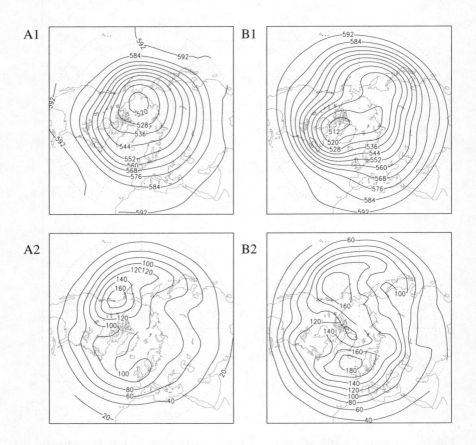

**Fig. 13.3** NH winter mean geopotential height at 500 hPa [dm] and the standard deviation of daily geopotential height values [m], both for the model (a1-a2) and for the NCEP-NCAR data (b1-b2).

rates the simulation of the jet strength and the stormtracks. In particular the latitudinal extent of the winter jet becomes much too broad and the jet maximum becomes 10 ms$^{-1}$ weaker.

The model has not been developed as a competitor to an atmospheric GCM, but rather as an alternative to the frequently used mixed boundary conditions for ocean models, so that oceanic long term variability can be studied taking to first order into account the atmospheric feed back processes to the ocean circulation. This means that we should demand that it generates reasonable surface fluxes. We will describe the simulation of these fluxes in detail in Opsteegh et al (1997). Here we briefly discuss the distribution of the precipitation and the zonal mean E-P fluxes. Fig. 13.4 displays the zonal mean E-P fluxes for the NH winter and summer both for the model and the NCEP-NCAR data. Given the simplicity of the model, the agreement with the observations is striking, especially for the summer hemispheres. In the winter hemispheres the positive E-P fluxes at approximately 20° N and 20° S are substantially weaker than observed. Fig. 13.5 displays the distribution of the NH winter mean precipitation for the model (a) and the NCEP-NCAR data (b). The simulation is quantitatively in very good agreement with the observations. The amount of rainfall in the NH stormtracks is a little too weak and the tropical ITCZ is partly missing, but the location of the areas of maximum precipitation is comparable to the results of much more complex models. The simulated sensible heat fluxes (not shown) compare similarly well with the NCEP-NCAR data.

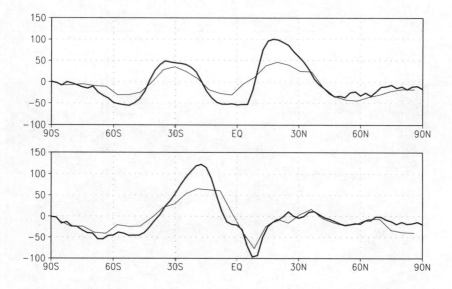

**Fig. 13.4** Zonal mean E-P fluxes [cm/year] for the NH winter (upper panel) and summer (lower panel) both for the model (thin solid line) and the NCEP-NCAR data (thick solid line).

## 13.4 Climate of the ocean model

To assess the ocean model's performance, sensitivity experiments and a spinup experiment were done with the ocean model only. We used the Hellerman and

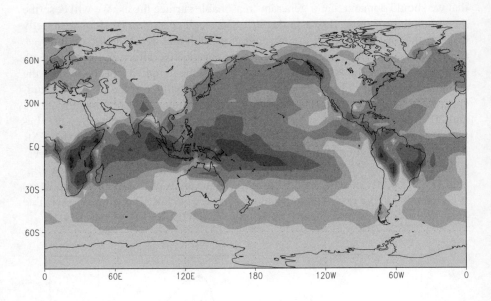

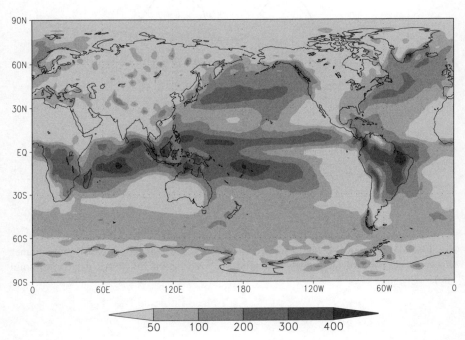

**Fig. 13.5** Distribution of the NH winter mean precipitation [cm/year] for the model (a) and the NCEP-NCAR data (b)

Rosenstein (1983) monthly mean surface stress fields, the sea surface salinity was relaxed towards Levitus (1982) salinities (yearly mean climatology) and the surface temperature was relaxed towards COADS (Woodruff et al 1987) SST or lower air temperature data (both monthly mean climatologies).   The 4000 year spinup demonstrates that the model can produce wind driven and thermohaline circulations that show a qualitative correspondence with 'state-of-the-art' OGCMs and observations. For example, the salinity along the GEOSECS (Bainbridge 1981) section in the West Atlantic (Fig. 13.6) shows relatively salty thermocline waters in the (sub)tropics, AAIW intruding from the South at 500 -- 1000 m, NADW intruding from the North between 1000 and 3000 m depth and a trace of AABW near the bottom.

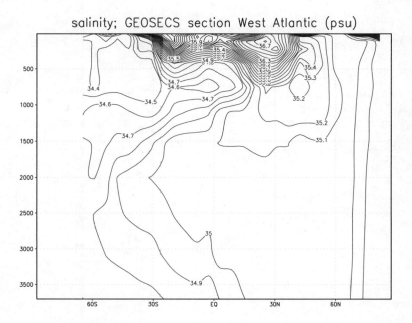

**Fig. 13.6** Salinity distribution along the GEOSECS section in ECBILT.

The model's climate, however, shows deficiencies related to the coarse resolution, flat bottom and possibly to the numerical techniques. The intermediate and deep waters at lower to mid latitudes are much too warm (up to about 5°C) and too salty (up to 0.15 psu) and meridional heat transport and gyre circulation are much too weak, especially in the Southern Hemisphere.

Though model deficiencies give rise to discrepancies in air/sea heat exchange and in meridional heat transport, these errors are not apparent everywhere. The strongest errors in this respect occur in the West Pacific and Southern Ocean and the model North Atlantic Ocean, on which we concentrate in this study, shows a reasonable surface heat budget. The amount (about 24 Sv) and source locations of the North Atlantic deep water (NADW) that is formed in the model are similar to those in e.g. state-of-the-art OGCMs. The annual mean surface heat exchange and meridional heat transport in the model North Atlantic Ocean are similar to those in the HOPE model (Drijfhout et al. 1996). Subsurface, however, the model is too warm, also in the North Atlantic. Much of the NADW is warmed by spurious horizontal diffusion of heat and most of the model NADW rises towards shallower depths at midlatitudes. The outflow of NADW into the South Atlantic Ocean is too little and too warm.

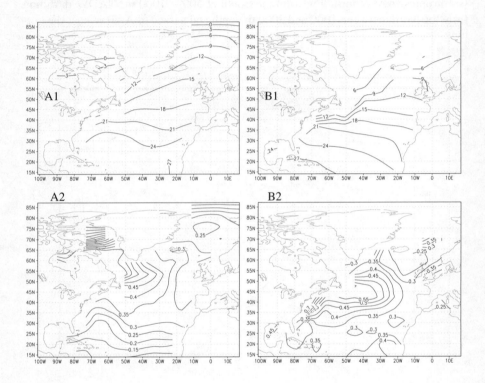

**Fig. 13.7** Mean (a1) and standard deviation (a2) of SST [°C] in the North-Atlantic basin based on winter half-year mean fields (ONDJFM) for the 1000 year model integration. (b1-b2) as (a1-a2) but for GISST data 1903-1993.

## 13.5  Climatology and variability of the coupled model

Starting from a state of rest and idealized temperature distribution, the coupled model was integrated forward in time until a statistical equilibrium was reached. Then some tuning experiments were performed in order to reduce the difference between the model and the observed climate. No flux correction was applied.

Details of the spin-up will be published elsewhere. In this section we will restrict ourselves to the description of the decadal variability in the North-Atlantic ocean in a 1000-year coupled run.

In order to compare the ocean climatology with observations, we plotted in Fig. 13.7 the mean and standard deviation of the SST's in the North-Atlantic basin as calculated from the 1000-year run and the 90-year GISST (Globa l Ice and Sea Surface Temperature) dataset provided by the British Meteorological Office, which comprises the years 1903-1993. The mean and standard deviation calculations were based on winter half-year mean fields (ONDJFM). Any linear trend in the observations was removed at each gridpoint. The model climate is warmer than observed ranging from 3 K in the subtropics until about 8 K in the Northern latitudes. The meridional temperature gradient between 40° and 60° North is substantially weaker. The model variability is slightly less than observed. Both observations and model show maximum variability located off the East coast of North-America, extending eastward between 40° and 60° North. The model maximum has shifted to the north by about 10°. In contrast to the observations the variability in the model decreases monotonically equatorward of 30° North. EOFs were calculated from the winter mean fields for both the GISST and the model data. The first four EOFs account for about 60% of the variance in both datasets. In Fig. 13.8, the first four EOFs are plotted for both the observations and the model. The first GISST EOF represents a basin wide warming or cooling, which is not reproduced in the model. GISST EOFs 2 and 3 are characterized by North-South oriented dipole structures as are model EOFs 1,2 and 3. GISST EOF 4 is characterized by an East-West oriented sequence of positive and negative anomalies and compares well with model EOF 4. Note however that due to the short observational record, the individual EOFs are not well resolved. This is particularly true for GISST EOFs 2 and 3 for which the eigenvalues are about the same.

In order to compare the climatology of the atmosphere, we plotted in Fig. 13.9 the mean and standard deviation of 800 hPa geopotential height for both the model and the observations. For the observations we used NCEP-NCAR reanalysis data from 1983-1996. The model jet is in about the right position, but weaker and more confined than in the NCEP-NCAR data. Especially in the eastern Atlantic, the model jet is confined below about 50° north, the diffluent part being displaced eastward over the European continent. The model variability is somewhat lower, but the pattern compares reasonably well with the NCEP-NCAR data. To compare modes of variability, EOFs were calculated and the first two are compared in Fig. 13.10. EOF 1 of the observations describes variations in the strength of the westerlies in the eastern Atlantic around 50° North and compares well with the north Atlantic oscillation (NAO) pattern of Wallace and Gutzler (1981, their Fig. 8). The second EOF describes north and southward displacements of the jet and compares well with the eastern Atlantic (EA) pattern of Wallace and Gutzler (1981, their Fig.12). Comparing the model EOFs with observations, one is tempted to identify the first model EOF with NAO and the second with EA. However, due to the fact that the model jet axis lies at about 45° north in the eastern Atlantic, it is

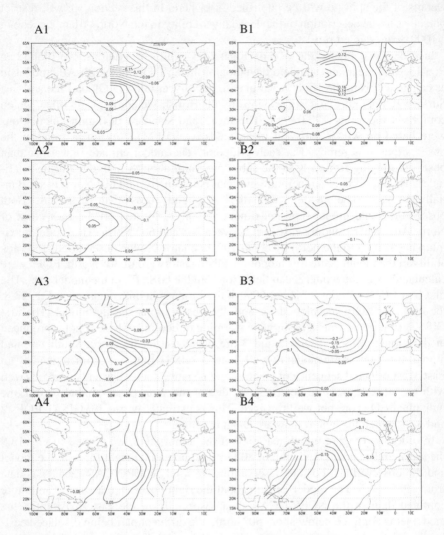

**Fig. 13.8** Panels A1-A4: EOF 1 to 4 of the 1000 year model integration. Panels B1-B4: as A1-A4 but of the GISST data.

the second model EOF that describes variations in the strength of the model westerlies around 45° north. The first model EOF describes primarily variations in the northward extension of the jet. Thus the second model EOF can be identified with a southward displaced NAO pattern, the first with a southward displaced EA pattern. The reason for discussing these EOFs thoroughly is that the 800 hPa winds are used to force the ocean through windstresses and latent and sensible heat fluxes. From previous studies (Zorita et al 1992, Saravanan and McWilliams 1996, Verbeek 1997) the picture has emerged that preferred spatial modes of surface winds force associated SST patterns through anomalous latent and sensible heat fluxes. In

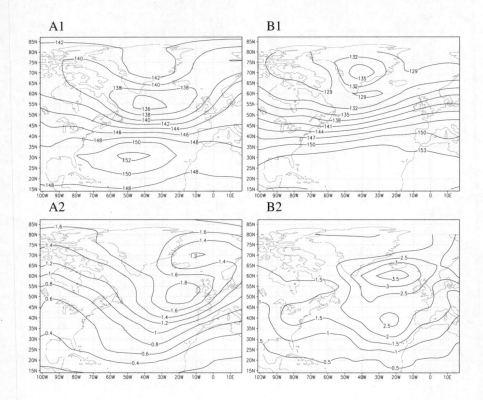

**Fig. 13.9** Mean (a) and standard deviation (b) of 800 hPa geopotential height [dm] in the North-Atlantic basin based on winter mean fields (DJF) of the 1000 year model integration. (B1-B2) as (A1-A2) but for NCEP-NCAR reanalysis data from 1983-1996.

studying air-sea interaction it is therefore relevant to know the preferred spatial modes of surface wind variability and how they compare with observations.

## 13.6  Decadal variability of the North-Atlantic.

To explore the connection between variations in the atmosphere and the ocean, a Singular Value Decomposition (SVD) was made of winter mean SST anomalies and 800 hPa geopotential height anomalies, denoted by $\Phi_{800}$. The second SVD pair is plotted in Fig. 13.11. A short timeseries of the amplitudes to give an impression of the temporal evolution is plotted in Fig. 13.12 along with a power spectrum estimate of these timeseries. The SST pattern (Fig. 13.11a) projects mainly onto SST EOFs 2 and 3, whereas the $\Phi_{800}$ anomaly pattern (Fig. 13.11b) is virtually EOF 2, the models equivalent of the NAO pattern. Consistent with the results of for instance Verbeek (1997), warmer (colder) than normal SSTs coincide with weaker

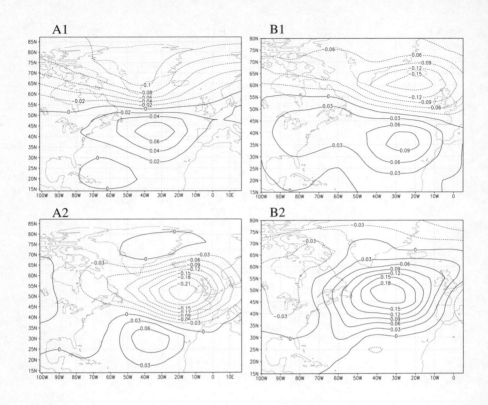

**Fig. 13.10** Panels A1-A2: EOF 1 and 2 of 800 hPa height of the 1000 year model integration. Panels B1-B2 as A1-A2 but for the NCEP-NCAR reanalyses data.

(stronger) than normal westerly winds. A clear decadal signal is visible in both the SST variance spectrum (Fig. 13.12b) and to a lesser extend in the $\Phi_{800}$ spectrum (Fig. 13.12c) at a timescale of about 18 years. The peaks are significantly different from a red noise spectrum at the 95% confidence level. The confidence limits were estimated from a Monte-Carlo simulation of 100000 spectra of AR(1) timeseries of the same length as the analysed timeseries (1000 points). The parameters of the AR(1) process were determined from the autocorrelation coefficient at lag 1 year and the total variance of the timeseries. In the timeseries (Fig. 13.12a) the 18 year period is clearly visible in both the SST variations (solid line) as well as in the $\Phi_{800}$ variations. The temporal correlation between both timeseries is 0.7. Note that no filtering was applied to remove the interannual variability that is also clearly present in the timeseries. The variability of the first and third SVD pairs (not shown) do not peak at decadal timescales; the SST variations are indistinguishable from red noise, whereas the variance spectra of the $\Phi_{800}$ variations are white. For that reason we will concentrate on the second SVD mode. This mode is similar to

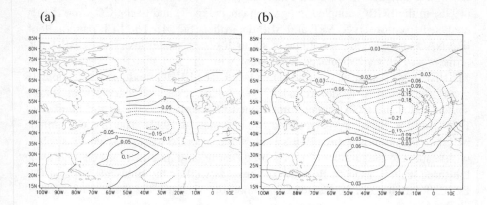

**Fig. 13.11** Second SVD mode of SST and $\Phi_{800}$ for the 1000-year coupled integration: (a) pattern SST, (b) pattern $\Phi_{800}$.

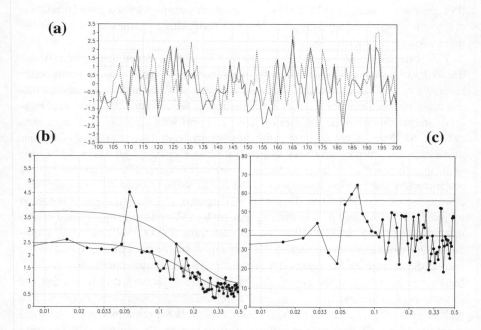

**Fig. 13.12** Second SVD mode of SST and $\Phi_{800}$ for the 1000-year coupled integration: (a) time series of SST [°C] (solid line) and $\Phi_{800}$ [5 m] (dotted line), (b) variance spectrum SST [°C] and (c) variance spectrum $\Phi_{800}$ [m$^2$]. The unmarked lines in (b) and (c) correspond to the spectra of a fitted AR(1) process and the 95% confidence limit determined from 100000 Monte-Carlo simulations.

the first Canonical Correlation Analysis (CCA) mode found by Grötzner et al. (1996) in the ECHO coupled integration (their Fig. 3) and to the CCA mode they analysed from observed SST and SLP data (their Fig. 4). The SLP patterns look like NAO and describe variations in strength of the westerlies. The ECBILT patterns are southward displaced, consistent with the fact that the position of the ECBILT NAO pattern is southward displaced. The ECHO mode peaks at about 18 years, the observed mode around 12 years. The results of a SVD analysis are shown, since the CCA modes turned out to be very sensitive to the number of EOF's used in the analysis (see also Bretherton et al 1992). The first CCA mode based on the dominant three EOF's is close the second SVD mode, which is robust with respect to the number of EOF's used (hardly any modifications beyond using the three leading EOF's). Both the patterns, as well as the timescale suggest that ECBILT simulates a similar decadal mode of variability as the ECHO coupled GCM.

In Grötzner et al (1996) it is hypothesized that this variability is due to unstable air-sea interaction: the SST anomaly excites an atmospheric response that reinforces the SST anomaly locally by enhanced or decreased net heat fluxes at the air-sea interface and enhanced or decreased entrainment of cool water from below. The SST anomaly is destroyed by a change in the heat transport by the gyre circulation due to the anomalous windstress. In order to verify this hypothesis, we performed three additional experiments.

First one might ask whether it is essential for the observed atmosphere-ocean decadal variability as characterized by the second SVD pair that the atmosphere responds to SST anomalies and feeds back on the SST anomalies. To answer this question, we performed the following experiment. We repeated the 1000-year coupled integration but in the summer of year 37, a year with virtually zero projection onto the SVD mode, we decoupled the atmosphere from the ocean. From that year onwards, we used the daily SST values and sea-ice cover of july year 37 to june year 38 as a lower boundary condition for the atmosphere. The ocean and sea-ice are forced by the varying atmosphere. The ocean is thus forced by "realistic" stochastic fluxes, that depend on the actual SST pattern, and the atmosphere is forced by fixed SST's. We will denote this experiment by OA and the fully coupled experiment by C. From year 50 to 1050, we calculated the SVD of SST and $\Phi_{800}$ anomalies. The patterns of the second SVD pair are virtually the same as those of the second SVD pair of the coupled integration. The temporal correlation is 0.77. Inspecting the variance spectra of SST and $\Phi_{800}$ (not shown), it is found that the spectral peak at around 18 years in the C experiment has disappeared in the SST as well as in the $\Phi_{800}$ spectrum of the OA experiment. What remains is a virtually red spectrum for the SST mode, and a white spectrum for the $\Phi_{800}$ mode. The response of the atmospheric circulation to SST anomalies appears indeed to be essential for the decadal variability in the model. Since a significant peak is detected in the $\Phi_{800}$ spectrum of the SVD mode of the coupled integration, and not in the OA integration, the conclusion can be drawn that the atmosphere responds to changes in the lower boundary condition at a preferred time scale of about 18 years by a change in the model's NAO index at that timescale. This reflects a slight shift in the frequency of occurrence of the weather regimes that can be associated with a positive and negative phase of the NAO index. This finding confirms the hypothesis of Palmer (1993) that the most likely atmospheric response to a change in diabatic forcing is a change in the population of weathe regimes.

In order to explore further the nature of the atmospheric response we selected a winter with a large positive phase of the second SVD pair and a winter with a large negative phase. The difference between the two winter mean SST patterns is plot-

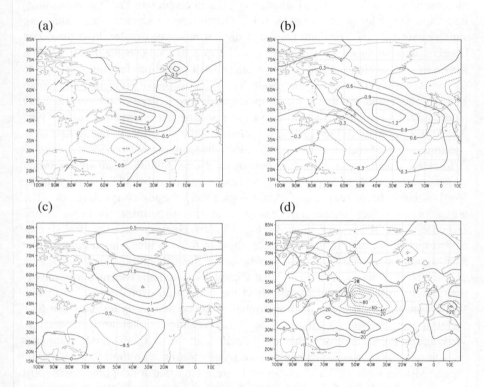

**Fig. 13.13** (a) winter mean SST difference between the two atmosphere only runs. (b) difference between DJF 30 year mean surface air temperature, (c) difference between DJF 30 year mean 500 hPa geopotential height [dm], (d) difference between DJF 30 year mean net surface heat fluxes [W/m²] with negative values indicating that the ocean is cooled

ted in Fig. 13.13a. The maximum SST difference is about 2 K and the pattern is close to the SST pattern of the second SVD pair (Fig. 13.11a). Of these two winters (from july preceding the winter until june next year), the daily global SST's were used as a lower boundary condition for two 40-year atmosphere only integrations. A climatological sea-ice cover, calculated from the first 100 years of the coupled run, was used in both atmosphere only integrations. The last 30 years were used to calculate the difference in climatology between both runs. We selected a period of 30 years, because the atmospheric response must show a clear signal on the decadal time scale in order to affect the variability on this time scale. The DJF atmospheric surface temperature, plotted in Fig. 13.13b, displays a response in the order of 1 K, about 10° downstream of the maximum SST anomaly. A negative response is found on top of the negative SST anomaly at 30° N, 50° W. A negative response is also seen over North-Western Europe. The local adjustment of the surface air temperature to the SST values diminishes the air-sea heat exchange as compared to the situation that the atmosphere would not respond. In a coupled integration, this will increase the persistence of SST anomalies. The difference in the DJF mean $\Phi_{500}$ field is in the order of 15 m Fig. 13.13c). Differences in surface wind values are smaller than 1.0 m/s. The main feature of the response, the dipole across the ocean basin implying reduced westerlies between 40° and 50° N, is visible during the first fifteen as well during the last fifteen years of the integration, lending credence to the statistical significance of the result. The $\Phi_{500}$ response projects mainly onto EOFs 2 and 3 and resembles the geopotential height pattern of the second SVD mode (Fig. 13.11b). Fig. 13.13d shows the resulting net anomalous surface heat flux for the anomaly in Fig. 13.13a. It is clear that the net effect of the surface temperature and wind response only act to reduce the damping of the SST anomaly.

To check whether SST anomalies outside the North-Atlantic basin are important or not, we repeated the experiment, this time restricting the SST anomaly to the North-Atlantic basin only, using climatological SST values everywhere else. The atmospheric surface temperature response of this experiment is close to the response to the global SST anomaly (Fig. 13.13b). However, no statistical significant response is found in the DJF mean $\Phi_{800}$ or $\Phi_{500}$ fields. Thus it appears that SST anomalies outside the North-Atlantic basin affect the atmospheric circulation over the North-Atlantic. It might be that Pacific SST anomalies set up a teleconnection pattern in the atmosphere that affect the circulation over the North-Atlantic. Wallace et al (1992) present observational evidence of a coupling between SST variability in the North Pacific and the North Atlantic on interannual timescales. We checked the variability in the North-Pacific and found a SVD mode that peaks at 18 years. There is some indication of a phase relation with the 18 year Atlantic mode. Also in Grötzner et al (1996) it is suggested that a coupling might exist between the North-Pacific and the North-Atlantic. In order to check how important the Pacific SST anomalies are for the presence of the peak in the power spectrum of the second SVD mode, we repeated the OA experiment for the Pacific only. In this experiment the Atlantic was fully coupled, but the atmosphere over the Pacific only feels the SST's of the year starting July of year 37 until June of year 38. The resulting power spectrum of the second SVD mode still shows the peak at approximately 18 years. In addition a small peak appears at a time scale of approximately 40 years. The latter peak is significant at the 95% level. Additional analyses need to be performed to unravel the coupling between Pacific and Atlantic.

In the foregoing evidence was presented that anomalous winds play a role in the observed decadal variability through anomalous surface heat fluxes. The last experiment that we describe was done to check whether the feed-back of wind anomalies on the ocean gyre circulation through anomalous windstresses is also of importance for the observed variability. So we repeated the 1000 year integration but from year 37 onwards we forced the ocean by the daily windstresses of that year, keeping everything else the same as in the fully coupled integration. We checked the variability in that run but found only small differences in the patterns between the second SVD mode of SST and $\Phi_{800}$ of this run and the fully coupled integration. The variance spectra of this mode are plotted in Fig. 13.14. Comparison with the fully coupled integration reveals small differences. The period of the spectral peak in both the SST as well as in the $\Phi_{800}$ spectrum has shifted slightly to a longer timescale of around 20 years. The peak in the $\Phi_{800}$ spectrum is no longer significant at the 95% confidence level. Thus it seems that anomalous windstresses are not crucial to explain the decadal variability in the model as described by the second SVD mode of SST and $\Phi_{800}$.

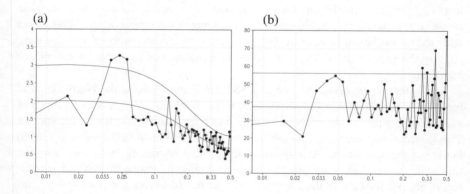

**Fig. 13.14** Second SVD mode of SST and $\Phi_{800}$ for the 1000-year coupled integration without windstress coupling: (a) variance spectrum SST [°C$^2$] and (b) variance spectrum $\Phi_{800}$ [m$^2$].

## 13.7 Conclusions

The North-Atlantic decadal variability in the ECBILT coupled integration compares qualitatively well with observations and a state-of-the-art coupled GCM, although its climatology is considerably warmer (note that no flux correction was applied). Because of its computational efficiency, additional experiments are possible to unravel cause and effect relationships. From the experiments performed so far, the following picture of decadal variability in the North-Atlantic has emerged. The atmosphere forces SST anomaly patterns due to the existence of preferred flow patterns that can dominate a whole winter by chance. For instance, a positive (negative) NAO pattern has reduced (increased) westerlies associated with it, leading to a reduced (enhanced) cooling of the ocean water and a positive (negative) SST anomaly results. To the south, an increased (decreased) northwesterly flow leads to increased (decreased) cooling and a negative (positive) SST anomaly emerges. This happens at the interannual timescale. If the atmosphere is forced by fixed SST's, the spectral characteristics of the NAO index are white. When ocean and atmosphere are coupled, a spectral peak at about 18 years appears. It is essential that the atmosphere is allowed to respond to SST changes for this peak to appear. The atmosphere responds in two ways. First, the surface air temperature adjusts to the SST values. Second, the statistics of occurrence of the atmospheric circulation regimes associated with the NAO anomaly pattern changes slightly. Both effects reduce the net air-sea heat exchange. These local thermodynamical and dynamical atmospheric responses allow for a longer life time of SST anomalies. There is some indication that the preferred time scale of 18 years is associated with the gyre circulation. The SST anomalies that are forced at the surface appear at the subsurface through the convection mechanism and are subsequently advected at subsurface levels by the gyre circulation. They partly survive the weak damping and reappear at the surface when they again enter the area where convection occurs. This effect adds to the stochastic forcing of SST anomalies by the atmosphere thus favoring slightly the generation of SST anomalies with the sign of the original subsurface anomalies.

So far the experiments indicate that SST anomalies outside the North-Atlantic basin influence the dynamical response over the North-Atlantic. A decadal mode of variability at a time scale of about 18 years has been identified in the North-Pacific. This coupling between the two basins was also hinted at in Grötzner et al (1996) but is not yet understood. Also the mechanisms that operate in the ocean to set the timescale of the variability are the subject of ongoing research. Several mechanisms are put forward in the literature such as the advective mechanism by Saravanan and Mc Williams (1997) and the idea of sustained convection by Lenderink and Haarsma (1994). In contrast to the results of Latif and Barnett (1994) the wind feedback on the ocean gyre circulation through anomalous windstresses appears not to be important. The weak response of surface winds to SST anomalies is in agreement with for instance Palmer and Sun (1985) and Saravanan and Mc Williams (1996). The very strong atmospheric response to a midlatitude Pacific SST anomaly found by Latif and Barnett might be due to the higher horizontal (T42) and vertical resolution (19 levels) used by these authors. Although the timescale and structure of the decadal variability in ECBILT is different from the one investigated by Saravanan and Mc Williams, our results support their idea that decadal variability originates from a damped oceanic mode. In our case reduced damping caused by the atmospheric response is, however, essential to detect this mode in the red background noise.

**Acknowledgments**. Qing Liu was supported by the EC in the context of the programme "Environment and Climate" under contract EV5V-CT94-0538 "Decadal and Interdecadal Climate Variability: Dynamics and Predictability Experiments"

(DICE). Selten was supported by the Dutch National Research Programme on Global Air Pollution and Climate Change, registered under nr. 951208, titled: "Climate Variability on Decadal Timescales".

# 14 A Theory For Interdecadal Climate Fluctuations

D. GU AND S.G.H. PHILANDER
*Atmospheric and Oceanic Sciences Program,*
*Princeton University,*
*Princeton, USA*

## 14.1 Introduction

The unexpected and prolonged persistence of warm conditions over the tropical Pacific during the early 1990's should be viewed, not as a prolonged El Niño, but as part of a decadal climate fluctuation that is governed by different physical processes. Whereas interannual fluctuations, including El Niño, amount primarily to a horizontal redistribution of warm surface waters within the tropics, interdecadal climate fluctuations involve changes in the properties of the equatorial thermocline because of an influx of water from higher latitudes. The influx affects equatorial sea surface temperatures and hence the tropical and extra-tropical winds that in turn affect the influx. Such processes can give rise to continual interdecadal oscillations.

Interactions between the ocean and atmosphere contribute to climate fluctuations over a broad spectrum of time-scales, from seasons to decades and longer. Studies of those interactions have thus far focused on the Southern Oscillation, which has its principal signature in the tropical Pacific sector, and which has a period of three to four years. Superimposed on this natural mode of the coupled ocean-atmosphere system, are lower frequency interdecadal fluctuations that contribute to the irregularity of the Southern Oscillation (Trenberth and Hurrell, 1994). The recent persistence of unusually warm conditions over the tropical Pacific during the early 1990's is an example (Latif et al 1996). In spite of theories and models that explain and simulate the Southern Oscillation (Neelin et al, 1995, Philander, 1990), and that correctly predicted the occurrence of its warm phase, El Niño, in 1987 and 1991 (Latif et al, 1994), the persistence of that recent warming came as a surprise (Ji et al, 1996). At present it has no explanation.

The Southern Oscillation, between complementary El Niño and La Niña states, involves an east-west redistribution of warm surface waters so that, during El Niño, the thermocline deepens in the eastern tropical Pacific while it shoals in the west. To a first approximation, neither the mean depth of the tropical thermocline nor the temperature difference across the thermocline changes. Many coupled ocean-atmosphere models of the Southern Oscillation exploit this result by having an ocean that is composed of two immiscible layers, a warm upper and cold deeper layer, that are separated by a thermocline whose mean depth is specified. During the persistent warming of the tropical Pacific in the early 1990's, isotherms in the equatorial thermocline deepened in the east but there is no evidence of a compensatory shoaling in the west. Fig. 14.1, which shows the measured change in the thermal structure along 110° W, just west of the Galapagos Islands, indicates that the warming was more pronounced at depth than at the surface. We next explore the implications of assuming that this warming was associated with an influx of warmer waters from the extra-tropics because of an earlier change in the prevailing westerly winds in the higher latitudes.

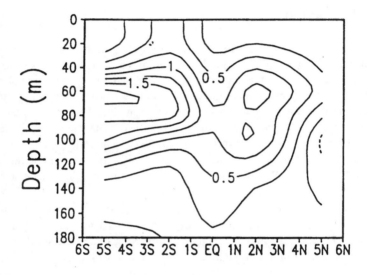

**Fig. 14.1** The change in temperature (in degrees centigrade) as a function of depth and latitude along 110°W, obtained by subtracting the mean temperature for the period 1985 to 1087 from that for the period 1990 to 1992, using data from Tropical Atmosphere Ocean (TAO) Array (14).

## 14.2  The influence of the extra-tropical winds

The influence of the extra-tropical winds extends through certain windows to the oceanic interior. The approximate location of the windows can be determined by following surfaces of constant density as they rise from the depth of the tropical thermocline to the ocean surface in the extratropics. If a change in atmospheric conditions causes the winds in the latter region to pump downward unusually warm water, then it is possible that, in due course, temperatures in the equatorial thermocline will rise. Liu and Philander (1994) and Liu et al, (1994), building on earlier results of Leyton et al, (1983), used an ocean model to locate the extratropical windows to the tropical thermocline, and to trace the routes that water parcels follow.

The subtropical regions where the surface winds drive convergent Ekman flow are potential windows to the deeper ocean, but not necessarily to the tropics. Water parcels that are forced downwards in the western side of an ocean basin join the subtropical gyres that include intense western boundary currents such as the Gulf Stream and Kuroshio Current. Water parcels that are forced downward in the central and eastern parts of the subtropical ocean basins are likely to travel westward and equatorward, to join the Equatorial Undercurrent that carries them eastward along the equator. Equatorial upwelling transfers these parcels to the surface layers whereafter poleward Ekman drift returns them to regions of subduction in the extratropics.

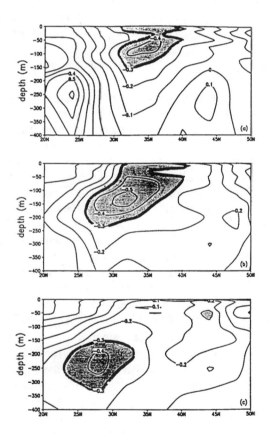

**Fig. 14.2** he equatorward and downward propagation of anomalous temperatures, averaged over longitudes 170W to 145W, during three periods: 1977 to 1981, 1982 to 1986, and 1987 to 1991.  (After Deser et al. 1996.)

Fig. 14.2 shows data that tentatively confirm the initial part of this journey. These results are from a study by Deser et al, (1996) of an interdecadal climate fluctuation that included a cooling of the surface waters in the western and central northern Pacific Ocean during the period 1976 to 1988. The reference thermal state is the time-average of temperatures as measured since 1900. The anomalous temperatures in °C are departures from the reference temperatures, during three different periods: 1977 to 1981, 1982 to 1986, and 1987 to 1991. Unusually cold water is seen to move downward and equatorward along the surface of constant density. The data are from the central Pacific -- they are averages for the longitudes 170°W to 145°W -- so that it is possible that some of the water subsequently joined the subtropical gyre while some continued equatorward. Tracer (tritium) data indicate that surface waters from the extra-tropics do indeed reach the Equatorial Undercurrent but further data analyses are necessary to determine the paths followed by water parcels. The results of Deser et al. indicate that the water parcels move along surfaces of constant density even though their temperatures are altered, a common assumption in theoretical studies. This is most likely if a cooling is accompanied by a decrease in salinities, a warming by an increase. (Stronger westerly winds that cause more evaporation and lower temperatures are usually accompanied by heavier rainfall that decreases salinity.)

Changes in extra-tropical winds can result in a subsequent changes in the structure of the equatorial thermocline, of the type seen in Fig. 14.1. Equatorial upwelling can translate a warming of the equatorial thermocline into a warming of the surface waters whereafter the unstable ocean-atmosphere interactions mentioned earlier amplify the warming. Tropical processes that establish the Southern Oscillation can reverse this trend but let us for the moment disregard those processes and instead turn to the extra-tropics for a mechanism that can also halt the warming trend in low latitudes. This mechanism depends on the link between atmospheric variability in the tropics and extra-tropics. On both interannual and interdecadal time-scales, the appearance of westerly winds (or relaxed easterly winds) over the western tropical Pacific, usually during periods when unusually warm surface waters cover large parts of the central and eastern tropical Pacific, are associated with an intensification and equatorward shift of the Subtropical Jet Stream, an eastward and equatorward extension of storm tracks across the Pacific, and with an intensification of the Aleutian low pressure system (Trenberth and Hurrell, 1994, Lau and Nath, 1994). A warming in the tropics is usually associated with more intense westerlies and colder surface waters (because of evaporation) in an extra-tropical region that happens to be a window to the equatorial thermocline. The cold water pumped downwards there in due course arrives in the tropical thermocline, halts the warming and sets the stage for cold conditions in the tropics. These arguments imply a continual, interdecadal climate fluctuation with a period that depends on the time it takes water parcels to travel from the extra-tropics to the equator. (See Latif and Barnett, (1994) for a discussion of decadal variability that does not involve the tropics).

## 14.3 A simple model of interdecadal variability

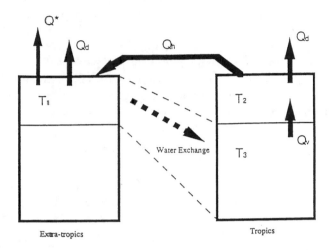

**Fig. 14.3** Sketch of the ocean box-model.

The arguments presented here can be quantified by means of the following ideal-ized model that intentionally suppresses interannual variations in order to focus on the interdecadal variations. The oceanic component of the model, shown in Fig. 14.3, consists of two tropical boxes, one at the surface at temperature $T_2$, the other immediately below it in the thermocline at temperature $T_3$, plus an extra-trop-ical surface box at temperature $T_1$. The tropical box covers the approximate region 20°S to 20°N, the extra-tropical box the region 25°N to 50°N (or 25°S to 50°S). The temperature of the surface box at the equator is determined by heat fluxes:

$$\frac{dT_2}{dt} = -Q_H + Q_V - Q_d \qquad (14.1)$$

Equatorial upwelling effects a vertical transport $Q_V$ that depends on the vertical temperature difference, and on a vertical velocity component. That vertical flow measures equatorial upwelling in response to zonal winds that drive divergent sur-face currents. The intensity of those wind, as argued earlier, depends on the temper-ature difference between the western and eastern equatorial Pacific. Temperature variations are far more modest in the west than in the east so that variations in the intensity of the wind, and hence in the intensity of upwelling, can be taken to be proportional to changes in $T_2$.

Hence $Q_v$ is proportional to $T_2(T_2 - T_3)$. If we assume all temperatures to be composed of a time-averaged value and a perturbation, and if we linearize, then this term has two components; one is proportional to the perturbation temperature difference $(T_2 - T_3)$, the other is proportional to the perturbation temperature $T_2$ and represents the positive feedback in which a change in temperature in the eastern equatorial Pacific intensifies the winds which in turn reinforces the change in temperature. $Q_h$ is the poleward atmospheric transport of heat out of the box and is assumed to depend on the temperature difference ($T_2 -- T_1$). The first term of its Taylor expansion can be written as $\gamma(T_2 -- T_1)$., where $\gamma$ is a constant. The term $Q_d$ in equation (1) represents damping that we take to be proportional to the cube of the perturbation temperature whose equation can therefore be written

$$\frac{dT_2}{dt} = -\gamma(T_2 - T_1) - \delta(T_2 - T_1) + \lambda_2 T_2 - \varepsilon T_2^3 \qquad (14.2)$$

where all the temperatures now refer to perturbation temperatures and $\delta$, $\lambda_2$, and $\varepsilon$ are positive constants. The terms on the right hand side represent respectively poleward heat transport, equatorial upwelling, the positive feedback term involving the zonal winds, that also stems from upwelling, and damping.

Similar arguments for the perturbation temperature of the extra-tropical box yield the equation

$$\frac{dT_1}{dt} = \alpha\gamma(T_2 - T_1) + \lambda_1 T_2 - \varepsilon T_1 + Q^* \qquad (14.3)$$

The terms on the right hand side represent respectively the fraction of the poleward atmospheric transport of heat that remains in the extra-tropical box, the effect of local winds that in turn depend on changes in sea surface temperatures in the tropics, damping, and stochastic forcing from weather systems unrelated to tropical temperature variations.

To link the extra-tropical and tropical oceans, we assume that at any time t, subsurface temperatures at the equator are the same as surface temperatures in the extra-tropics at an earlier time:

$$T_3(t) = T_1(t-d) \qquad (14.4)$$

Functional (or delay) differential equations such as (14.2) or (14.3) have been studied extensively and are known to have unstable (growing) oscillatory solutions (Gyori and Ladas, 1991). The presence of damping terms in our equations ensures bounded solutions so that we focus on oscillatory solutions and their sensitivity to changes in the parameters. Our reference case corresponds to the following numerical values for the constants. The time-scale for poleward heat transport, $1/\gamma$, is a year; the time-scale associated with the positive feedback in the tropics, $1/\lambda_2$, is approximately 100 days, and that associated with upwelling, $1/\delta$, is 60 days. The negative constant $\gamma_1$ is assigned a value equal to half of $\lambda_2$. The delay time $d$ for the connection between the tropics and extra-tropics is taken to be 20 years.

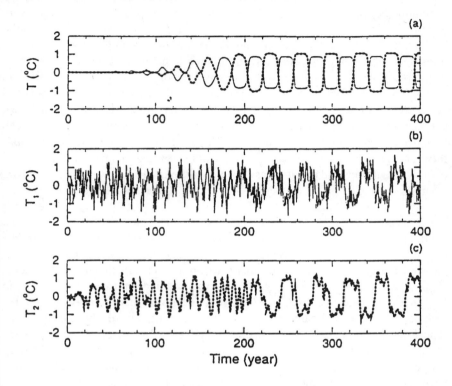

**Fig. 14.4** (a) The interdecadal oscillations obtained by solving equations (2)-(4) for the case of no random forcing (Q* = 0), and with parameters assigned their reference values given in the text. In (b) the random white noise forcing has a normal distribution, a zero mean value, and a rms of 2.

   The results in Fig. 14.4(a) show how an initial small perturbation in the surface layers at the equator slowly amplifies while generating extratropical temperature fluctuations before settling down to an oscillation with a period of 45 years and an amplitude of about 1° C. The tropical and extratropical fluctuations are out of phase. A cycle starts with a rapid increase in equatorial temperatures (because of the local positive feedback).The developments in the tropics cause an intensification of the extratropical westerly winds, and hence enhanced evaporation and a drop in surface temperatures of the extratropics. The latitudinal temperature difference created in this manner gives rise to a poleward transport of heat that halts both the warming of the tropics and the cooling of the extratropics, thus establishing an equilibrium state that persists for a considerable time before coming to an end when the cool conditions in the extratropics affect first the equatorial thermocline and then the surface layer at the equator. The decrease in equatorial surface temperatures influences extratropical winds in such a manner as to increase extratropical temperatures. The resultant effect on the poleward heat transport   in due course leads to the complementary phase of the interdecadal oscillation. In Fig. 14.4(a) the oscillation is perfectly periodic and transitions from one phase to the other are very abrupt.

The introduction of stochastic forcing in the extratropics, in panel (b), causes the transition to be more gradual and the oscillation to be more irregular. It also leads to the presence of oscillations with a relatively short period. Such bifurcation to oscillations with different periods also occur when the values of certain parameters are altered. A change in the parameter "$\gamma$" which determines the rate at which heat is transported poleward has no effect on the period of the oscillation which is seen to depend linearly on the delay time "$d$". The parameter "$\gamma$" does affect the amplitude of the oscillation because the magnitude of the temperature difference between the tropics and extratropics depends on the rate of poleward heat transport. Changes in the feedback parameters "$\lambda_1$" and "$\lambda_2$" alter the period of the oscillation.

The results presented here serve two purposes. The one is to demonstrate that continual, interdecadal climate fluctuations with a time-scale that depends on the time it takes extratropical atmospheric disturbances to influence the equatorial thermocline, are indeed possible. The other is to motivate observational and theoretical studies that explore the validity of the proposed mechanisms. The results in Fig. 14.1 are suggestive of equatorward propagation but it is conceivable that the path of the water parcels subsequently curved poleward, towards the Kuroshio Current. Liu and Philander (1994) calculated the trajectories of parcels that proceed equatorward but did so for idealized oceanic circulation driven by idealized winds. The calculations need to be repeated with realistic winds and special attention needs to be paid to asymmetries, relative to the equator, of windows from the extratropics to the equatorial thermocline. The interactions between the ocean and atmosphere need to be explored with models that have greater realism.

# 15 Decadal Variability of the Thermohaline Ocean Circulation

STEFAN RAHMSTORF
*Potsdam Institute for Climate Impact Research,*
*Potsdam, Germany*

## 15.1 The thermohaline circulation as component of the climate system

In contrast to the wind-driven part of the ocean circulation, the thermohaline circulation is driven by density gradients and extends from the surface to the abyssal ocean. It is organized in basin-scale coherent current systems, which are globally connected through the Southern Ocean (see e.g. Schmitz 1995 for a review and Macdonald and Wunsch 1996 for a recent quantitative estimate). The flow is driven by surface heat and freshwater (i.e. buoyancy) fluxes. The heat fluxes dominate in that the densest surface waters are found in cold high-latitude regions, not in the salty subtropics; the flow is thus thermally direct. It is the downward penetration of buoyancy through diffusion which in the long run ultimately engages all ocean depths in the circulation, a fact first recognized by Sandström (1908). On shorter time scales, intense buoyancy loss (due to heat loss) in a few localized areas of deep convection has a controlling influence on the thermohaline ocean circulation: if all deep convection is interrupted in models (e.g. through a surface freshwater cap) the thermohaline circulation breaks down in a matter of years. On a time scale of millennia, diffusion then erodes the stratification until convection restarts and a long-term heat balance between downward diffusion and convection is re-established. The role of the thermohaline circulation in this balance is to transport heat horizontally towards the convection sites. The deep flow is constrained by continental boundaries and the bottom topography, so that the flow through deep channels and over sills plays an important role in the spreading of deep waters. Due to the Earth's rotation the flow is strongly concentrated in boundary currents attached to continental slopes or mid-ocean ridges. An overview of key localities and mechanisms for the thermohaline circulation is given in Fig. 15.1. Note that deep water renewal of the world ocean through convection occurs only in a few locations: in the North Atlantic in the Greenland and Labrador Seas, and in the Southern Ocean near the Antarctic continent.

The thermohaline circulation plays a crucial role in the climate system because of its large meridional heat transport. It roughly equals that of the atmosphere but is much more localized: in the present climate, about 1.2 PW (1 PW = $10^{15}$ W) is transported towards the North Atlantic across 24°N by the ocean (Roemmich and Wunsch, 1985). A simple calculation illustrates this: about 17 Sv (1 Sv = $10^6$ m$^3$/s)

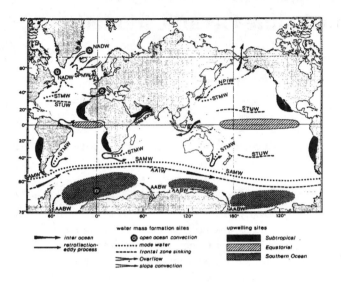

**Fig. 15.1** Location of generation and transformation areas of the major water masses: North Atlantic Deep Water (NADW), Antarctic Bottom Water (AABW), Subtropical Mode Water (SAMW), Subantartctic Mode Water (STMW), Subtropical Underwater (STUW), Subpolar Mode Water (SPMW) and North Pacific Intermediate Water (NPIW). Also indicated are interocean and intergyre exchange and retroflection processes. (From CLIVAR Scientific Steering Group 1995, p. 84.)

of North Atlantic Deep Water (NADW) flows southward across 24°N at a temperature of ~2.5°C, and is replaced by northward flow of near-surface waters which are some 16 degrees warmer; this implies a heat transport of 1.1 PW. In contrast, the wind-driven cell of about 23 Sv Florida current and 6 Sv Ekman transport recirculates towards the south at barely lower temperature, giving a heat transport of only 0.1 PW (Roemmich and Wunsch 1985). The wind-driven gyres are furthermore of limited latitudinal extent, in contrast to the deep boundary currents which provide an almost continuous flow from the Arctic to the Southern Ocean. The climatic effect of this thermohaline flow pattern is often illustrated by comparing the surface temperatures of the northern Atlantic with comparable latitudes of the Pacific; the former are 4-5°C warmer. It is thus plausible that variations of the thermohaline circulation could lead to multi-year sea surface temperature (SST) anomalies in the subpolar North Atlantic; this was first suggested in the 1960's by Bjerknes (1964).

The climatic effects are not limited to the North Atlantic, however. Model studies suggest that a climatic equilibrium without NADW formation would be significantly warmer in most ocean regions outside the North Atlantic (Manabe and Stouffer 1988; Rahmstorf and Willebrand 1995), as the ocean circulation moves less heat from other regions into the Atlantic. A recent coupled simulation of glacial climate (Ganopolski et al. 1998), however, shows that a change in the latitude of NADW formation may also reduce the *global mean* temperature, as sea-ice advances southward in the Atlantic and the planetary albedo is increased. After a shutdown of deep convection the ocean can temporarily act as a global heat sink while the deep ocean warms. This can cause a transient global cooling as in a simulation of Fanning and Weaver (1997a; perhaps aided by unrealistically strong thermal coupling of northern and southern hemispheres in their diffusive energy balance model).

The thermohaline circulation also plays an important part in the global carbon cycle, by coupling the atmosphere to the huge oceanic carbon reservoir and by providing a connection to the carbon sink in the ocean sediments. Recent model simulations (Mikolajewicz 1997) show that a breakdown of NADW formation leads to a large transient increase in atmospheric $CO_2$ content, as the oceanic uptake of $CO_2$ in high latitudes and the outgassing in low latitudes get out of balance. A simulation of anthropogenic climate change which involves a reduction of NADW formation also shows that this leads to reduced oceanic carbon uptake and increased atmospheric $CO_2$ levels, providing a positive feedback which amplifies the anthropogenic global warming (Sarmiento and Le Quéré, 1996).

## 15.2  Evidence for decadal variability of the thermohaline circulation

The previous section has highlighted the active role of the thermohaline ocean circulation in the climate system; this suggests that variations in the circulation could have important climatic repercussions. However, direct measurements of decadal or longer term variability of the deep ocean circulation are not available. No deep boundary currents were even discovered in the Pacific until 1967 and in the Indian Ocean until 1970 (B. Warren, personal communication). The ongoing World Ocean Circulation Experiment (WOCE) is expected to give new estimates of the deep flow at key locations, but at present large uncertainties surround even the present mean transport of the deep western boundary currents, let alone their long-term variability. If the variance spectrum is red, measuring a true mean may not be possible (Wunsch 1992). Because of the problem of defining a valid "level of no motion" even repeated hydrographic sections may not yield reliable information on deep current variability. Wunsch (1992) further highlights the difficulty of interpreting variability, even if we manage to observe it.

The situation is best in the North Atlantic. Several moored current meter arrays (typically lasting one year) have measured the overflow of NADW from Denmark Strait and found a strikingly constant transport, possibly due to hydraulic control of the flow over the sill (Dickson, Gmitrowicz, and Watson 1990; Dickson and Brown 1994). However, this does not rule out variations on longer time scales. Very recently, Bacon (1998) has compiled hydrographic sections off Greenland and has concluded that there is good evidence for decadal variations in overflow from the Nordic Seas (Fig. 15.2). He linked these overflow changes to the North Atlantic

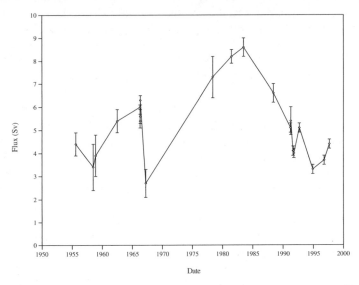

**Fig. 15.2**  Transport of the deep western boundary current flowing southward off south-east Greenland, derived from various hydrographic sections since 1955. (From Bacon 1998.)

Oscillation (NAO) index.

More information is available about convection variability. Convection not only plays a role in driving the deep flow but is also the prime mechanism for releasing heat from the deep ocean, so that convection variability has a direct influence on surface climate. Because deep convection events in the ocean occur only sporadically under harsh winter conditions, they have not been observed directly until recently. However, convection homogenizes temperature and salinity in the water column, so that much information about past convective activity can be deduced from vertical temperature and salinity profiles and other tracers. For the Greenland Sea, Schlosser et al. (1991) concluded from tracer measurements that deep convection had been reduced by 80% during the 1980's. Strong convective variability was also found in the Labrador Sea. Regular observations from Ocean Weather Station Bravo (Lazier 1980) show that deep convection occurred almost every winter until 1967, but then stopped for several years until convective activity resumed in 1972. The interruption of convection was a consequence of the so-called Great Salinity Anomaly (Dickson et al. 1988). A similar event occurred in the early eighties (Lazier 1995). Model studies suggest that a change in convection leads to a change in deep flow within a few years (Rahmstorf 1995b), but direct evidence is lacking. In the Mediterranean, the effects of an individual convection event have been observed in the deep current downstream (Send, Font, and Mertens 1996); a pulse of cold water was seen passing several weeks after a convection event, but did not lead to a pulse in deep current transport.

Further indirect evidence for thermohaline circulation variability is provided by observed changes in water mass characteristics. Hydrographic data from different periods of the past decades have been compared, using either comprehensive data compilations (Levitus 1989a; Levitus 1989b; Levitus and Antonov 1995), individual repeated sections (Roemmich and Wunsch 1984; Bryden et al. 1996) or long-term time series (e.g. the one maintained near Bermuda, Joyce and Robbins 1996). As the signals are small, the former pose a significant quality control challenge. To interpret hydrographic changes, it is useful to distinguish changes in isopycnal depth from temperature and salinity changes on a given isopycnal (Bindoff and McDougall 1994). Isopycnals may move up or down in response to basin-scale wind changes, while an increase in thickness of a particular density layer may indicate an increased production of the water mass concerned. Changes of the θ-S (potential temperature vs. salinity) signature on an isopycnal, on the other hand, reflect the buoyancy forcing in the source region and should change only on a slower (advective) time scale. Increasing salinity from 1957 to 1992 in the intermediate water at 24°N in the Atlantic has been tentatively interpreted as a reduction of Antarctic Intermediate Water inflow (Lavin et al. 1994; Bryden et al. 1996). Deep water changes at the same latitude from 1981 to 1992 may reflect reduced flow of Labrador Sea Water (possibly a consequence of the interruption of convection discussed above) and a freshening of deep waters formed in the Greenland-Iceland-Norwegian Sea (Bryden et al. 1996). Further north, between 40°N and 50°N, repeat sections have revealed a significant cooling of Labrador Sea water since 1988 and an eastward spreading of the colder water (Sy et al. 1997). Changes in the properties of the Upper North Atlantic Deep Water layer near Bermuda appear to be linked to changes in Labrador Sea convection six years earlier (Curry, McCartney, and Joyce 1998).

Finally, paleoclimatic proxy data show that profound changes in the thermohaline circulation must have occurred in the past, some of which apparently happened on decadal time scales. During the last glacial maximum, the rate of NADW export from the Atlantic was probably similar to today's (Yu, Francois, and Bacon 1996), but convection sites were shifted to the south and NADW flow was shallower, while Antarctic Bottom Water (AABW) penetrated further northward (Sarnthein et al. 1994; Sarnthein et al. 1995). This picture is confirmed by the so far only coupled ocean-atmosphere simulation of glacial climate (Ganopolski et al. 1998). Several times during the glacial period the NADW circulation appears to have collapsed rapidly (within about a decade), probably due to an inflow of meltwater or a surge of icebergs (Heinrich event) into the North Atlantic (MacAyeal 1993; Sarnthein et al. 1994; Sarnthein et al. 1995; Björck and al. 1996).

## 15.3 Physics of the thermohaline circulation

Before we discuss variability mechanisms of the thermohaline circulation, it is useful to recall some of its basic physical characteristics (see also the review papers of Weaver and Hughes 1992 and Rahmstorf, Marotzke, and Willebrand 1996). As stated above, the flow is driven by thermally dominated density gradients which are ultimately a result of the uneven heating of our planet by the sun. It would thus be natural to conclude that the thermohaline circulation is driven by the density gradients between low and high latitudes, and that it acts to transport heat from low to high latitudes. This is indeed the picture chosen by Stommel (1961) in his famous conceptual model of a one-basin, one-hemisphere circulation, which was designed to demonstrate the salinity feedback (discussed below). However, the complex geography of the world ocean and the interaction and vertical layering of deep waters of North Atlantic and Antarctic origin make the situation more complex. For the Atlantic thermohaline circulation, Rahmstorf (1996b) has recently argued that it behaves like a true interhemispheric flow system driven by density gradients between the northern deep convection regions and the southern end of the enclosed Atlantic (south of 30°S), rather than the tropics. In this view, which is based on GCM experiments, the thermohaline circulation primarily transports heat from the South Atlantic to the North Atlantic, rather than from low to high latitudes. When NADW formation is shut down, the entire North Atlantic surface cools and the South Atlantic surface warms, with the zero line near the equator. Such a cross-hemispheric flow differs in its stability from Stommel's one-hemisphere circulation. For example, inflow of freshwater in the tropics, which in Stommel's model would decrease tropical density and thus enhance NADW flow, was found to weaken NADW flow in the GCM experiments. And as we will see below, the magnitude of the temperature difference driving the flow is an important parameter for the stability of the system.

Stommel (1961) was the first to recognize the fundamental salinity feedback which is one factor responsible for the non-linear behavior of the thermohaline circulation. This positive feedback is easy to understand: the circulation transports salt towards the deep water formation region, thereby increasing the salinity and density there and enhancing the circulation. It is again arguable whether this salt transport is essentially from the tropics to the high latitudes as in Stommel's simple model, or whether salt is transported from the South Atlantic (or even from outside the Atlantic) to the North Atlantic - for equilibrium conditions, the GCM results show a salinity reduction even in the Atlantic tropics in the absence of NADW flow. In either case, Stommel's salinity feedback makes the thermohaline circulation a partly self-sustaining system which can have multiple equilibria.

A schematic stability diagram (Fig. 15.3) illustrates this. The parabolic curves are solutions of Stommel's conceptual model, or more precisely of Rahmstorf's (1996b) modified version for cross-hemispheric flow in the Atlantic (the solutions of the two differ only in the values and interpretation of their parameters). Note that in a certain parameter regime multiple equilibria exist: essentially equilibria with NADW "on" and "off", and a third unstable branch (dotted line) in between. For the equilibrium marked "NADW off", Stommel's model would predict a reverse, haline driven flow. In GCMs Antarctic Bottom Water (AABW) dominates the deep Atlantic after the cessation of NADW formation, but the dynamics governing

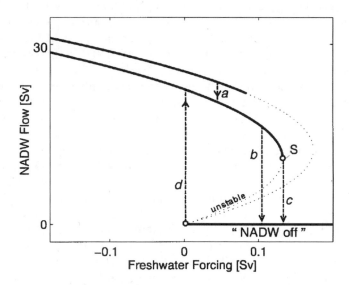

**Fig. 15.3** Schematic bifurcation diagram for North Atlantic Deep Water (NADW) flow, as derived from GCM results and theoretical models. Solid lines show stable equilibrium states; unstable states are dotted. The upper two solid lines correspond to equilibria with different NADW formation sites, the bottom one to an equilibrium without NADW formation. The uppermost stable branch is not drawn out to its saddle-node bifurcation, indicating that it becomes convectively unstable before reaching the advective stability limit. Examples of the four basic transition mechanisms are shown by dashed lines: (**a**) local convective instability, a rapid shutdown or startup of a convection site, (**b**) polar halocline catastrophe, a total rapid shutdown of North Atlantic convection, (**c**) advective spindown, a slow spindown process triggered when the large-scale freshwater forcing exceeds the critical value at Stommel's saddle-node bifurcation S while convection initially continues, (**d**) startup of convection in the North Atlantic from a "NADW off" state. The convective transitions a, b, d can be triggered either when a gradual forcing change pushes the system to the end of a stable branch or by a brief but sufficiently strong anomaly in the forcing.

AABW flow into the Atlantic are not well understood and it is not included in the diagram. The origin of the freshwater axis divides a flow regime where both salinity and temperature gradients drive the flow (on the left), from a regime where temperature drives and salinity inhibits the flow (on the right). Only in the latter the salinity feedback is positive and the two basic equilibria exist. Beyond Stommel's saddle-node bifurcation S the freshwater input is so strong that no NADW formation can be sustained.

For a GCM the meaning of "zero freshwater input" is not obvious - the experiments were performed with a fully two-dimensional surface freshwater flux field (fixed in time), with anomalies added in certain regions. Zero freshwater input is in this case defined as the point where the regime of the "no NADW" equilibrium begins - i.e. the point where freshwater input starts to inhibit NADW flow, and where the outflow of NADW becomes fresher than the northward near-surface return flow. The theory predicts that at this point the NADW flow for the "on" state is twice the flow at Stommel's bifurcation, which is the minimum sustainable NADW flow. This is confirmed in the GCM experiments. The GCM results and the fitted conceptual model further show a ratio of NADW flow to freshwater input at Stommel's bifurcation which is equal to $m_{crit}/F_{crit} = 53$. This dimensionless number depends only on the thermal and haline expansion coefficients and the temperature difference driving the flow; the latter was found to be 5.9°C (Rahmstorf 1996b). This is clearly not compatible with the idea that it is the equator-to-pole temperature difference which drives NADW flow.

Another consequence of the stability diagram is that the stability of the circulation will depend strongly on the mean NADW formation rate; the location of a model on the stability map is crucial for understanding its stability behavior. A model with very strong NADW formation (more than 25 Sv, say) will be situated in the left half of the diagram, where no 'NADW off' state exists, and it will require a very large freshwater anomaly to suppress NADW formation (in excess of 0.2 Sv). This situation is found e.g. in the ECHAM3/LSG model (Schiller, Mikolajewicz, and Voss 1997), which forms 26 Sv of NADW in the control run.

In the North Pacific, the net freshwater input is so large that it is most likely beyond the limit up to where deep water formation could be sustained, even with the help of the salinity increase that such a flow would bring. For a recent review of reasons for the difference between Atlantic and Pacific in this respect, see Rahmstorf, Marotzke and Willebrand (1996).

A second factor responsible for the non-linear behavior of the thermohaline circulation is convection. The integral effect of convection is essentially a vertical homogenization of the water column with little net vertical transport (Send and Marshall 1995). The net sinking motion associated with deep water formation is a secondary effect, which does not necessarily have to all occur in the same location as convection (see Ernst 1995 for Lagrangian deep water trajectories from an eddy-resolving model). Even in the absence of feedback, convection is strongly non-linear in its behavior: if the surface buoyancy flux is gradually varied, convection is turned 'on' or 'off' as the buoyancy flux changes sign. In addition there is a positive feedback, which can be described as follows: reduced surface density reduces vertical convection, which can lead to an accumulation of freshwater near the surface, which reduces surface density even further. This feedback can be illustrated with a simple conceptual model of two vertically stacked boxes (Welander 1982; Lenderink and Haarsma 1994). Like Stommel's salt advection feedback, the convective feedback can lead to multiple equilibria if temperature and salinity are coupled at different time scales. An example of such coupling are mixed boundary conditions, where surface temperature is relaxed towards some chosen value and a fixed surface salt (or freshwater) flux is prescribed. Assume a convection patch where cooling exceeds the input of freshwater in its effect on buoyancy, keeping convection going. If convection is interrupted, the surface cools to the prescribed relaxation temperature; after that there is no more heat loss. The freshwater input continues, however, assuring that convection remains permanently switched off. Convection thus behaves like a 'flip-flop' - in a certain parameter regime it can stabilize itself as being either 'on' or 'off'. In a three-dimensional ocean model, this can lead to the existence of many different, more or less stable convection patterns (Lenderink and Haarsma 1994; Rahmstorf 1995b) and transitions between them (indicated on Fig. 15.3 by the two upper branches connected with a convective transition, $a$). Such a convective state transition has recently also been found in the Hadley Centre coupled model, where Labrador Sea convection started up after 700 years of the control run, causing a sudden increase in NADW flow (Tett, Johns, and Mitchell 1997).

A number of further feedbacks affect the thermohaline circulation. The most important is the temperature advection feedback, which is analogous to the salinity advection feedback discussed above, except that it is negative: the advection of warm water from the south by the thermohaline circulation reduces density in the convection region, thereby weakening the flow. Both simple estimates and GCM experiments (Rahmstorf 1995b; Rahmstorf and Willebrand 1995; Rahmstorf, Marotzke, and Willebrand 1996) demonstrate that this feedback is powerful enough to have a strongly stabilizing effect on the flow rate of NADW. The traditional restoring boundary condition on temperature (Haney 1971) largely suppresses this important feedback, and if combined with fixed freshwater fluxes (allowing for the positive salt advection feedback) leads to a much too unstable thermohaline circulation in models.

Other feedbacks operate through the atmosphere. The first of these is intimately connected to the temperature feedback discussed above and limits it: if the surface of the ocean is warmed through increased oceanic advection of heat, the atmosphere will absorb some of this extra heat and both transport it away and radiate it to space. Atmospheric transport is much more effective for small scale temperature anomalies (Bretherton 1982). On the smallest scales (e.g. those of convective heat release by the ocean) the damping of SST anomalies is limited only by the efficiency of local air-sea coupling (typically ~50 W/m$^2$K but strongly dependent on atmospheric conditions - it can be especially large during the winter-time convection season). A large-scale warming of North Atlantic SST due to heat advection in the thermohaline circulation, however, is damped by the atmosphere at only about 10 W/m$^2$K: at this rate, a heat transport of 1.2 PW across 24°N leads to an average SST warming of 4°C north of this latitude. Simple energy balance models of the atmosphere can be used to account for this feedback (e.g. Power et al. 1995; Rahmstorf and Willebrand 1995), and recent comparisons have shown that these compare well with coupled models in studies of thermohaline circulation sensitivity (Capotondi and Saravanan 1997; Weber 1997).

Another feedback is the change in atmospheric vapor transport in response to thermohaline circulation (i.e. SST) changes (Nakamura, Stone, and Marotzke 1994). However, this feedback is probably weak and not even its sign is known with certainty (Tang and Weaver 1995; Hughes and Weaver 1996; Weber 1997). Finally, there is the possibility of wind feedback (Fanning and Weaver 1997a;

Mikolajewicz 1997; Schiller, Mikolajewicz, and Voss 1997) which deserves further study. If this feedback primarily acts on small scale convection regions, the resolution and realism of present models is probably not adequate to draw firm conclusions.

The same is true for all feedbacks acting locally on convection, rather than on the large-scale forcing of the thermohaline circulation. Atmospheric models do not yet reliably forecast regional precipitation changes. Another problem is that present global ocean models are unable to simulate the physics of the overflow of NADW over the Greenland-Iceland-Scotland sill, and usually form most of their NADW through convection south of Greenland or Iceland. NADW formation in these models is well removed from the sea ice edge, reducing interaction with ice and ice-albedo-feedback (Ganopolski et al. 1998). This could be one of the reasons why the significant reduction of NADW formation found in most greenhouse scenario experiments (e.g. Manabe and Stouffer 1994; Cubasch et al. 1995) only leads to a regional SST cooling south of Greenland, but has surprisingly little effect on Northern European air temperature (Rahmstorf 1997).

## 15.4 Basic variability mechanisms

This section will provide a brief general discussion of possible mechanisms of thermohaline circulation variability. More specific examples of variability modes found in circulation models will follow in later sections. One may group the variability mechanisms into the following five categories:

- Hasselmann's mechanism: integration of white noise forcing
- stochastic excitation of damped oceanic internal modes
- self-sustained internal oscillations
- externally forced variability (including state transitions)
- coupled modes.

Other ways to classify variability are of course equally possible - see e.g. Stocker (1996) and Sarachik et al. (1996) for useful reviews. We will now give a brief discussion of some characteristics of the variability types.

### 15.4.1 Hasselmann's mechanism

In this mechanism the ocean plays but a passive role, integrating the effect of weather fluctuations in the overlying atmosphere, which essentially act like a random forcing on the ocean. Consider a temperature response of the form

$$\frac{dT}{dt} = Q - \lambda T \qquad (15.1)$$

where Q is a stochastic thermal forcing and $-\lambda T$ a damping term, then a Fourier transformation of this equation leads to the temperature variance spectrum

$$S_T(\omega) = \frac{S_Q(\omega)}{\omega^2 + \lambda^2} \qquad (15.2)$$

where $S_Q(\omega)$ is the heat flux variance. For a white noise forcing ($S_Q$=const.) the response is red roughly up to the damping frequency $\lambda$ of temperature anomalies; for longer frequencies the spectrum becomes flat. If the memory of the ocean ($1/\lambda$) is long enough, this mechanism can lead to climate variability at decadal time scales. This mechanism does not directly involve the thermohaline circulation, but could force thermohaline circulation variability as a result of persistent density anomalies in the upper ocean (as in the model of Weisse, Mikolajewicz, and Maier-Reimer 1994).

### 15.4.2 Internal variability modes of the ocean

Internal variability modes come in two flavors: modes that are self-sustaining due to non-linear feedback and modes which are damped, so that they would die out if not excited by some forcing. Simple conceptual models are useful to demonstrate how thermohaline feedbacks can lead to internal oscillations; an instructive "gallery" of such oscillators was presented by Welander (1986). A prototype of advective oscillations is the Howard-Malkus loop oscillator (Welander 1967), where flow in a differentially heated ring of fluid is driven by temperature differences. In this flow loop, a warm anomaly can be sustained if it passes through the region of cooling more quickly than through the heating region. If the flow responded instantaneously to the temperature changes, this would not be the case - the key to make the oscillation self-sustaining is a delay in the flow response. The oscillation arises because the same anomaly alternates between decelerating or accelerating the flow in different phases of the oscillation cycle; the period of the oscillation is thus set by the advection time around the loop. If we follow the anomaly around the loop, it experiences a positive feedback which maintains it, but if we look at one particular place we see a delayed negative feedback: the warm anomaly creates a current which removes it. Because of the delayed response, the reaction overshoots and a cold anomaly is generated instead of the previous warm anomaly, and an oscillation starts. This is a simple thermal/advective oscillation, characterized by feedback between heating and advection. A linear box model where circulation anomalies are related out of phase to temperature anomalies, and which therefore can oscillate, has been suggested by Huck, Colin de Verdière and Weaver (1997). In this model the feedback is not dominated by the advection of a thermal anomaly in the mean flow ($\bar{v}T'$) as in Welander's loop oscillation, but by the advection of the mean thermal gradient due to a flow anomaly ($v'\bar{T}$)

Other oscillations are truly thermohaline, i.e. depend on the different response of temperature and salinity; anomalies of temperature and salinity are damped at different time scales. Welander (1986) discusses a thermohaline/advective version of the loop oscillator which does not require any delay in the dynamic response in order to oscillate. A direct oceanic analogy to the loop oscillation would be a density anomaly travelling with the thermohaline overturning in a loop around the

North and South Atlantic; indeed such oscillations are found in ocean circulation models on centennial time scales. However, more localized decadal advective oscillations, in which density anomalies are not advected around the whole loop but arise from flow anomalies ($v'\overline{T}$, $v'\overline{S}$) are more common in circulation models. A simple linear thermohaline oscillator model has been proposed by Griffies and Tziperman (1995); this shows damped oscillations where temperature anomalies in the sinking region (providing negative feedback) lag behind salinity anomalies (providing positive feedback). This characteristic phase relation is similar to that found in the coupled model of Delworth, Manabe and Stouffer (1993).

Yet other oscillations involve oceanic convection instead of advection, e.g. the thermohaline/convective oscillator of Welander (1982). Here, convection is periodically switched on and off. When it is off, salt slowly accumulates at the surface until convection starts. However, once the salt has been removed by convection, strong surface heating stops convection until the slower, but ultimately dominant salt accumulation triggers another convection event. Note that this particular oscillation works only for haline (not thermally) driven convection. However, a similar, purely thermal convective oscillation can be obtained due to the temperature nonlinearity in the equation of state (Pierce, Barnett, and Mikolajewicz 1995; Cai and Chu 1997).

Finally, diffusive 'flushes' are another type of internal thermohaline variability (reviewed e.g. by Weaver 1995), but their long time scales are beyond the scope of this paper.

To understand a particular internal oceanic oscillation, we thus have to ask which is the essential feedback loop which causes the oscillation and sets its time scale, and which feedbacks act to dampen or maintain the oscillation. The basic ingredients of internal decadal oscillations will be temperature and salinity anomalies, advection, convection and wave propagation, and in some cases interaction with sea ice (Yang and Neelin 1993).

### 15.4.3 Forced variability

As forced variability we consider such variations in thermohaline circulation which are directly forced by the atmosphere or from the land. For example, if there are internal decadal variations in atmospheric circulation, these could force decadal variations in thermohaline circulation. The ocean would simply be a slave to periodic variations in surface conditions.

An interesting case of forced variability are state transitions in the thermohaline circulation. Due to the advective and convective feedbacks discussed earlier, the thermohaline circulation may have several different equilibrium states, and changes in surface conditions may lead to a highly non-linear response in the ocean, such as a shutdown of NADW formation or a shift in convection sites (Rahmstorf 1995a). An example is the spectacular collapse of NADW flow during the so-called Heinrich Events (MacAyeal 1993; Sarnthein et al. 1994) mentioned above. Simulations of anthropogenic climate change in the next century show a reduction or in some cases complete cessation of NADW formation (Cubasch et al. 1992; Manabe and Stouffer 1994).

Both types of feedback - advective and convective - have their own critical thresholds and time scales for state transitions (Rahmstorf 1995b; Rahmstorf 1995a; Rahmstorf 1996a; Rahmstorf, Marotzke, and Willebrand 1996). The advective spindown mechanism has a time scale of several centuries; its critical threshold is Stommel's bifurcation shown in Fig. 15.3. Convective transitions are much faster and can occur within a few years; they are also more complex and difficult to predict as convection is a very localized process depending on regional details, which are not usually well represented in models. Convective transitions include regional shifts in convection sites, but also the so-called "polar halocline catastrophe" (Bryan 1986), i.e. a rapid and total shutdown of convection in the high latitudes of the North Atlantic.

### 15.4.4 Coupled modes

There is no universally agreed definition of what constitutes a coupled ocean-atmosphere mode. It is clear that oceanic variability arising in an ocean model driven by fixed surface fluxes is not a coupled mode. Likewise, stochastic surface forcing does not make a mode coupled, as it does not respond to oceanic changes. Variability of an atmosphere model driven by prescribed SST at its lower boundary could also not be called coupled. The equivalent boundary condition for the ocean is to prescribe fixed atmospheric conditions (temperature and other variables such as wind speed), which leads to a restoring boundary condition (Haney 1971; Rahmstorf and Willebrand 1995); ocean variability arising under such a completely unchanging atmosphere could hardly be called a coupled mode, even though surface fluxes are not fixed. We will therefore consider an oceanic mode uncoupled if it arises either with prescribed fluxes or under a restoring boundary condition (i.e. a purely local damping of anomalies), or a combination of the two (mixed boundary conditions). This is in line with common terminology calling a model driven by any of these boundary conditions an ocean-only model, not a coupled model. We will speak of coupled modes only if a *non-local feedback* via the atmosphere is important in generating the variability, as in simple ENSO models (Cane and Zebiak 1985) for wind feedback or in diffusive or advective energy balance models (Power et al. 1995; Rahmstorf and Willebrand 1995), which simulate atmospheric heat transport changes. Conversely, with this definition atmospheric models with a local swamp or slab ocean boundary are uncoupled atmosphere-only models, and variability arising in these models is internal atmospheric variability.

The classic example of a coupled mode is ENSO, where changes in atmospheric winds trigger a dynamical response of the oceanic mixed layer and equatorial waves, which ultimately lead to a delayed negative feedback on the initial anomaly (Suarez and Schopf 1988; Enfield 1989; Cane 1992). Another coupled mode has recently been suggested for the mid-latitude North Pacific (Latif and Barnett 1994). This involves the wind-driven gyre circulation. The best place to look for a coupled mode involving the thermohaline circulation is the North Atlantic; as early as 1964

Bjerknes proposed that variations in the ocean's meridional heat transport could explain decadal climate variations there. This hypothesis is finding increasing support from both data analysis and models (Bryan and Stouffer 1991; Kushnir 1994; Hall and Manabe 1997). The link between the North Atlantic Oscillation (NAO) pattern in the atmosphere and the ocean circulation has become an important research topic in recent years. Decadal temperature anomalies can be followed along the North Atlantic Current in step with the NAO (McCartney, Curry, and Bezdek 1996). Likewise, the latitude of the Gulf Stream correlates with the NAO index (Taylor 1996). Wohlleben and Weaver (1995) have analysed observed data and proposed a coupled subpolar North Atlantic mode which involves Labrador Sea convection. They suggest the following feedback loop: high Labrador SST $\rightarrow$ high Greenland sea level pressure $\rightarrow$ transport of ice/freshwater through Fram Strait and Denmark Strait into Labrador Sea $\rightarrow$ reduced convection $\rightarrow$ cooling of Labrador Sea. Given the delays in the system this leads to a 20-year oscillation. While the existence of a coupled decadal climate mode involving the thermohaline circulation is thus not yet firmly established, there are some interesting results pointing in this direction. In the remainder of this paper we will discuss decadal variability of the thermohaline circulation as it is found in models.

## 15.5 Models with realistic geography

The first interdecadal oscillation of the thermohaline circulation in an ocean-atmosphere GCM was described by Delworth, Manabe and Stouffer (1993, hereafter DMS93) in the GFDL coupled model. A time series of the maximum of the meridional transport stream function in the Atlantic is reproduced in Fig. 15.4. Spectral analysis of this oscillation revealed a fairly broad peak centered at a period of ca. 50 years. DMS93 point out similarities between the associated SST pattern and the observed interdecadal SST variability in the Atlantic (Kushnir 1994). The current and dynamic height anomaly pattern associated with this oscillation is shown in Fig. 15.5. The oscillation is centered on the regions to the east and south of Newfoundland in a dipole pattern. The mechanism of the oscillation was described essentially as a delayed negative feedback between oceanic advection and density. The feedback loop has an additional twist, though, involving a horizontal gyre circulation in quadrature with the vertical overturning. Consider the weak phase of the thermohaline flow: northward heat transport in the Atlantic is reduced, and a pool of cold dense water accumulates in the central North Atlantic. This then causes an anomalous cyclonic flow which transports salty water into the deep water formation region of the model, enhancing the thermohaline circulation and starting the warm phase of the oscillation. The atmosphere appears to play little role in this oscillation, except perhaps by exciting it and making it irregular due to its stochastic buoyancy forcing (DMS93). Subsequent work has linked this variability with salinity and temperature variability in the Greenland Sea and fluctuations in the East Greenland current (Delworth, Manabe, and Stouffer 1997). These

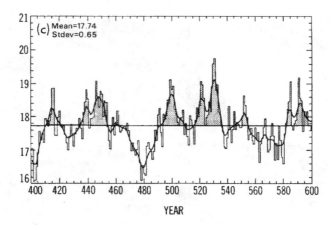

**Fig. 15.4** Time series of 200 years of the Atlantic meridional overturning in the coupled model of Delworth, Manabe, and Stouffer 1993.

results have spawned a number of further studies attempting to reproduce the oscillations of the coupled model with simplified models, some of which will be discussed below. Griffies and Tziperman (1995) have proposed a linear damped oscillator in the form of a simple box model which reproduces aspects of the DMS93 oscillation, in particular the phase relations between salinity, temperature and flow anomalies.

A thousand-year integration of the Hamburg ECHAM1/LSG coupled model has also revealed thermohaline oscillations (von Storch et al. 1997). The first EOF of the variability of the Atlantic meridional overturning stream function is shown in Fig. 15.6 (lower panel). It shows a pattern of anomalous downwelling north of 25°N accompanied by anomalous upwelling south of this latitude, with little connection to the outflow across 30°S. This pattern is different from the mean flow pattern, so it is not a loop oscillation with the whole Atlantic "conveyor belt" accelerating and decelerating, but rather a more localized advective mode. Consistent with this is the period of the oscillation of broadly 50 years. The overturning pattern associated with the mode of DMS93 is strikingly similar (Fig. 15.6, upper panel). A uniform upwelling anomaly of 1 Sv across the Atlantic between the equator and 25°N would produce a vertical excursion of isopycnals by about 15 m if maintained for 10 years, so that isopycnal movements associated with decadal variability would be quite small unless the upwelling was highly localized.

In a periodically coupled 700-year simulation of the ECHAM3/LSG model (a successor of the above), Timmermann et al. (1998) found a 35-year oscillation of the Atlantic thermohaline circulation. It resembles the DSM93 mode in the ocean component, but in contrast to DSM93 it was identified as a coupled ocean-atmosphere mode in which the atmospheric response to SST anomalies provides the negative feedback on the thermohaline overturning. When the thermohaline circu-

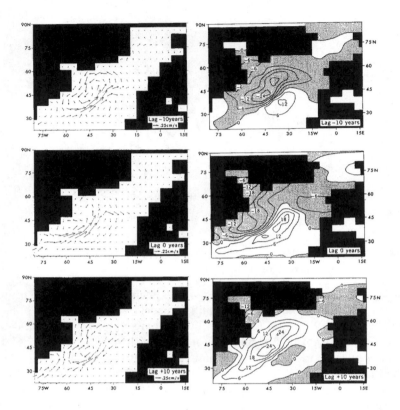

**Fig. 15.5** Regression between the Atlantic overturning intensity (see Fig. 15.4) and the currents at 170 m depth (left panel), and the dynamic topography relative to 915 dbar (in $10^2$ $cm^2s^{-2}$, right panel). (From Delworth, Manabe, and Stouffer 1993.)

lation is strong, a warm SST anomaly develops in the subpolar North Atlantic which leads to a high NAO index in the atmosphere. This in turn leads to a positive surface freshwater input, lowering surface salinity and density and weakening the thermohaline circulation.

In the coupled CSM model of NCAR, Capotondi and Holland (1998) have found a rich decadal climate variability. This model is distinguished by a higher resolution than the above models (T42 atmosphere with 18 levels, 2.4° ocean resolution with 45 levels) and was consequently integrated for only 300 years. The model uses no flux adjustments. A 50-year mode of the meridional overturning stands out, which is similar to a mode found in an ocean-only version of the model. The authors therefore interpret the mode as a stochastically forced internal ocean mode, similar to DSM93.

Decadal oscillations of the thermohaline circulation were also found in Rahmstorf's (1995a) hybrid coupled model, using an ocean model configuration

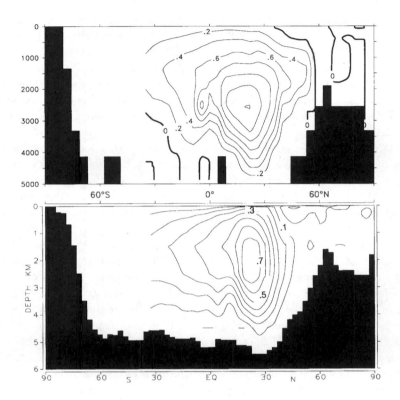

**Fig. 15.6** First EOF (empirical orthogonal function) of the meridional mass transport stream function variability in the Atlantic for the two coupled model experiments of Delworth, Manabe, and Stouffer 1993 (top) and von Storch et al. 1997 (bottom).

similar to that of DMS93 (the main difference being the use of isopycnal diffusion in the latter), combined with a simple energy balance model and fixed freshwater fluxes. This model produces a steady circulation when spun up for present-day climatological conditions, but a minor increase in precipitation in the high-latitude North Atlantic leads to a transition to an oscillatory regime through a Hopf bifurcation (without stochastic forcing). Only a brief discussion of this oscillation was included in Rahmstorf (1995a); results of a further analysis are summarized here. The oscillation shows many similarities to the one of DMS93, e.g. in the phase relations: sea surface salinity (averaged over the northern North Atlantic) leads the overturning peak by a year or two while SST lags it by a similar amount. For depth-averaged quantities, density leads (as in Figure 8 of DMS93) while salinity and temperature both lag (as in their Figure 9). Dynamic height and current anomalies (Fig. 15.7, compare to Fig. 15.5) show two distinct patterns: a large-scale flow pattern in the form of a band of northeastward flow across the interior Atlantic, very similar to that of DMS93, and a strong anomaly at the convecting grid point off Newfoundland which is in opposite phase and dominates the vertical overturning response. The large-scale dipole pattern in dynamic topography is dominated by temperature anomalies just as in DMS93 (their Figures 18 and 19). We suggest that a non-linear self-sustained oscillation localized at the convection site off Newfoundland is what drives this oscillation; the moment that this convection is stopped (through a further prescribed increase in precipitation) the oscillation also ends. The dynamics of a single convection point next to a land boundary are unrealistic (as analysed in Rahmstorf 1995b); this oscillation probably behaves like a low-order box model involving a few grid cells (this idea is consistent with the clean Hopf bifurcation and the phase relationships). This local oscillator then forces the large-scale flow variations just as stochastic noise excites them in DMS93, and in a very similar pattern. In this interpretation, the large-scale oscillation would be a robust internal ocean mode which is essentially the same as in DMS93, except that it is excited by a local grid-cell instability instead of stochastic surface forcing.

It is a major advantage of hybrid coupled models that they can include specified climatic feedbacks without internal variability of the atmosphere; they can produce a 'pure' form of thermohaline oscillations (almost sinusoidal in Rahmstorf, 1995a) which does not need to be extracted from noise by statistical techniques and is thus easier to analyse. To make full use of this advantage, hybrid models should be constructed which exactly match coupled GCM's in their ocean component and their climatological surface fluxes, in order to emulate the variability found in the GCM. This would not only help to isolate the oscillation mechanism, but would also allow sensitivity studies, such as testing where stochastic noise is necessary to excite the oscillation, or how changes in surface climatology alter it. At present many modelers have used their rather different models to produce oscillations and then speculate how similar to DMS93 they are. A perhaps more productive approach would be to start with the very model of DMS93 and progressively 'strip it down' to establish the essence of the oscillation. A good step in this direction has been taken by Weaver and Valcke (1997), albeit using fixed surface fluxes rather than a feedback model. They showed that with fixed climatological surface fluxes taken from DMS93, with or without stochastic component, the oscillation of DMS93 cannot be reproduced.

Weisse et al. (1994) describe decadal variability in the North Atlantic in a stochastically forced experiment with the LSG model. They show that this mode is due to the formation of salinity anomalies in the Labrador Sea and their subsequent advection into the deep water formation sites of the model south of Greenland. The salinity anomalies are generated by Hasselmann's mechanism (see section 15.4.1) and arise because of the absence of convection in the Labrador Sea in this model. This mechanism is thus rather different from the advective feedback described in DMS93, and the spectrum is red rather than showing a decadal spectral peak.

The same model also shows a thermohaline oscillation with a period of 320 years, which at first sight looked similar to Welander's loop oscillation (Mikolajew-

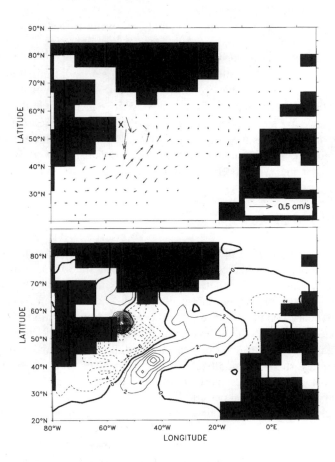

**Fig. 15.7** Current anomalies at 170 m (top) and dynamic height anomalies (as in Fig. 15.5, bottom) at the time of minimum overturning in the decadal oscillation of Rahmstorf 1995a. The deep convection point off Newfoundland is marked by 'X'.

icz and Maier-Reimer 1990). Subsequent work (Mikolajewicz and Maier-Reimer 1994) showed that with a more realistic, weaker thermal damping (and thus a more effective negative temperature feedback) the oscillation disappears unless atmospheric noise is increased. Pierce et al. (1995) and Osborn (1997) have shown that this oscillation is caused by a convective flip-flop oscillator in the Southern Ocean; it may be a model artifact related to the spurious widespread convection which most models display in this region (an exception are models with sophisticated eddy mixing parameterisations, e.g. Danabasoglu, McWilliams, and Gent 1994). Pierce et al. (1996) have demonstrated that the oscillation can also occur in a hybrid coupled model (with atmospheric energy balance model), unless suppressed by a flux correction. Such century-scale overturning variability is however not the main theme of this review; for further discussion see e.g. Weaver (1995).

A common feature of the variability modes described in this section is that they are centered on the regions of convection and deep water formation. It is therefore of some concern that in these models these convection regions are not well resolved, nor are they necessarily in the right places (see Fig. 15.1). Typically the models form their NADW south-east of the tip of Greenland, in the kind of short-cut thermohaline circulation that probably existed during the last glacial maximum 21,000 years ago. The formation of lower NADW by convection in the Greenland-Norwegian Seas and subsequent overflow over the Greenland-Scotland sill is not properly represented. It is therefore not yet clear how realistic the variability patterns and feedback mechanisms in these coarse models can be, and to what extent they can be compared with observed regional patterns.

## 15.6  Idealized models

A rapidly growing number of idealized modelling studies has been performed to pin down the essential ingredients of thermohaline oscillations and to investigate their parameter sensitivity. Oscillations have been found in idealized ocean models (usually square basins spanning one or both hemispheres) driven by fixed fluxes, restoring boundary conditions, mixed boundary conditions or atmospheric energy balance models, both with three-dimensional or zonally averaged circulations. It is difficult to gain an overview over the various oscillation modes and to determine to what extent they share essentially the same mechanisms and where they differ. However, most oscillations appear to be due to a delayed negative advective feedback and have a multi-decadal period.

At first sight one might expect fixed surface fluxes to be most conducive to variability, as they do not damp any anomalies. However, when both salinity and temperature effects are considered, their effect on density (and thus flow) is often partly cancelling. A warm anomaly (e.g. due to advection from the subtropics) is often accompanied by high salinity. In these cases mixed boundary conditions are more prone to variability, as they damp the temperature anomaly but leave the salinity anomaly intact. For the effect on thermohaline overturning this means that the negative thermal feedback is suppressed, allowing the positive haline feedback to dominate.

Decadal or interdecadal oscillations under mixed boundary conditions have been found in many models (e.g. Weaver and Sarachik 1991; Weaver, Sarachik, and Marotzke 1991; Weaver et al. 1993; Chen and Ghil 1995; Yin and Sarachik 1995). These were mainly due to various advective mechanisms. The studies identified certain surface forcing conditions which are conducive to oscillations: strong freshwater forcing (as compared to the thermal forcing) helps to destabilize the thermally direct thermohaline flow, and a minimum in P-E in mid-latitudes combined with high-latitude freshening supports decadal variability because it leads to a northward salt advection. Chen and Ghil (1996) obtained a 40-50 year oscillation in a hybrid model using an atmospheric energy balance model, with no salinity included in the model. Greatbatch and Zhang (1995) describe a decadal oscillation driven only by constant surface heat flux. Again, the mechanism is due to a delayed negative advective feedback. High-latitude temperatures control density and over-turning, and are in turn controlled by the intensity of advection from low latitudes. Several authors stress that surface fluxes derived from a spinup with restoring boundary conditions may be unique in giving the ocean the fluxes it needs for a steady solution, and thus tend to suppress variability (Cai 1995; Pierce, Barnett, and Mikolajewicz 1995). Even under a thermal restoring boundary condition (again without salt) oscillations can in some circumstances be found (Cai and Chu 1997). Capotondi and Holland (1997) performed a sensitivity study with respect to surface forcing, and found in their two-hemisphere square basin that decadal variability (period around 24 years) arose with weak thermal restoring, an energy balance model or fixed fluxes, but irregular 'flushes' were triggered when mixed boundary conditions with strong thermal restoring were used. They also investigated the role of stochastic forcing, and found essentially the same variability mode irrespective of the pattern of stochastic forcing, or whether the stochastic component was in the heat or freshwater fluxes. However, large scale stochastic forcing patterns were more effective in exciting the variability (leading to larger amplitude) than small scale forcing.

An interesting question is the potential role of waves in decadal variability. Greatbatch and Peterson (1996) propose that decadal oscillations can arise because the adjustment of the flow to density changes, occurring through Kelvin waves, is so slow that the north-south density gradient and overturning vary out of phase (overturning is in phase with east-west density gradient). The almost absent stratifi-cation at their convective polar model boundary slowed the Kelvin waves there enough to give a decadal oscillation, which thus would be determined by a wave propagation rather than advective time scale. A similar mechanism was invoked by Cai and Chu (1997). On the other hand, Huck, Colin de Verdière and Weaver (1997) performed a thorough sensitivity study of decadal variability in a square basin and found that it did not depend on boundary waves; the variability survived a removal of the boundaries (by extending the basin far away from the region of variability) except for the western boundary. They further confirmed Winton's (1996) earlier finding that the oscillations also occur in an $f$-plane model, eliminat-ing Rossby waves from the possible mechanisms.

A weakness of idealized models is that they don't include topography, which appears to have an important influence on decadal thermohaline variability. Moore and Reason (1993) report oscillations in the North Atlantic in a global model asso-ciated with deep convection, which arose only with flat bottom. Likewise, Weaver et al. (1994) have found a convective-advective oscillation centered on periodic convection in the Labrador Sea in an Atlantic model, which was suppressed when topography was introduced into the model. Winton (1997) suggests that some of the variability found in flat-bottom models is caused by currents impinging on weakly stratified coasts (i.e. vertical walls), and shows that variability is suppressed when a bowl-shaped bottom topography is included. Another limitation of most models is their coarse resolution; Fanning and Weaver (1997b) show that coarse resolution and high diffusion both tend to suppress variability.

Even two-dimensional models have shown advective decadal oscillations (Saravanan and McWilliams 1995). Taking Welander's conceptual models again as guidance, we expect that a model will be more prone to advective oscillations if there is either a delayed response of the flow to density changes or different damping time scales for temperature and salinity perturbations. Indeed the model of Winton (1996) with an instantaneous diagnostic momentum balance and purely thermal forcing does not support decadal oscillations. However, such oscillations can be easily induced in a two-dimensional model if an artificial delay is added to the diagnostic momentum balance by restoring the velocities to their equilibrium values on a 10-year time scale (A. Ganopolski, pers. comm.).

## 15.7  Conclusions

Decadal variability of the thermohaline circulation probably exists. Both observational evidence and the prevalence of decadal oscillations in all types of ocean models point to this conclusion, and it would indeed be surprising if the complex turbulent flow in the ocean would not vary on decadal time scales (Wunsch 1992). There is as yet no consensus on the exact mechanism of this variability and models show a range of possibilities. The simplest mechanism, and the one which seems to play a role in most models, is an advective oscillation centered on a deep water formation site and the current feeding it. By necessity deep water formation sites are points of maximum surface density and water of lesser density is advected towards these regions near the surface. If deep water formation increases, more light water is advected into the convection site, tending to decrease deep water formation. This is a negative feedback. Add a delay in the response of the flow to the density change and the basic ingredients for an oscillation are there. This advective oscillator is captured in essence by the conceptual model of Huck, Colin de Verdière and Weaver (1997) or similar simple models and can be considered the null hypothesis for a decadal oscillation involving the ocean's deep circulation. It works for purely thermal forcing and does not require salinity effects. However there is also a different type of advective oscillation which depends on the different damping of temperature and salinity anomalies, but does not require a delay in the flow response. A non-linear, self-sustaining prototype has been described by Welander (1986), while Griffies and Tziperman (1995) have used a linear damped version. The oscillation of DMS93 shows the tell-tale phase relation between salinity and temperature anomalies (Griffies and Tziperman 1995), but also a 3-year lag in the flow response. A further indication that the fast damping of temperature (compared to salinity) anomalies may be important is the fact that the oscillation of DMS93 could not be reproduced with fixed fluxes (Weaver and Valcke 1997).

The details of the advective pathways, the interaction with the wind-driven circulation, the topography and the air-sea coupling differ between idealized and more geographically explicit models, and therefore a detailed scenario that can be compared with observations can only come from the most realistic coupled models. An in-depth analysis and sensitivity study of a particular oscillation mode found in such a coupled model can probably best be achieved by fitting a simple atmospheric feedback model to the coupled model, with the same mean climate but controllable noise and feedback.

Some models suggest a possibility for other oscillation mechanisms which are fundamentally different from the simple advective oscillators described above, e.g. the convective oscillation of Weaver et al. (1994), where a time scale for cooling the Labrador Sea by convection is important in setting the period, or the Hasselmann mechanism in Weisse et al. (1994). An as yet controversial issue is the role of Kelvin waves in the oscillation mechanism (Greatbatch and Peterson 1996). Finally, there is also the possibility of a sudden shut-down (or start-up) of a particular convection site; this has been studied in idealized models (Lenderink and Haarsma 1994; Rahmstorf 1995b), a hybrid model (Rahmstorf 1995a) and was found to occur in a full coupled model (Tett, Johns, and Mitchell 1997). It is possible - indeed likely - that thermohaline variability in the real ocean may not be explained by one simple mechanism but by a combination of several factors.

Given the existence of thermohaline circulation oscillations, do they offer an opportunity for long-term prediction of climatic conditions over the Atlantic and Europe? Griffies and Bryan (1997) have analysed ensembles of coupled model integrations and found predictability up to ~20 years ahead, particularly during periods of strong circulation variability. The behavior of the model could be explained by damped harmonic oscillations of the thermohaline circulation excited by stochastic weather 'noise', with an oscillation period and damping time scale of around 40 years. These results suggest that there are real prospects for multi-year climate forecasts in the North Atlantic region. Modelling progress will probably continue apace in the next decade if computer resources continue to increase. A greater challenge will be to put a monitoring system in place to provide the necessary input data for forecast models.

**Acknowledgments.** The raw data for Fig. 15.6 (top panel) were kindly provided by Ron Stouffer, while the EOF analysis shown in this Figure was performed by Jin-Song von Storch. Andrey Ganopolski and two anonymous reviewers provided helpful reviews of the manuscript.

# References

Aagaard K, Carmack EC (1989) On the role of sea ice and other fresh water in the Arctic circulation. *J. Geophys. Res.* , 96, 14485-14498

Adem, J. and W. J. Jacob, 1968: One year experiment in numerical predictions of monthly mean temperature in the atmosphere-ocean-continent system, *Mon. Wea. Rev.*, 96, 714-719.

Alexander, M.A. and C. Deser, 1995: A mechanism for the recurrence of winter-time midlatitude SST anomalies. *J. Phys.Oceanogr.*, 25, 122-137.

Allan, R, Lindesay, J. and Parker, D.E., 1996: *El Niño Southern Oscillation and Climatic Variability.* CSIRO Publishing, 416pp.

Allen, M.R. and L.A. Smith, 1996, Monte Carlo SSA: Detecting irregular oscillations in the presence of coloured noise. *J. Climate*, 3373-3404.

Annamalai, H., 1995: Intrinsic problem in the seasonal prediction of the Indian summer monsoon rainfall. *Meteorol. Atmos. Phy.*, 55, 61-76.

Anderson, D. L. T. and J. Willebrand, 1996: Decadal climate variability: Dynamics and Predictability. NATO ASI Series, Series I: Global Environmental Change, Vol. 44, Springer, 493 pp

Arnault, S.A., Tropical Atlantic surface currents and ship drifts, *J. Geophys. Res.*, 92, 5076-5088, 1987.

Bacher, A., J. Oberhuber, E. Roeckner, ENSO dynamics and seasonal cycle in the tropical Pacific as simulated by the ECHAM4/OPYC3 coupled general circulation model. in press, *Clim. Dyn.*

Bacon, S., 1998: Evidence for decadal variability in the outflow from the Nordic Seas. *Nature*, submitted.

Bainbridge, A.E., 1981: GEOSECS Atlantic expedition. Vol.2: Sections and profiles. U.S. Govt. Printing Office, Washington, D.C., Stock No. 038-000-00435-2.

Balmaseda, M. A., M. K. Davey, and D. L. T. Anderson, 1995: Decadal and seasonal dependence of ENSO prediction skill. *J. Climate*, 8, 2705-2715.

Barnett, T. P., 1995: Monte Carlo Climate Forecasting. *J. Climate*, 8, 1005-1022.

Barnett, T. P., M. Latif, N. Graham, M. Flügel, S. Pazan, and W. B. White, 1993: ENSO and ENSO-related predictability. Part I: Prediction of equatorial Pacific sea surface temperature with a hybrid coupled ocean-atmosphere model. *J. Climate*, 6, 1545-1566.

Barnett, T.P. and R.W. Preisendorfer, 1987: Origins and levels of monthly and seasonal forecast skill for United States surface air temperatures determined by canonical correlation analysis. *Mon. Wea. Rev.*, 115, 1825-1850.

Barnston, A.G. and R.E. Livezey, 1987: Classification, seasonality and persistence of low-frequency atmospheric circulation patterns. *Mon. Wea. Rev.*, 115, 1083-1126.

Barnston, A., H. van den Dool, S. Zebiak, T. Barnett, M. Ji, D. Rodenhuis, M. Cane, A. Leetmaa, N. Graham, C. Ropelewski, V. Kousky, E. O'Lenic, and R. Livezey, 1994: Long-lead seasonal forecasts - where do we stand? *Bull. American Met. Soc.*, 75, 2097-2114.

Barsugli, J.J. and D.S. Battisti, 1997: The basic effects ofatmosphere-ocean thermal coupling on midlatitude variability. *J. Atmos.Sci.*, to appear.

Basnett, T.A. and Parker, D.E., 1997: Development of the Global Mean Sea Level Pressure Data Set GMSLP2. *Climate Research Technical Note 79*, Hadley Centre, Meteorological Office, Bracknell, U.K.

Battisti, D.S., U. Bhatt, and M.A. Alexander, 1995:A modeling study of the interannual variability in the wintertime North Atlantic ocean. *J.Climate*, 8, 3067-3083.

Bengtsson, L., 1974: A three parameter model for limited area forecasting.Techn. Paper No.5-74. Naval Postgraduate School, Monterey, California 93940.

Bjerknes, J., 1964: Atlantic air-sea interaction. *Adv. in Geophys.*, Academic Press, 10, pp 1-82.

Bjerknes, J., 1966: A possible response of the Hadley circulation to equatorial anomalies of ocean temperature, *Tellus*, 18, 820-829.

Bjerknes, J., 1969: Atmospheric teleconnections from the equatorial Pacific, *Mon. Wea. Rev.*, 97, 163-172.

Björck, S. and e. al., 1996: Synchronized terrestrial-atmospheric deglacial records around the North Atlantic. *Science,* **274**, 1155-1160.

Bjornsson H, Mysak LA, Brown RD (1995) On the interannual variability of precipitation and runoff in the Mackenzie drainage basin. *Clim. Dyn.,* 12: 67-76

Bindoff, N. L. and T. J. McDougall, 1994: Diagnosing climate change and ocean ventilation using hydrographic data. *J. Phys. Oceanogr.,* **24**, 1137-1152.

Bladé, I., 1997: The influence of mid-latitudecoupling on the low-frequency variability of a GCM. Part I: No tropical SSTforcing. *J. Climate*, 10, 2087-2106.

Blandford, H. F., 1884: On the connection of the Himalaya snowfall with dry winds and seasons of drought in India. *Proc. Roy. Soc.*, 37, 3-22.

Bottomley, M., Folland, C.K., Hsiung, J., Newell, R.E. and Parker, D.E., 1990: *Global Ocean Surface Temperature Atlas (GOSTA)*. Joint Meteorological Office and Massachusetts Institute of Technology Project, supported by US Dept. of Energy, US National *Science* Foundation and US Office of Naval Research and funded by UK Depts of the Environment and Energy. HMSO, London. 20 + iv pp and 313 Plates.

Brankovic C, Palmer TN, Ferranti L (1994). Predictability of seasonal atmospheric variations. *J Climate*, 7,217-237.

Bretherton, F. P., 1982: Ocean climate modeling. *Progr. Oceanogr.,* **11**, 93-129.

Bretherton, C.S., C. Smith and J.M. Wallace, 1992: An intercomparison of methods for finding coupled patterns in climate data. *J. Climate.,* 5, 541-560.

Bryan, F., 1986: High-latitude salinity and interhemispheric thermohaline circulation. *Nature,* **323**, 301-304.

Bryan, K. and R. Stouffer, 1991: A note on Bjerknes' hypothesis for North Atlantic variability. *J. Mar. Syst.,* **1**, 229-241.

Bryden, H. L., M. J. Griffiths, A. M. Lavin, R. C. Millard, G. Parilla and W. M. Smethie, 1996: Decadal changes in water mass characteristics at 24 N in the subtropical North Atlantic ocean. *J. Clim.,* **9**, 3162-3186.

CabosNarvaez, W. D., M. OrtizBevia and J. Oberhuber, The tropical Atlantic variability, *J. Geophys. Res.* 103, 7475-7489, 1998.

Cai, W., 1995: Interdecadal variability driven by mismatch between surface flux forcing and oceanic freshwater/heat transport. *J. Phys. Oceanogr.,* **25**, 2643-2666.

Cai, W. and P. C. Chu, 1997: A thermal oscillation under a restorative forcing. *Q. J. R. Met. Soc.,* submitted.

Cane, M. A., 1992: Tropical Pacific ENSO models: ENSO as a mode of the coupled system. In *Climate system modeling,* edited by K. E. Trenberth, Cambridge University Press, Cambridge, pp. 583-614.

Cane, M. A. and S. Zebiak, 1985: A theory for El Niño and the Southern Oscillation. *Science,* **228**, 1085-1087.

Capotondi, A. and W. R. Holland, 1997: Decadal variability in an idealized ocean model and its sensitivity to surface boundary conditions. *J. Phys. Oceanogr.,* 27, 1072-1093.

Capotondi, A. and W. R. Holland, 1998: Thermohaline circulation variability in the CSM coupled integration and its influence on the North Atlantic model climate.

Capotondi, A. and R. Saravanan, 1997: Sensitivity of the thermohaline circulation to surface buoyancy forcing in a two-dimensional ocean model. *J. Phys. Oceanogr.,* 27, 1072-1093.

Carton, J. A. and B. Huang, 1994: Warm events in the tropical Atlantic. *J. Phys. Oceanogr.,* 24, 888-893.

Carton, J. A., X. Cao, B. S. Giese, and A. M. da Silva, 1996: Decadal and interannual SST variability in the tropical Atlantic Ocean. *J. Phys. Oceanogr.,* 26, 1165-1175.

Cayan, D.R., 1992: Latent and sensible heat flux anomalies over thenorthern oceans: driving the sea surface temperature. *J. Phys.Oceanogr.,* 22, 859-881.

Chang, P., L. Ji, and H. Li, 1996: A decadal climate variation in the tropical Atlantic Ocean from thermodynamic air-sea interactions. *Nature,* 385, 516-518.

Chao, Y. and S.G.H. Philander, On the structure of the southern oscillation, *J. Climate,* 6, 450-469, 1993.

Chapman WL, Walsh JE, 1993: Recent variations of sea ice and air temperature in high latitudes. Bull. Amer. Met. Soc. 74, 33-47

Charney, J.G., 1975: Dynamics of deserts and drought in the Sahel. Quart. J. Roy. Meteor. Soc., 101, 193-202.

Charney, J.G. and Shukla, J., 1981: Predictability of monsoons. In: Monsoon Dynamics, Eds. J Lighthill and R.P. Pearce, Cambridge University Press, pp99-109.

Chen, D., S. E. Zebiak, M. A. Cane, and A. J. Busalacchi, 1996: On the initialization and predictability of a coupled ENSO forecast model. *Mon. Wea. Rev.*, Special Issue on "Coupled ocean-atmosphere models", in press.

Chen, F. and M. Ghil, 1995: Interdecadal variability of the thermohaline circulation and high-latitude surface fluxes. *J. Phys. Oceanogr.*, **25**, 2547-2568.

Chen, F. and M. Ghil, 1996: Interdecadal variability in a hybrid coupled ocean-atmosphere model. *J. Phys. Oceanogr.*, **26**, 1561-1578.

Cheng, X., G. Nitsche, and J. M. Wallace, 1995: Robustness of low-frequency circulation patterns derived from EOF and rotated EOF analyses. *J. Climate*, 8, 1709-1713.

CLIVAR Scientific Steering Group, 1995: *CLIVAR Science Plan.* Vol. 89, *WCRP Report.* World Climate Research Programme, Geneva.

Cockcroft, M.J., Wilkinson, M.J. and Tyson, P.D., 1987: The application of a present-day climatic model to the late Quaternary in southern Africa, *Climatic Change*, 10, 161-181

Cohen, A.L. and Tyson, P.D., 1995: Sea-surface temperature fluctuations during the Holocene off the south coast of Africa: implications for terrestrial climate and rainfall. *The Holocene*, **5**, 304-312.

Colony R, Thorndike AS (1984) An estimate of the mean field of Arctic sea ice motion. *J. Geophys. Res*, 89: 10623-10629

Cox, M. D., 1984: A primitive equation, three dimensional model of the ocean. GFDL Ocean Group Technical Report 1, GFDL.

Cubasch, U., G. C. Hegerl, R. Voss, J. Waszkewitz, and T. J. Crowley, 1996: Simulation with an OAGCM of the influence of variations of the solar constant on the global climate. *Geophys. Res. Lett.*, submitted.

Cubasch, U., K. Hasselmann, H. Höck, E. Maier-Reimer, U. Mikolajewicz, B. D. Santer and R. Sausen, 1992: Time-dependent greenhouse warming computations with a coupled ocean-atmosphere model. *Clim. Dyn.*, **8**, 55-69.

Cubasch, U., G. C. Hegerl, A. Hellbach, H. Höck, U. Mikolajewicz, B. D. Santer and R. Voss, 1995: A climate change simulation starting from 1935. *Clim. Dyn.*, **11**, 71-84.

Cullen, M. J. P., 1993: The Unified forecast/climate model. *Met. Magazine*, 122, 81-94.

Curtis, S. and S. Hastenrath, Forcing of anomalous sea surface temperature evolution in the tropical Atlantic during Pacific warm events, *J. Geophys. Res.* 100, 15835-15847, 1995.

Currie, R.G., 1996: Variance contribution of luni-solar (Mn) and solar cycle (Sc) signals to climate data. *Int. J. Climatol.*, 16, 1343-1364.

Curry, R. G., M. S. McCartney and T. M. Joyce, 1998: Oceanic transport of sub-polar climate signals to mid-depth subtropical waters. *Nature,* **391**, 575-578.

Danabasoglu, G., J. C. McWilliams and P. R. Gent, 1994: The role of mesoscale tracer transports in the global ocean circulation. *Science,* **264**, 1123-1126.

Darby MS, Mysak LA (1993) A Boolean delay equation model of an interdecadal Arctic climate cycle. *Clim. Dyn.* 8: 241-246

Darby MS, Willmott AJ (1993) A simple time-dependent coupled ice-ocean model with application to the Greenland-Norwegian Sea. *Tellus* 45A: 221-246

da Silva, A. M., C. C. Young, and S. Levitus, 1994: Atlas of surface marine data 1994, Volumes 1 and 3. NOAA Atlas NESDIS 6 and 8. Available from NOCD, NOAA/NESDIS E/OC21, Washington, D.C. 20235, USA.

Davey, M. K., S. Ineson, and M. A. Balmaseda, 1994: Simulation and hindcasts of tropical Pacific Ocean interannual variability. *Tellus*, 433-477.

Delecluse, P. and J. Servain, C. Levy, K. Arpe and L. Bengtsson, On the connection between the 1984 Atlantic warm event and the 1982-1983 ENSO, *Tellus* A 4, 448-464, 1994.

Delworth, T.L., 1996: North Atlantic interannual variability in a coupled ocean-atmosphere model. *J. Climate*, 9, 2356-2375.

Delworth, T., S. Manabe and R. J. Stouffer, 1993: Interdecadal variations of the thermohaline circulation in a coupled ocean-atmosphere model. *J. Climate,* **6**, 1993-2011.

Delworth, T. L., S. Manabe and R. J. Stouffer, 1997: Multidecadal climate variability in the Greenland Sea and surrounding regions: a coupled model simulation. *Geophys. Res. Let.,* **24**, 257-260.

Deser, C., and M.L. Blackmon, 1993: Surface climate variations over the North Atlantic Ocean during winter 1900-1989. *J. Climate*, 6, 1743-1753.

Deser, C., and M. L. Blackmon, 1995: On the relationship between tropical and North Pacific Sea surface temperature variations. *J. Climate*, 8, 1677-1680.

Deser, C., and M.S. Timlin, 1997: Atmosphere-ocean interactions on weekly-time scales in the North Atlantic and Pacific. *J. Climate* , 10, 393-408.

Deser, C., M. A. Alexander, and M. S. Timlin, 1996: Upper-ocean thermal variations in the North Pacific during 1970-1991. *J. Climate*, 9, 1840-1855.

Diaz, H., and V. Markgraf, 1992: El Niño. Historical and paleoclimatic aspects of the Southern Oscillation. Cambridge University Press. 471pp.

Dickson, R. R., 1984: Eurasian snow cover versus Indian monsoon rainfall - An extension of the Hahn-Shukla results. *J. Climate* Appl. Meteor., 23, 171-173.

Dickson RR, Lamb HH, Malmberg SA, Colebrook JM,1975: Climatic reversal in the northern North Atlantic. *Nature* 256: 479-482

Dickson, R. R. and J. Brown, 1994: The production of North Atlantic Deep Water: Sources, rates, pathways. *J. Geophys. Res.,* **99**, 12319-12341.

Dickson, R. R., E. M. Gmitrowicz and A. J. Watson, 1990: Deep water renewal in the northern North Atlantic. *Nature,* **344**, 848-850.

Dickson, R. R., J. Meincke, S. A. Malmberg and A. J. Lee, 1988: The "Great Salinity Anomaly" in the northern North Atlantic, 1968-82. *Progr. Oceanogr.,* **20**, 103-151.

Dirmeyer, P.A. and J. Shukla, 1996: The effect on regional and global climate of expansion of the world's deserts. Quart. J. Roy. Meteor. Soc., 122, 451-482.

Drijfhout, S., C. Henze, M. Latif and E. Maier-Reimer, 1996: Mean Circulation and Internal Variability in an Ocean Primitive Equation Model. *J. Phys. Oceanogr.,* 26, 559-580.

Dietrich G, Kalle W, Krauss W, Siedler G (1975) "General Oceanography 2nd Edition". John Wiley & Sons, New York 626pp

Dolman, A. J., and D. Gregory, 1992: The parametrization of rainfall interception in GCMs. Quart. Journ. R. Met. Soc., 118, 455-468.

Duchene, C. and C. Frankignoul, Sensitivity and realism of wind driven ocean models, J. Marin. Sys. 1, 97-117, 1990.

Duchon, C.E., 1979: Lanczos filtering in one and two dimensions. *J. Appl. Meteor.*, 18, 1016-1022.

Dunbar M, Wittman W (1963) Some features of ice movement in the Arctic Basin. In "Proceedings of the Arctic Basin Symposium, October 1962" pp90-108 (The Arctic Institute of North America, Washington, DC 1962)

Duxbury AC, Duxbury AB, 1994:An Introduction to the World's Oceans 4th Edition. WC Brown Publishers, Dubuque, Iowa

Eckert, C. and M. Latif, 1996: Predictability of a stochastically forced hybrid coupled model of El Niño. *J. Climate*, accepted.

Enfield, D. B., 1989: El Niño, past and present. *Rev. Geophys.,* **27** (2), 159-187.

Enfield, D. B. and D. A. Mayer, Tropical Atlantic sea surface temperature and its relations with El Niño -Southern Oscillation, *J. Geophys. Res.*, 102 929-949, 1997.

Ernst, U., 1995: Lagrangesche Analyse der Bildung und Ausbreitung von Nordatlantischem Tiefenwasser im CME-Modell. Diplom Thesis, Institut für Meereskunde, Christian Albrechts Universität, Kiel.

Fanning, A. F. and A. J. Weaver, 1997a: Temporal-geographical meltwater influences on the North Atlantic conveyor: implications for the Younger Dryas. *Paleoceanography,* **12**, 307-320.

Fanning, A. F. and A. J. Weaver, 1997b: Thermohaline variability: the effects of horizontal resolution and diffusion. *J. Climate.*

Ferranti, L., J. M. Slingo, T. N. Palmer and B. J. Hoskins, 1997: Relations between interannual and intraseasonal monsoon variability as diagnosed from AMIP integrations. *Q. J. R. Meteorol. Soc.*, 123, 1323-1357.

Findlater, J., 1969: A major low-level air current near the Indian Ocean during the northern summer. *Q. J. R. Meteorol. Soc.*, 95, 362-380.

Fissel DB, Melling H (1990) Interannual variability of oceanographic conditions in the southeastern Beaufort Sea. Can Contractor Rep of Hydrography and Ocean *Sciences* No 35, 102pp (+6 microfiche)

Folland, C.K., Colman, A., Salinger, M.J. and Parker, D.E., 1993: Eigenvectors of Global Sea Surface Temperature and night marine air temperature, 1901-90, and relationships with New Zealand Air temperature, 1871-1992. *Proc. Fourth Int. Conf. on Southern Hem. Meteorology and Oceanography, Hobart, Tasmania, 29 Mar- 2 Apr 1993*, 322-323.

Folland, C.K., Owen, J., Ward, M.N. and Colman, A., 1991: Prediction of seasonal rainfall in the Sahel region using empirical and dynamical methods. *J. Forecasting,* **10,** 21-56.

Folland, C.K., Palmer, T.N., and Parker, D.E., 1986: Sahel rainfall and worldwide sea temperatures, 1901-85. *Nature*, 320, 602-607.

Folland, C.K. and Parker, D.E., 1995: Correction of instrumental biases in historical sea surface temperature data. *Quart. J. Roy. Meteorol. Soc.*, **121**, 319-367.

Folland, C.K., Parker, D.E., Ward, M.N., and Colman, A.W., 1988: Sahel rainfall, Northern Hemisphere circulation anomalies and worldwide sea temperature changes. Study Week on "Persistent Meteo-oceanographic anomalies and teleconnections", Pontifical Academy of *Sciences*, Rome, 23-27 Sep 1986. Ed: C Chagas and G Puppi. Pontificiae Academiae Scientiarum Scripta Varia, 69, pp393-436. Also forms Long Range Forecasting and Climate Memorandum No. 7a. Available from the National Meteorological Library, Meteorological Office, Bracknell, Berkshire, UK.

Folland, C.K., Parker, D.E. and Kates, F.E., 1984: Worldwide marine temperature fluctuations 1856-1981. *Nature*, **310**, 670-673.

Folland, C.K. and Salinger, M.J., 1995: Surface temperature trends and variations in New Zealand and the surrounding ocean, 1871-1993. *Int. J. Climatol.*, **15**, 1195-1218.

Folland C.K., Salinger, M.J. and Rayner, N., 1997: Annual South Pacific island air temperatures and ocean surface temperatures. *N.Z. J. Weather and Climate,* **17(1),** *23- 42*.

Fontaine, G. and S. Bigot, 1993: West African rainfall deficits and sea surface temperatures. *Int. J. Climatol.*, 13, 271-285.

Frankignoul, C., 1981: Low frequency temperature fluctuations off Bermuda. *J. Geophys. Res.*, 86, 6522-6528.

Frankignoul, C., 1985: Sea surface temperature anomalies, planetary wavesand air-sea feedback in the middle latitudes. *Rev. Geophys.*, 23, 357-390.

Frankignoul, C, A. Czaja and B. L'Heveder, 1998: Air-sea feedback in the North Atlantic and surface boundary conditions for ocean models. *J. Climate*, 11, 2310-2324.

Frankignoul, C. and K. Hasselmann, 1977: Stochastic climate models. Part II: Application to sea surface temperature variability and thermocline variability. *Tellus*, 29, 284-305.

Frankignoul, C., P. Müller and E. Zorita, 1997: A simple model of the decadal response of the ocean to stochastic windforcing. *J. Phys. Oceanogr.*, 27, 1533-1546.

Frankignoul, C., and R. W. Reynolds, 1983: Testing a dynamical model for mid-latitude sea surface temperature anomalies. *J. Phys. Oceanogr.*, 13, 1131-1145.

Gadgil, S. and G. Asha, 1992: Intraseasonal variation of the summer monsoon. I: Observational aspects. J. Meteorol. Soc. Japan, 70, 517-527.

Gadgil, S., A. Guruprasad, D. R. Sikka and D. K. Paul, 1992 : Intraseasonal variation and the simulation of the Indian summer monsoon. Simulation of Interannual and Intraseasonal Monsoon Variability, WCRP-68, World Meteorological Organization, Geneva.

Ganopolski, A., S. Rahmstorf, V. Petoukhov and M. Claussen, 1998: Simulation of modern and glacial climates with a coupled global model of intermediate complexity. *Nature,* **391**, 350-356.

Gates, W.L., 1993: AMIP: The atmospheric model intercomparison project. Bull Amer Meteo Soc, 73, 1962-1970.

Ghil M, Mullhaupt A (1985) Boolean delay equations. II. Periodic and aperiodic solutions. J Stat Phys 41: 125-173

Ghil, M. and Vautard, R., 1991: Interdecadal oscillations and the warming trend in global temperature time series. *Nature*, **350**, 324-327.

Gill, A. E., 1980: some simple solutions for heat-induced tropical circulation, Quart. J. Roy. Meteor. Soc., 106,447-462.

Glantz, M. H., R. W. Katz, and N. Nicholls, 1991: Teleconnections linking worldwide climate anomalies. Cambridge University Press, United Kingdom, 535 pp.

Goldenburg, S.B. and L.J. Shapiro, 1996: Physical mechanisms for the association of El Niño and West African rainfall with Atlantic major hurricane activity. *J. Climate*, 9, 1169-1187.

Goddard, L. and N. E. Graham, 1997: El Niño in the 1990s. *J. Geophys. Res.*, submitted.

Graham, N. E., T. P. Barnett, and R. Wilde, 1994: On the roles of tropical and mid-latitude SSTs in forcing interannual to interdecadal variability in the winter northern hemisphere circulation. *J. Climate*, 7, 1416-1441.

Graham, N. E., T. P. Barnett, R. Wilde, M. Ponater and S. Schubert, 1994: Low-frequency variability in the winter circulation over the Northern Hemisphere: On the relative role of tropical and mid-latitude sea surface temperatures. *J. Climate*, 7, 1416-1442.

Graham, N.E., 1994: Decadal-scale variability in the tropical and North Pacific during the 1970s and 1980s: observations and model results. *Clim. Dyn.*, **10**, 135-162.

Gray, W.M., J.D. Sheaffer and C.W. Landsea, 1997: Climate trends associated with multi-decadal variability of Atlantic hurricane activity. Proc. Workshop on Atlantic hurricane variability on decadal timescales: *Nature*, causes and socio-economic impacts. NOAA, in press.

Greatbatch, R. J. and K. A. Peterson, 1996: Interdecadal variability and oceanic thermohaline adjustment. *J. Geophys. Res.,* **101**, 20467-20482.

Greatbatch, R. J. and S. Zhang, 1995: An interdecadal oscillation in an idealised ocean basin forced by constant heat flux. *J. Clim.,* **8**, 81-91.

Gregory, D., and P. R. Rowntree, 1990: A mass flux convection scheme with repre sentation of cloud ensemble characteristics and stability-dependent closure, *Mon. Wea. Rev.,* 118(7), 1483-1506.

Gregory, D., and R. N. B. Smith, 1990: Canopy, surface and soil hydrology. Unified Model documentation paper 25, Meteorological Office, London Rd, Bracknell, Berkshire, RG12 2SY.

Griffies, S. M. and K. Bryan, 1997: Predictability of North Atlantic multidecadal climate variability. *Science,* **275**, 181-184.

Griffies, S. M. and E. Tziperman, 1995: A linear thermohaline oscillator driven by stochastic atmospheric forcing. *J. Climate.,* **8**, 2440-2453.

Grötzner, A., M. Latif and T.P. Barnett, 1996: A decadal climate cycle in the North Atlantic ocean as simulated by the ECHO coupled GCM. Max-Planck-Institut für Meteorologie. Report No. 208

Grötzner, A., M. Latif and T.P. Barnett, 1998: A Decadal Climate Cycle in the North Atlantic Ocean as Simulated by the ECHO Coupled GCM. *J. Climate,* 11, 831-847.

GSP Group (1990) Greenland Sea Project: A venture toward improved understanding of the oceans' role in climate. Eos Trans Amer Geophys Union 71: No 24 (June 12) 750-754

Gu, D. and S. G. H. Philander, 1996: A theory for decadal climate fluctuations, *Science,* **275**, 805-807.

Gu, D., and S. Philander, 1995: Secular changes of annual and interannual variability in the tropics during the last century. *J. Clim.,* 8, 864-876.

Gu, D, S G H Philander, and M McPhaden, 1997: The seasonal cycle and its modulation in the eastern tropical pacific ocean, *J. Phys.Oceanogr,* 27, 2209-2218.

Gyori, L, and G Ladas, 1991: Oscillation theory of delay differential equations.Oxford University Press, Oxford, 368pp.

Haarsma, R.J., F.M. Selten. J.D. Opsteegh, G. Lenderink and Q. Liu, 1997: ECBILT: A coupled atmosphere ocean sea-ice model for climate predictability studies. KNMI technical report TR-195, De Bilt, The Netherlands.

Haigh, J.D., 1996: The impact of solar variability on climate. *Science,* **272**, 981-984.

Hall, A. and S. Manabe, 1997: Can local linear stochastic theory explainsea surface temperature and salinity variability?, *Clim. Dyn.,*13, 176-180.

Halliwell, G.R., Jr., P. Cornillon and D.A. Byrne, 1991:Westward-propagating SST anomaly features in the Sargasso sea. *J. Phys.Oceanogr.,* 21, 635-649.

Halliwell, G.R., Jr. and D.A. Mayer, 1996: Frequency response properties offorced climatic SST anomaly variability in the North Atlantic. *J.Climate*, 9, 3575-3587.

Häkkinen S (1993) An Arctic source for the Great Salinity Anomaly: a simulation of the Arctic ice-ocean system for 1955-1975. *J. Geophys. Res.*, 98: 16397-16410

Hanawa, K., Y. Yoshikawa, and T. Watanabe, 1989a: Composite analyses of wintertime wind stress vector fields with respect to SST anomalies in the western North Pacific and the ENSO events. Part I: SST composites. J. Meteor. Soc. Japan, 67, 385-400.

Hanawa, K., Y. Yoshikawa, and T. Watanabe, 1989b: Composite analyses of wintertime wind stress vector fields with respect to SST anomalies in the western North Pacific and the ENSO events. Part II: ENSO composites. J. Meteor. Soc. Japan, 67, 833-865.

Hanawa, K, S. Ishizaki, and Y. Tanimoto, 1996: Examination of the strengthening of wintertime mid-latitude westerlies over the North Pacific in the 1970s. J. Meteor. Soc. Japan, 74, 715-721.

Haney, R. L., 1971: Surface thermal boundary condition for ocean circulation models. *J. Phys. Oceanogr.*, **1**, 241-248.

Hansen, D.V. and H.F. Bezdek, 1996: On the decadal anomalies in NorthAtlantic sea surface temperature. *J. Geophys. Res.*, 101,8749-8758.

Harrisson, E. F., P. Minnis, B. R. Barkstrom, V. Ramanathan, R. D. Cess, and G. G. Gibson, 1990: Seasonal variation of cloud radiative forcing derived from the Earth radiation budget experiment. *J. Geophys. Res.*, 95(D11), 18687-18703.

Harzallah, A., and R. Sadourny, 1995: Internal versus SST-forced atmospheric variability as simulated by an atmospheric general circulation model. *J. Climate*, 8, 474-495.

Harzallah A, Rocha de Aragao JO, Sadourny R (1996) Interannual rainfall variability in North-East brazil : observation and model simulation. *Int. J. Climatol.*, 16, 861-878.

Hasselmann, K., 1976: Stochastic climate models. Part I: Theory. *Tellus*, 28, 473-485.

Hastenrath, S., 1984: Interannual variability and annual cycle: Mechanisms of circulation and climate in the tropical Atlantic sector. *Mon. Wea. Rev.*, 112, 1097-1107.

Hastenrath, S., 1990: Decadal-scale changes of the circulation in the tropical Atlantic sector associated with Sahel drought. *Int. J. Climatol.*, 10, 459-472.

Haston, L. and J. Michaelson, 1994: Long-term central coastal California precipitation variability and relationships to El Niño. *J. Climate*, 7, 1373-1387.

Held, I.M. and M.J. Suarez, 1978: A two level primitive equation atmosphere model designed for climate sensitivity experiments. *J.Atmos. Sci.*, 35, 206-229.

Hellerman, S. and M. Rosenstein, 1983: Normal monthly wind stress over the World Ocean with error estimates. *J. Phys. Oceanogr.*, 13, 1093-1104.

Hewitt, C. D., and J. F. B. Mitchell, 1996: GCM simulations of the climate of 6k B.P. : mean changes and inter-decadal variability. *J. Climate.*, 9, 3505-3529.

Hibler III W.D., Johnsen SJ (1979) The 20-yr cycle in Greenland ice core records. *Nature* 280: 481-483

Horel, J. D. and J. M. Wallace, 1981: Planetary-scale atmospheric phenomena associated with the Southern Oscillation. *Mon. Wea. Rev.*, 109, 813-829.

Hoskins, B.J., and D.J. Karoly, 1981: The steady linear response of a spheri-cal atmosphere to thermal and orographic forcing. *J. Atmos. Sci.*, 38, 1179-1196.

Houghton, R. W., 1993: The relationship of sea surface temperature to thermo-cline depth at annual and interannual time scales in the Tropical Atlantic Ocean, *J. Geophys. Res.*, 96 15173-15185.

Houghton, R.W., 1996: Subsurface quasi-decadal fluctuations in the NorthAt-lantic. *J. Climate*, 9, 1363-1373.

Houghton, R. W. and Y. M. Tourre, 1992: Characteristics of low-frequency sea surface temperature fluctuations in the tropical Atlantic. *J. Climate*, 5, 765-771.

Huang, B., J. A. Carton and J. Shukla, 1995: A numerical simulation of the vari-ability in the Tropical Atlantic Ocean, *J. Phys. Oceanogr.*, 25 , 835-864.

Huck, T., A. Colin de Verdière and A. J. Weaver, 1997: Decadal variability of the thermohaline circulation in ocean models. *J. Phys. Oceanogr.*, submitted.

Hughes, T. M. C. and A. J. Weaver, 1996: Sea surface temperature - evaporation feedback and the ocean's thermohaline circulation. *J. Phys. Oceanogr.*

Hulme, M., 1994: Validation of large-scale precipitation fields in General Circu-lation Models. In *Global precipitations and climate change*, M. Desbois and F. Désalmand (eds), Springer-Verlag, pp387-405.

Hulme M., 1991: An intercomparison of model and observed global precipita-tion climatologies. *Geophys Res Letters* 18 : 1715-1718.

Hurrell, J.W., 1995: Decadal trends in the North Atlantic Oscillation:Regional temperatures and precipitation. *Science*, 269, 676-679.

Ineson, S., and M. K. Davey, 1997: Interannual climate simulation and predict-ability in a coupled TOGA GCM. *Mon. Wea. Rev.*, 125, 721-741.

Jacobs, G. A., H. E. Hurlbert, J. C. Kindle, E. J. Metzger, J. L. Mitchell, W. J. Teague, and A. J. Wallcraft, 1994: Decade-scale trans-Pacific propagation and warming effects of an El Niño anomaly. *Nature*, 370, 360-363.

James, I. N. and P. M. James, 1989: Ultra-low-frequency variability in a simple atmospheric model. *Nature*, 342, 53-55.

Janicot, S., 1992: Spatio temporal variability of West African rainfall. Part II: Associated surface and airmass characteristics. *J. Climate*, 5, 499-511.

Janowiak, J.E., 1988: An investigation of interannual rainfall variability in Africa. *J. Climate*, 1, 240-255.

344

Janowiak, J. E. and P. A. Arkin, 1991: Rainfall variations in the tropics during 1986-1989, as estimated from observations of cloud-top temperature. *J. Geophys. Res.*, 96, 3359-3373.

Ji, M., A. Leetmaa, and V. E. Kousky, 1997: Coupled model forecasts of ENSO during the 1980s and 1990s at the National Meteorological Center. *J. Climate*, Special Issue "Stan Hayes", in press.

Ji, M., and T. M. Smith, 1995: Ocean model response to temperature data assimilation and varying surface wind stress: intercomparisons and implications for climate forecast. Mon.Wea.Rev., 123, 1811-1821.

Ji, M., A. Leetmaa, and J. Derber, 1995: An ocean analysis system for seasonal to interannual climate studies. *Mon. Wea. Rev.*, 123, 460-481.

Jiang, S., F.-F. Jin, and M. Ghil: Multiple equilibria, periodic, and aperiodic solutions in a wind-driven, double gyre, shallow-water model. *J. Phys. Oceanogr.*, 25, 764-786.

Jin, F.-F., J. D. Neelin, and M. Ghil, 1994: El Niño on the devil's staircase: Annual subharmonic steps to chaos. *Science*, 264, 70-72.

Johns, W.E., T.N. Lee, F.A. Schott, R.J. Zantopp and R.H. Evans, 1990: The North Brazil current retroflection: seasonal structure and eddy variability, *J. Geophys. Res.*, 95 22103-22120..

Johns, T. C., R. E. Carnell, J. F. Crossley, J. M. Gregory, J. F. B. Mitchell, C. A. Senior, S. F. B. Tett, and R. A. Wood, 1997: The second Hadley Centre coupled ocean-atmosphere GCM: Model description, spinup and validation. *Clim. Dyn.*, 13, 103-134.

Jolliffe, I.T., 1986: Principal component analysis. Springer-Verlag, Berlin.

Jones, P. D., S. C. B. Raper, R. S. Bradley, H. F. Diaz, P. M. Kelly and M. L. Wigley, 1986: Northern Hemisphere surface air temperature variations, 1851-1984. *J. Clim. Appl. Meteor.*, 25, 161-179.

Joseph, P. V., J. K. Eischeid and R. J. Pyle, 1994 : Interannual variability of the onset of the Indian summer monsoon and its association with atmospheric features, El Niño and sea surface temperature anomalies. *J. Climate.*, 7, 81-105.

Joyce, T.M. and P.Robbins, 1996: The long-term hydrographic record atBermuda. *J. Climate,* 9, 3122-3131.

Ju, J. and J. M. Slingo, 1995 : The Asian Summer Monsoon and ENSO. *Q. J. R. Meteorol. Soc.*, 121, 1133-1168.

Kachi, M. and Nitta, T., 1997: Decadal variations of the global atmosphere-ocean system. *J. Met. Soc. Japan*, **75**, 657-675.

Kalnay,E., M. Kanamitsu, R. Kistler, W. Collins, D. Deaver, L. Gandin, M. Iredell, S. Saha, G. White, J. Woollen, Y. Zhu, A. Leetma, R. Reynolds, M. Chelliah, W. Ebisuzaki, W. Higgins, J. Janowiak, K.C. Mo, C. Ropelewski, J. Wang, R. Jenne, 1996: The NCEP/NCAR 40-year reanalysis project. Bull. Am. Met. Soc., March 1996.

Kauker F. and J.M. Oberhuber 1997.'A regional version of the ocean general circulation model OPYC with open boundaries and tides'. Accepted by *Tellus*.

Katz, R.W., and Glantz, M.H., 1986: Anatomy of a rainfall index. *Mon. Wea. Rev.*, 114, 764-771.

Kawabe, M., 1995: Variations of current path, velocity, and volume transport of the Kuroshio in relation with the large meander. *J. Phys. Oceanogr.*, 25, 3103-3117.

Kawamura, R., 1994: A rotated EOF analysis of global sea surface temperature variability with interannual and interdecadal time scales. J. *Phys. Oceanog.*, **24**, 707- 715.

Kellogg WW (1983) Feedback mechanisms in the climate system affecting future levels of carbon dioxide. J Geophys Res 88: 1263-1269

Kelly P.M., Goodess C.M., Cherry B.S.G. , 1987: The interpretation of the Icelandic sea ice record. J Geophys Res 92: 10835-10843

Kelly, K.A. and B. Qiu, 1995: Heat flux estimates for the western NorthAtlantic. Part II: The upper-ocean heat balance. *J. Phys. Oceanogr.*,25, 2361-2373.

Kim, K-Y., 1996: Temporal and spatial subsampling errors for global empirical orthogonal functions: Applications to surface temperature field. *J. Geophys. Res.*, **101,** 23433-23446.

Kitoh, A., A. Noda, Y. Nikaidou, T. Ose, and T. Tokioka, 1995: AMIP simulations of the MRI GCM. Pap. Meteor. Geophys., 45, 121-148.

Kitoh, A., 1991a: Interannual variations in an atmospheric GCM forced by the 1970-1989 SST. Part I: Response of the tropical atmosphere, J. Meteor. Soc. Japan, 69, 251-269.

Kitoh, A., 1991b: Interannual variations in an atmospheric GCM forced by the 1970-1989 SST. Part II: Low frequency variability in the wintertime Northern Hemisphere, J. Meteor. Soc. Japan, 69, 271-291

Kleeman, R., Colman, R.A., Smith, N.R. and Power, S.B., 1996: A recent change in the mean state of the Pacific basin climate: observational evidence and atmospheric and oceanic responses. *J. Geophys. Res.*, **101**, 20483-20499.

Knox, J.L., K. Higuchi, A. Shabbar and N.E. Sargent, 1988: Secular variation of Northern hemisphere 50 kPa geopotential height. *J. Climate*, 1, 500-511.

Knutson, T.R. and Manabe, S., 1997: Model assessment of decadal variability and trends in the tropical Pacific Ocean. Submitted to *J. Climate*.

Koch L (1946) The east Greenland ice. Medd om Grønland 130: 1-374

Kodera, K., 1998: Consideration of the origin of the different midlatitude atmospheric res-ponses among El Niño events. J. Meteor. Soc. Japan, 76, 347-361.

Kodera, K, and K. Yamazaki, 1990: Long-term variation of upper stratospheric circulation in the Northern Hemisphere in December. J. Meteor. Soc. Japan, 68, 101-105.

Koide, H., and K. Kodera, 1997: Characteristics of the recent long-term wintertime variability in the atmosphere and ocean. Tenki, 44, 535-550.

346

Konig, W., E. Kirk and R. Sausen, 1990: Sensitivity of an atmosphere general circulation model to interannually varying SST. Ann. Geophys., 8, 829-844.

Kraus, E. B., and J. S. Turner, 1967: A one dimensional model of the seasonal thermocline. Part II. Telus, 19, 98-105.

Krishnamurti, T. N. and H. N. Bhalme, 1976: Oscillations of a monsoon system. Part I: Observational aspects. J. Atmos. Sci., 33, 1937-1954.

Kumar A, Hoerling MP (1995) Prospects and limitations of seasonal atmospheric GCM predictions. Bull Amer Meteo Soc 74 : 335-345.

Kushnir, Y., 1994: Interdecadal variations in North Atlantic sea surface temperature and associated atmospheric conditions. J. Climate., 7, 142-157.

Kushnir, Y. and I. M. Held, 1996: Equilibrium atmospheric response to NorthAtlantic SST anomalies. J. Climate, 9, 1208-1220.

Kushnir, Y., and N.-C. Lau, 1992: The general circulation model response to a North Pacific SST anomaly. J. Climate, 5, 271-283.

Kutsuwada, K., 1991: Quasi-periodic variabilities of wind-stress fields over the Pacific Ocean related to ENSO events. J. Meteor. Soc. Japan, 69, 687-700.

Labitzke, K., 1987: Sunspots, the QBO, and the stratospheric temperature in the north polar region. Geophys. Res. Lett., 14, 535-537.

Lamb HH (1977) 'Climate: Present, Past and Future, 2'. Methuen, London

Lamb, P.J., 1978a: Case studies of tropical Atlantic surface circulation patterns during recent sub-Saharan weather anomalies: 1967 and 1968. Mon. Wea. Rev., 106, 482-491.

Lamb, P.J., 1978b: Large-scale tropical Atlantic surface circulation patterns associated with Subsaharan weather anomalies. Tellus, A30, 240-251.

Lamb, P.J., 1982: Persistence of sub-Saharan drought. Nature, 299, 46-48.

Lamb, P.J. and R.A. Peppler, 1987: North Atlantic Oscillation: Concept and application. Bull. Am. Met. Soc., 68, 1218-1225.

Lamb, P.J. and R.A. Peppler, 1991: West Africa. Chapter 5 in Teleconnections Linking Worldwide Climate Anomalies: Scientific Basis and Societal Impact (M.H. Glantz, R.W. Katz, and N. Nicholls, Eds.), Cambridge University Press, 121-189.

Lamb, P.J. and R.A. Peppler, 1992: Further case studies of tropical Atlantic surface and atmospheric patterns associated with Subsaharan drought. J. Climate, 5, 476-488.

Lamb, P.J., M. El Hamly and D.H. Portis, 1997a: North Atlantic Oscillation. Geo Observateur, 7, 103-113.

Lamb, P.J., M. El Hamly, M.N. Ward, R. Sebbari, D.H. Portis and S. El Khatri, 1997b: Experimental Precipitation Prediction for Morocco for 1997-98. Report for Moroccan Ministry of Public Works, Cooperative Institute for Mesoscale Meteorological Studies, University of Oklahoma, Norman. 10 pp.

Landsea, C.W. and W.M. Gray, 1992: The strong association between western Sahelian monsoon rainfall and intense Atlantic hurricanes. J. Climate, 5, 435-453.

Latif, M. and T. P. Barnett, 1994: Causes of decadal climate variability over the North Pacific and North America. *Science,* 266, 634-637.

Latif M. and T.P. Barnett, Interactions of the Tropical Oceans, 1995: *J. Climate,* 8 952-964.

Latif, M, T. P. Barnett, M. A. Cane, M. Flugel, N.E. Graham, H. Von Storch, J-S Xu, and S E Zebiak, 1994: *Clim. Dyn.* , 9, 167.

Latif, M. and T. P. Barnett, 1996: Decadal variability over the North Pacific and North America: Dynamics and predictability. *J. Climate,* **9**, 2407-2423.

Latif, M., J. Biercamp, H. von Storch, M. J. McPhaden and E. Kirk, 1990: Simulation of ENSO-related surface wind anomalies with an atmospheric GCM forced by observed SST, *J. Climate,* 3, 509-521.

Latif, M., A. Sterl, E. Maier-Reimer, and M. M. Junge, 1993a: Climate variability in a coupled GCM. Part I: The tropical Pacific. *J. Climate.,* 6, 5-21.

Latif, M., A. Sterl, E. Maier-Reimer, and M. M. Junge, 1993b: Structure and predictability of the El-Niño/Southern Oscillation phenomenon in a coupled ocean-atmosphere general circulation model. *J. Climate.,* 6, 700-708.

Latif, M., T. Stockdale, J. Wolff, G. Burgers, E. Maier-Reimer, and M. M. Junge, 1994: Climatology and variability in the ECHO coupled GCM. *Tellus,* 46A, 351-366.

Latif, M. and N.E. Graham, How much predictive skill is contained in the thermal structure of an OGCM? *J. Phys. Oceanogr.,* 22 951-962, 1992.

Latif, M., A. Grotzner, and H. Frey, 1996a: El Hermanito: El Niño's overlooked little brother in the Atlantic. *Nature,* submitted.

Latif, M. A. Grotzner, M. Mtnnich, E. Maier-Reimer, S. Venzke, and T. P. Barnett, 1996b: A mechanism for decadal climate variability. In Decadal climate variability: Dynamics and Predictability. NATO ASI Series, Series I: Global Environmental Change, Vol. 44, Springer, pp 263-292.

Latif, M., R. Kleeman, and C. Eckert, 1997a: Greenhouse warming, decadal variability, or El Niño? An attempt to understand the anomalous 1990s. *J. Climate,* 10, 2221-2239.

Latif, M., ,D. L. T. Anderson, T. P. Barnett, M. A. Cane, R. Kleeman, A. Leetmaa, J. J. O'Brien, A. Rosati, and E. K. Schneider, 1997b: TOGA review paper: "Predictability and Prediction". *J. Geophys. Res.,* submitted.

Latif, M., A. Grotzner, A. Timmermann, S. Venzke, and T. P. Barnett, 1997c: Dynamics of decadal climate variability over the Northern Hemisphere. Proceedings of the JCESS/CLIVAR workshop on Decadal Climate Variability, April 22-24, 1996, Columbia MD, USA. Available from CLIVAR Office, c/o Max-Planck-Institut für Meteorologie, Bundesstr. 55, D-20146 Hamburg, Germany, in prep.

Lau, N-C, 1985: Modeling the seasonal dependence of the atmospheric response to observed El Niños in1962-1976, *Mon. Wea. Rev.,* 113, 1970-1996.

Lau, N.-C., S. G. H. Philander, and M. J. Nath, 1992: Simulation of ENSO-like phenomena with a low-resolution coupled GCM of the global ocean and atmosphere. *J. Climate.,* 5, 284-307.

Lau, N-C, M..J. Nath,1990: A general circulation model study of the atmospheric response to extratropical SST anomales observed during 1950-79, *J Climate*, 3, 965-989.

Lau, N-C., 1997: Interactions between global SST anomalies and the midlatitude atmospheric circulation. *Bull. Amer. Meteorol. Soc.*, **78**, 21-33.

Lau, N.-C. and M..J. Nath, 1994: A modeling study of the relative roles of tropical and extratropical SST anomalies in the variability of the global atmosphere-ocean system. *J. Climate*, 7,1184-1207.

Lavin, A., H. L. Bryden, M. J. Garcia and G. Parrilla, 1994: *Decadal time changes in the circulation at 24 N in the Atlantic Ocean, ICES Report, Session S*, 16 pp.

Lazier, J. R. N., 1980: Oceanographic conditions at Ocean Weather Ship *Bravo*, 1964-74. *Atmos.-Ocean,* **18**, 227-238.

Lazier, J. R. N., 1995: The salinity decrease in the Labrador Sea over the past thirty years. In *Natural climate variability on decade-to-century time scales*, edited by D. G. Martinson and e. al, Natl. Res. Council, Washington DC, pp. 295-305.

Lean, J., J. Beer, and R. Bradley, 1995: Reconstruction of solar irradiance since 1600: Implications for climate change. *Geophys. Res. Lett.*, 22, 3195-3198.

Leetmaa, A., and M. Ji, 1989: Operational hindcasting of the tropical Pacific. Dynamics of Atmospheres and Oceans, 13, 465-490.

Leith, C. E., 1974: Theoretical skill of Monte Carlo forecasts, *Mon. Wea. Rev.*, 102, 409-418.

Lenderink, G. and R.J. Haarsma, 1994: Variability and multiple equilibria of the thermohaline circulation associated with deep water convection. *J. Phys. Oceanogr.*,24,1480-1493.

Lenderink, G. and R.J. Haarsma, 1996: Rapid convective transitions in the presence of sea-ice. *J. Phys. Oceanogr.*, 26, 1448-1467.

Lenderink, G. and R. J. Haarsma, 1994: Variability and multiple equilibria of the thermohaline circulation, associated with deep water formation. *J. Phys. Oceanogr.*, **24**, 1480-1493.

Levitus, S., 1982: Climatological Atlas of the World Ocean, NOAA Prof. Paper No. 13, U.S. Govt. Printing Office, 173 pp.

Levitus, S. and J. Antonov, 1995: Observational evidence of interannual to decadal-scale variability of the subsurface temperature-salinity structure of the World Ocean. *Clim. Change,* **31**, 495-514.

Levitus, S., 1989a: Interpentadal variability of temperature and salinity atintermediate depths of the North Atlantic ocean, 1970-74 versus 1955-59.*J. Geophys. Res.*, 94, 6091-6131.

Levitus, S., 1989b: Interpentadal variability of temperature and salinity in the deep North Atlantic, 1970-1974 versus 1955-1959. *J. Geophys. Res.,* **94**, 16125-16131.

Levitus, S., J.I. Antonov and T.P. Boyer, 1994: Interannual variability oftemperature at depth of 125 meters in the North Atlantic Ocean. *Science*, 266, 96-99.

Levitus, S., T. P. Boyer, and J. Antonov, 1994: World Ocean Atlas 1994. Volume 5: Interannual variability of upper ocean thermal structure. U.S. Department of Commerce, NOAA, Washington, D. C.

Liu, Z., 1993: Thermocline forced by varying Ekmanpumping, Part II: Annual and decadal Ekman pumping. *J. Phys. Oceanogr.*, 23, 2523-2540.

Liu, Z., 1996: Thermocline vriability in different dynamic regions. *J.Phys. Oceanogr.*, 26, 1634-1645.

Liu, Z. and S. G. H. Philander, 1995: How different wind stress patterns affect the tropical subtropical circulations of the upper ocean. *J. Phys. Oceanogr.*, 25, 449-462.

Liu, Z., S. G. H. Philander, and R. C. Pacanowski, 1994: A GCM study of the tropical-subtropical upper-ocean water exchange. *J. Phys. Oceanogr.*, 24, 2606-2623.

Livezey, R., C. Folland, J. Davies, P. Glecker, D. Karoly, D. Parker, M. Sugi and J. Walsh, 1995: AGCM intercomparison diagnostics: Key time series. In: Workshop on simulations of the climate of the twentieth century using GISST, Eds. C.K. Folland and D.P. Rowell, pp40-49. Climate Research Technical Note No. 56. Available from the National Meteorological Library, London Road, Bracknell, UK.

Lohmann, G., 1996: Stability of the thermohaline circulation in analytical and numerical models. PhD thesis. Available from Alfred-Wegener-Institut für Polar- und Meeresforschung, D-27568 Bremerhaven, Germany.

Lough, J.M., 1986: Tropical Atlantic sea surface temperatures and rainfall variations in Subsaharan Africa. *Mon. Wea. Rev.*, 114, 561-570.

Lu, P. and J. P. McCreary, 1995: Influence of the ITCZ on the flow of thermocline water from the subtropical to the equatorial Pacific Ocean. *J. Phys. Oceanogr.*, 25, 3076-3088.

Luksch, U., 1996: Simulation of North Atlantic low-frequency SST variability. *J. Climate*, 9, 2083-2092.

Luyten, J R, J Pedlosky, and H Stommel, 1983: The ventilated thermocline.*J. Phys. Oceanogr*, 3, 292-309.

MacAyeal, D. R., 1993: Binge/purge oscillations of the Laurentide ice sheet as a cause of the North Atlantic's Heinrich events. *Paleoceanography,* 8, 775-784.

Macdonald, A. and C. Wunsch, 1996: An estimate of global ocean circulation and heat fluxes. *Nature,* 382, 436-439.

Malmberg SA (1969) Hydrographic changes in the waters between Iceland and Jan Mayen in the last decade. Jokull 19: 30-43

Manabe, S., and A. J. Broccoli, 1985: A comparison of climate model sensitivity with data from the last glacial maximum, J. Atmos. Sci., 42, 2643-2651.

Manabe, S. and R. J. Stouffer, 1988: Two stable equilibria of a coupled ocean-atmosphere model. *J. Climate.*, 1, 841-866.

Manabe, S. and R. J. Stouffer, 1994: Multiple-century response of a coupled ocean-atmosphere model to an increase of atmospheric carbon dioxide. *J. Climate.*, 7, 5-23.

Manabe, S., R. J. Stouffer, M. J. Spelman, and K. Bryan, 1991: Transient respon ses of a coupled ocean-atmosphere model to gradual changes of atmospheric CO2. Part I: Annual mean response. *J. Climate.*, 4(8), 785-818.

Manabe, S. and R. J. Stouffer, 1996: Low-frequency variability of surface air-temperature in a 1000-year integration of a coupled atmosphere-ocean-land surface model. *J. Climate*, 9, 376-393.

Manak DK, Mysak LA (1989) On the relationship between Arctic sea ice anomalies and fluctuations in northern Canadian air temperature and river discharge. *Atmosphere-Ocean,* 27: 682-691

Mann, M. E. and J. Park, 1994: Global-scale modes of surface temperature variability on interannual to century timescales. *J. Geophys. Res.*, 99, 25819-25833.

Mann, M.E. and Park, J., 1996: Joint spatio-temporal modes of surface temperature and sea level pressure variability in the Northern Hemisphere during the last century. *J. Climate*, **9**, 2137-2162.

Mantua, N. J., S. R. Hare, Y. Zhang, J. M. Wallace, and R. C. Francis, 1997: A Pacific interdecadal climate oscillation with impacts on salmon production. Bull. Amer. Meteorol. Soc., 78, 1069-1079..

Marotzke J, Welander P, Willebrand J (1988) Instability and multiple steady states in a meridional-plane model of thermohaline circulation. *Tellus* 40A: 162-172

Marshall, J. and F. Molteni, 1993: Toward a dynamic understanding of plane-tary-scale flow regimes. *J.Atmos. Sci.*, 50, 1792-1818.

Martinson DG, Killworth RD, Gordon AL (1981) A convective model for the Weddell polynya. J Phys Oceangr 11: 466-488

May, W. and L. Bengtsson, 1996: On the impact of theEl Niño/Southern Oscilla-tion phenomenon on the atmospheric circulation inthe northern hemisphere extratropics. Max-Planck-Institute fürMeteorologie Report 224, 61 pp.

McCartney, M. S., R. G. Curry and H. F. Bezdek, 1996: North Atlantic's trans-formation pipeline chills and redistributes subtropical water. *Oceanus,* **39** (2), 19-23.

Mechoso, C. R., A. W. Robertson, N. Barth, M. K. Davey, P. Delecluse, P. R. Gent, S. Ineson, B. Kirtman, M. Latif, H. L. Treut, T. Nagai, J. D. Neelin, S. G. Philander, J. Polcher, P. S. Schopf, T. Stockdale, M. J. Suarez, L. Terray, O. Thual, and J. J. Tribbia, 1995: The seasonal cycle over the tropical Pacific in coupled ocean-atmosphere general circulation models. *Mon. Wea. Rev.*, 123, 2825-2838.

Meehl, G. A., 1987: The annual cycle and interannual variability in the tropical Pacific and Indian ocean regions. Monthly Weather Review, 115, 27-50.

Meehl, G. A., 1990: Seasonal cycle forcing of El-Niño-Southern Oscillation in a global , coupled ocean-atmosphere GCM. *J. Climate.*, 3, 72-98.

Meehl, G. A., 1993: A coupled air-sea biennial mechanism in the tropical Indian and Pacific regions: Role of the ocean. *J. Climate.*, 6, 31-41.

Meehl, G. A., 1996: Characteristics of decadal variability in a global coupled GCM. Proceedings of the JCESS/CLIVAR workshop on Decadal Climate Variability, April 22-24, 1996, Columbia MD, USA, Available from CLIVAR Office, c/o Max-Planck-Institut für Meteorologie, Bundesstr. 55, D-20146 Hamburg, Germany, in prep.

Meehl, G. A., M. Wheeler, and W. M. Washington, 1994: Low-frequency variability and CO2 transient climate change. part 3. Intermonthly and interannual variability. *Clim. Dyn.*, 10, 277-303.

Mehta, V. M., 1997: Variability of the tropical ocean surface temperatures at decadal-multidecadal time scales. Part I: The Atlantic Ocean. *J. Climate*, submitted.

Mehta, V. M. and T. Delworth, 1995: Decadal variability of the tropical Atlantic ocean surface temperature in shipboard measurements and in a global ocean-atmosphere model. *J. Climate*, 8, 172-190.

Mikolajewicz, U., 1997: A meltwater induced collapse of the 'conveyor belt' thermohaline circulation. In *Tracer Oceanography*, edited by P. Schlosser, W. M. Smethie and R. Toggweiler.

Mikolajewicz, U. and E. Maier-Reimer, 1990: Internal secular variability in an ocean general circulation model. *Clim. Dyn.,* 4, 145-156.

Mikolajewicz, U. and E. Maier-Reimer, 1994: Mixed boundary conditions in OGCMs and their influence on the stability of the model's conveyor belt. *J. Geophys. Res.,* 99, 22633-22644.

Miller, A. J., D. R. Cayan, T. P. Barnett, N. E. Graham and J. M. Oberhuber, 1994: Inter-decadal variability of the Pacific Ocean: Model response to observed heat flux and wind stress anomalies. Climate Dyn., 9, 287-302.

Miller, A.J., D.R. Cayan, T.P. Barnett, N.E. Graham and J.M. Oberhuber, 1994: The 1976-77 Climate Shift of the Pacific Ocean. Oceanography, 7, 21-26.

Miller, A, D. R. Cayan and W. B. White, 1998: A westward-intensified decadal change in the North Pacific thermocline and gyre-scale circulation. *J. Climate*,11, in press.

Minobe, S., 1997: A 50-70 year climatic oscillation over the North Pacific and North America. *Geophys. Res. Lett.*, 24, 683-686.

Mitchell, J. F. B., T. C. Johns, J. M. Gregory, and S. F. B. Tett, 1995: Climate response to increasing levels of greenhouse gases and sulphate aerosols. *Nature*, 376, 501-504.

Mizuno, K. and W. B. White, 1983: Annual and interannual variability in the Kuroshio current system. *J. Phys. Oceanogr.*, 13, 1847-1867.

Moore, A. M. and C. J. Reason, 1993: The response of a global ocean general circulation model to climatological surface boundary conditions for temperature and salinity. *J. Phys. Oceanogr.,* 23, 300-327.

Moron, V., S. Bigot and P. Roucou, 1995: Rainfall variability in sub-equatorial America and Africa and relationships with the main SST modes (1951-90). *Int. J. Climatol.*, 15, 1297-1322.

Moron, V., A. Navarra, M.N. Ward and E. Roeckner, 1998: Skill and reproducibility of seasonal rainfall patterns over tropical land areas in GCM simulations with prescribed SST. *Clim. Dyn.*, 14, 83-100.

Moura, A. D. and J. Shukla, 1981: On the dynamics of drought in northeast Brazil: observations, theory, and numerical experiments with a general circulation model. J. Atmos. Sci., 38, 2653-2675.

Molchanov, S.A., O.I. Piterbarg and D.D. Sokolov,1987: Generation of large-scale ocean temperature anomalies by short-periodatmospheric processes. *Izvestiya, Atmos. Ocean Phys.*, 23, 405-409.

Münnich, M., M. A. Cane, and S. E. Zebiak, 1991: A study of self-excited oscillations in a tropical ocean-atmosphere system. Part II: Nonlinear cases. J. Atmos. Sci., 48, 1238-1248.

Münnich, M., M. Latif, and E. Maier-Reimer, 1997: Decadal oscillations in a simple coupled model. *J. Phys. Oceanogr.*, submitted.

Murphy, J. M., and J. F. B. Mitchell, 1995: Transient response of the Hadley Centre coupled ocean-atmosphere model to increasing carbon dioxide. Part II: Spatial and temporal structure of response. J. Clim., 8, 57-80.

Mutai, C.C., M.N. Ward and A.W. Colman, 1998: Towards the Prediction of the East Africa Short Rains based on SST-Atmosphere coupling. *Int. J. Climatol.*, in press.

Mysak LA, Manak DK (1989) Arctic sea-ice extent and anomalies, 1953-84. Atmosphere-Ocean 27: 376-405

Mysak, L.A. D.K. Manak, and D.F. Marsden, 1990: Sea-ice anomalies observed in the Greenland and Labrador Seas during 1901-1984 and their relations to an interdecadal Arctic climate cycle. *Clim. Dyn. ,* 5,111-133.

Mysak LA, Power SB (1991)  Greenland Sea ice and salinity anomalies and interdecadal climate variability. *Climatol Bull* 25: 81-91

Mysak LA, Power SB (1992)  Sea-ice anomalies in the western Arctic and Greenland-Iceland Sea and their relation to an interdecadal climate cycle. Climatol Bull 26: 147-176

Mysak LA, Manak DK, Marsden RF (1990) Sea-ice anomalies observed in the Greenland and Labrador Seas during 1901-1984 and their relation to an interdecadal Arctic climate cycle. *Clim. Dyn.* 5: 111-133

Mysak LA, Ingram RG, Wang J, van der Baaren A (1996) The anomalous sea-ice extents in Hudson Bay, Baffin Bay and the Labrador Sea during three simultaneous NAO and ENSO episodes. Atmosphere-Ocean 34: 313-343

Nagai, T., T. Tokioka, M. Endih, and Y. Kitamura, 1992: El Niño-Southern Oscillation simulated in an MRI atmosphere-ocean coupled general circulation model. *J. Climate.*, 5, 1202-1233.

Nakamura, M., P. H. Stone and J. Marotzke, 1994: Destabilization of the thermohaline circulation by atmospheric eddy transports. *J. Climate,* **7**, 1870-1882.

Nakamura, H., 1996: Year-to-year and interdecadal variability in the activity of intraseasonal fluctuations in the Northern Hemisphere wintertime circulation. *Theor. Appl. Climatol.*, 55, 19-32.

Nakamura, H, G. Lin and T. Yamagata, 1997: Decadal climate variability in the North Pacific during the recent decades, *Bull. Amer. Meteor. Soc.*, 78, 2215-2225.

Namias, J., and R. M. Born, 1970: Temporal coherence in the North Pacific sea-surface temperature patterns. *J. Geophys. Res.*, 75, 5952-5955.

Namias, J.,1959: Recent seasonal interactions between the North Pacific waters and the overlying atmospheric circulation, *J. Geophys Res.*, 94,631-646.

Namias, J., 1963: Large scale air-sea interactions over the North Pacific from summer 1962 through the subsequent winter, *J. Geophys Res.*, 68, 6171-6186.

Navarra, A., N. Ward e N. Rayner, A stochastic model of the SST for climate simulation experiments, *Clim. Dyn.*, in press, 1998.

Navarra, A., and M. N. Ward, The Forced Manifold, IMGA Technical Report 1/1998.

Neelin, J. D. and W. Weng, 1997: Analytical prototypes for ocean-atmosphere interaction at midlatitudes. Part I: deterministic coupled feedbacks in presence of stochastic forcing. *J. Climate*, to be submitted.

Neelin, J. D., M. Latif, and F.-F. Jin, 1994: Dynamics of coupled ocean-atmosphere models: The tropical problem. Ann. Rev. Fluid Mech., 26, 617-659.

Neelin, J. D., M. Latif, M. A. F. Allart, M. A. Cane, U. Cubasch, W. L. Gates, P. R. Gent, C. Gordon, C. R. Mechoso, G. A. Meehl, J. M. Oberhuber, S. G. H. Philkander, P. S. Schopf, K. R. Sperber, A. Sterl, T. Tokioka, J. Tribbia, and S. E. Zebiak, 1992a: Tropical air-sea interaction in general circulation models. *Clim. Dyn.*, 7, 73-104.

Newell, N.E., Newell, R.E., Hsiung, J. and Wu, Z-X., 1989: Global marine temperature variation and the solar magnetic cycle. *Geophys. Res. Lett.*, **16**, 311-314.

Ng, C.N., and Young, P.C., 1990: Recursive estimation and forecasting of non-stationary time series. *J. Forecasting*, 9, 173-204.

Nicholson, S.E., 1980: The *nature* of rainfall fluctuations in subtropical West Africa. *Mon. Wea. Rev.*, 108, 473-487.

Nicholson, S.E., 1988: Land-surface atmosphere interaction: physical processes and surface changes and their impact. Progr. Phys. Geogr., 12, 36-65.

Nicholson, S.E., 1996: Rainfall in the Sahel during 1994. *J. Climate*, 9, 1673-1676.

Nicholson, S.E. and I.M. Palao, 1993: A re-evaluation of rainfall variability in the Sahel. Part I. Characteristics of rainfall fluctuations. *Int. J. Climatol.*, 4, 371-389.

Nicholson, S.E. and J. Kim, 1997: The relationship of the El Niño-Southern Oscillation to African Rainfall. *Int. J. Climatol.*, 17, 117-135.

354

Nitta, T., and Kachi, M., 1994: Interdecadal variations of precipitation over the tropical Pacific and Indian Oceans. *J. Met. Soc. Japan*, **72**, 823-831.

Nitta, T., and S. Yamada, 1989: Recent warming of tropical sea surface temperature and its relationship to the northern hemisphere circulation. *J. Meteor. Soc. Japan.*, 67, 375-383.

Oberhuber, J. M., 1988: An atlas based on the COADS data set: the budgets of heat, buoyancy and turbulent kinetic energy at the surface of the global ocean, MPIM Rep. 15,Hamburg.

Oberhuber, J. M., 1993: Simulation of the Atlantic circulation with a coupled sea ice-mixed layer-isopycnal general circulation model. Part I: Model description, *J. Phys. Oceanogr.* 23 808-829.

Oberhuber, J. M., 1993: Simulation of the Atlantic circulation with a coupled sea ice-mixed layer-isopycnal general circulation model. Part II: Model experiment,*J. Phys. Oceanogr.*, 23, 830-844.

Opsteegh, J.D., R.J. Haarsma and F.M. Selten,1997: An atmospheric climate model of intermediate complexity: a suitable alternative to mixed boundary conditions in ocean models. Will be submitted to *Clim. Dyn.*

Osborn, T. J., 1997: Thermohaline oscillations in the LSG OGCM: propagating anomalies and sensitivity to parameterizations. *J. Phys. Oceanogr.*, **27**, 2233-2255.

Ostrovskii, A., and L. Piterbarg, 1995: Inversion of heat anomaly transport from sea surface temperature time series in the northwest Pacific. *J. Geophys. Res.*, 100, 4845-4865.

Pacanowski, R. C., and S. G. Philander, 1981: Parameterization of vertical mixing in numerical models of tropical oceans. *J. Phys. Oceanogr.*, 11, 1443-1451.

Palmer, T.N., 1993: A nonlinear dynamical perspective on climate change. *Weather*, 48, 313-348

Palmer, T. N., 1994: Chaos and predictability in forecasting the monsoon,*Proc. Indian nat. Sci. Acad.*, 60, No. 1, 57-66.

Palmer, T.M. and Z. Sun, 1985: A modelling and observational study of therelationship between sea surface temperature in the north-west Atlantic andthe atmospheric general circulation. *Quat. J. Roy. Meteor. Soc.*, 111,947-975.

Palmer, T.N., 1986: Influence of the Atlantic, Pacific and Indian Oceans on Sahel rainfall. *Nature*, 322, 251-253.

Palmer, T.N., 1993: Extended-range atmospheric prediction and the Lorenz model. *Bull. Amer. Met. Soc.*, 74, 49-65.

Palmer, T.N., Brankovic, C., Viterbo, P., and Miller, M.J., 1992: Modeling interannual variations of summer monsoons. *J. Climate*, 5, 399-417.

Pant, G. B. and K. Rupa Kumar, 1997: Climates of South Asia. Wiley, 320pp..

Pan, Y. H., and A. H. Oort, 1990: Correlation analysis between sea surface temperature anomalies in the eastern equatorial Pacific and the world ocean. *Clim. Dyn.*, 4, 191-205.

Parker, D.E., 1983: Documentation of a Southern Oscillation index. *Meteorological Magazine*, 112, 184-187.

Parker, D.E., Jones, P.D., Folland, C.K. and Bevan, A., 1994: Interdecadal changes of surface temperature since the late nineteenth century, *J. Geophys. Res.* **99**, 14373- 14399.

Parker, D.E., Jackson, M. and Horton, E.B., 1995: The GISST2.2 sea surface temperature and sea-ice climatology. *Climate Research Technical Note 63*, Hadley Centre, Meteorological Office, Bracknell, U.K., 35pp.

Parker, D.E., Folland, C.K. and Jackson, M., 1995a: Marine surface temperature: observed variations and data requirements. *Climatic Change*, **31**, 559-600.

Parker, D.E. and Folland, C.K., 1991: Worldwide surface temperature trends since the mid-19th century. In *Greenhouse-Gas-Induced Climatic Change: A Critical Appraisal of Simulations and Observation*, M.E.Schlesinger (ed.), Elsevier, pp 173-193.

Parker, D.E., Jackson, M. and Horton, E.B., 1995b: The GISST2.2 sea surface temperature and sea-ice climatology. *Climate Research Technical Note 63*, Hadley Centre, Meteorological Office, Bracknell, U.K., 35pp.

Parker, D.E., Folland, C.K., Bevan, A., Ward, M.N., Jackson, M. and Maskell, K. 1995c: Marine surface data for analysis of climatic fluctuations on interannual to century time scales. In: *"Natural Climate Variability on Decade-to-Century Time Scales"* Eds: D.G. Martinson, K. Bryan, M.Ghil, M.M. Hall, T.R. Karl, E.S. Sarachik, S.Sorooshian and L.D.Talley. National Acad. Press, Washington, DC. pp241-250. Colour figs pp222-228.

Parthasarathy, B., A. A. Munot and D. R. Kothawale, 1994 : All-India monthly and seasonal rainfall series: 1871-1993. *Theoretical and Applied Climatology*, 49, 217-224.

Parthasarathy, B., A. A. Munot and D. R. Kothawale, 1995 : Monthly and seasonal rainfall series for All-India homogeneous regions and meteorological subdivisions: 1871-1994. Research Report No. RR-065, Indian Institute of Tropical Meteorology, Pune, 113pp.

Peng S, Mysak LA (1993) A teleconnection study of interannual sea surface temperature fluctuations in the northern North Atlantic and precipitation and runoff over western Siberia. *J. Climate* 6: 876-885

Peng, S., L.A. Mysak, H. Richtie, J. Derome, and B. Dugas, 1995: The differences between early and midwinter atmospheric responses to seasurface temperature anomalies in the Northwest Atlantic. *J. Climate*, 8, 137-157.

Peng, S., W. A. Robinson, M. P. Hoerling, 1997: The modeled atmospheric response to mid-latitude SST anomalies and its dependence on background circulation states. *J. Climate*, 10, 981-987.

Philander, S. G. H. and A.D. Seigel, 1985: Simulations of El Niño of 1982-83. In "Coupled Ocean-Atmosphere Models", J. Nihoul ed., pp. 517-541, Elsevier, Amsterdam.

Philander, S. G. H., N. C. Lau, R. C. Pacanowski, and M. J. Nath, 1989: Two different simulations of the Southern Oscillation and El-Niño with coupled ocean-atmosphere general circulation models. *Phil. Trans. R. Soc. Lond.* A, 329, 1 67- 178.

Philander, S. G. H., R. C. Pacanowski, N.-C. Lau, and M. J. Nath, 1992: Simulation of ENSO with a global atmosphere GCM coupled to a high-resolution tropical Pacific ocean GCM. *J. Climate.*, 5, 308-329.

Philander S.G. H. and R. C: Pacanowski, 1986: A model of the seasonal cycle of the tropical Atlantic Ocean, *J. Geophys. Res.*, 91 14192-14206.

Philander, S. G. H., 1990: El Niño, La Niña, and the Southern Oscillation. Academic Press, Inc., San Diego, 293 pp.

Pierce, D. W., T. B. Barnett and U. Mikolajewicz, 1995: On the competing roles of heat and fresh water flux in forcing thermohaline oscillations. *J. Phys. Oceanogr.*, 25, 2046-2064.

Pierce, D. W., K.-Y. Kim and T. P. Barnett, 1996: Variability of the thermohaline circulation in an ocean general circulation model coupled to an atmospheric energy balance model. *J. Phys. Oceanogr.*, 26, 725-738.

Pollard RT, Pu S (1985) Structure and circulation of the upper Atlantic Ocean northeast of the Azores. *Prog. Oceanogr.* 14: 443-462

Power, S. B., R. Kleeman, R. A. Colman and B. J. McAvaney, 1995: Modelling the surface heat flux response to long-lived SST anomalies in the North Atlantic. *J. Clim.*, 8, 2161-2180.

Qui, B., W. Miao and P. Müller, 1997: Propagation and decay of forced and free baroclonic Rossby waves in off-equatorialo ceans. *J. Phys. Oceanogr.*, to appear.

Rahmstorf, S., 1995a: Bifurcations of the Atlantic thermohaline circulation in response to changes in the hydrological cycle. *Nature*, 378, 145-149.

Rahmstorf, S., 1995b: Multiple convection patterns and thermohaline flow in an idealised OGCM. *J. Clim.*, 8, 3028-3039.

Rahmstorf, S., 1996a: Comment on "Instability of the thermohaline circulation with respect to mixed boundary conditions". *J. Phys. Oceanogr.*, 26, 1099-1105.

Rahmstorf, S., 1996b: On the freshwater forcing and transport of the Atlantic thermohaline circulation. *Clim. Dyn.*, 12, 799-811.

Rahmstorf, S., 1997: Risk of sea-change in the Atlantic. *Nature*, 388, 825-826.

Rahmstorf, S., J. Marotzke and J. Willebrand, 1996: Stability of the thermohaline circulation. In *The warm water sphere of the North Atlantic ocean*, edited by W. Krauss, Borntraeger, Stuttgart, pp. 129-158.

Rahmstorf, S. and J. Willebrand, 1995: The role of temperature feedback in stabilizing the thermohaline circulation. *J. Phys. Oceanogr.*, 25, 787-805.

Roemmich, D. and C. Wunsch, 1984: Apparent changes in the climatic state of the deep North Atlantic Ocean. *Nature*, 307, 447-450.

Rasmusson, E.M. and T. H. Carpenter, 1982: Variations in tropical sea surface temperature and surface wind fields associated with the Southern Oscillation/ El Niño. *Mon. Weath. Rev.*, 110, 354-384.

Rasmusson, E. M. and T. H. Carpenter, 1983 : The relationship between eastern equatorial Pacific sea surface temperatures and rainfall over India and Sri Lanka. *Mon. Weath. Rev.*, 111, 517-528.

Rasmusson, E. M. and J. M. Wallace, 1983: Meteorological aspects of the El Nio/Southern Oscillation. *Science*, 222, 1195-1202.

Rasmusson, E. M., X. Wang, and C. F. Ropelewski, 1990: The biennial component of ENSO variability. Journal of Marine Systems, 1, 71-96.

Rayner, N.A., Horton, E.B., Parker, D.E., Folland, C.K. and Hackett, R.B., 1996: Version 2.2 of the Global sea-Ice and Sea Surface Temperature data set, 1903-1994. *Climate Research Technical Note CRTN74*, Hadley Centre, Meteorological Office, Bracknell, U.K., 35pp.

Ratcliffe, R.A.S. and S. Murray, 1970: New lag association between North Atlantic sea surface temperature and European pressure applied to long-range weather forecasting. *Quart. J. R. Met.Soc.*, 96, 226-246.

Reverdin, G., D. Cayan and Y. Kushnir, 1997: Decadal variability of hydrography in the upper northern North Atlantic in 1948-1990. *J. Geophys. Res.*, 102, 8505-8531.

Reynolds, R.W., 1979: A stochastic forcing model of sea surface temperature anomalies in the North Pacific and North Atlantic. *Climate Research Institute Report No. 8,* Oregon State University, 23pp.

Reynolds, R. W., 1988: A real-time global sea surface temperature analysis. J. of Climate, 1, 75-86.

Reynolds, R.W. and T.M. Smith, 1994: Improved global sea surface temperature analyses using optimum interpolation. *J. Climate*, 9,2958-2972.

Richardson P.L. and G. Reverdin, Seasonal cycle of velocity in the Atlantic North Equatorial countercurrent as measured by surface drifters, currentmeters and shipdrifts , *J. Geophys. Res.*, 92 3691-3708, 1987.

Richman, M. B., 1986: Rotation of principal components. *J. of Climatology*, 6, 293-335.

Robertson, A. W., 1996: Interdecadal variability over the North Pacific in a multi-century climate simulation. *Clim. Dyn.*, 12, 227-241.

Robertson, A. W., C.-C. Ma, M. Ghil, and C. R. Mechoso, 1995: Simulation of the tropical Pacific climate with a coupled ocean-atmosphere general circulation model. Part II: Interannual variabilty. *J. Climate.*, 8, 1199-1216.

Robitaille DY, Mysak LA, Darby MS (1995) A box model study of the Greenland Sea, Norwegian Sea, and Arctic Ocean. Climate Dyn 11: 51-70

Robock, A. D. and J. Mao, 1995: The volcanic signal in surface temperature observations. *J. Climate*, 8, 1086-1103.

Rodwell, M. J., 1997: Breaks in the Asian Monsoon: The influence of the southern hemisphere weather systems, *J. Atmos. Sci.*, 54, 2597-2611.

Rodwell, M. J. and B. J. Hoskins, 1996: Monsoons and the dynamics of deserts. *Q. J. R. Meteorol. Soc.*, 122, 1385-1404.

Roeckner E ,Arpe K, 1995: AMIP experiments with the new Max Planck Institute Model ECHAM4. Proceedings of the first international AMIP scientific conference, Monterrey, WRCP-92, WMO/TD No 732, 307-312.

Roeckner E, Arpe K, Bengtsson L, Christoph M, Claussen M, Dumenil L, Esch M, Giorgetta M, Schlese U, Schulzweida U, 1996: The atmospheric general cicrulation model ECHAM-4 : model description and simulation of present-day climate. Report number 218. MPI, Hamburg.

Roeckner E., J. M. Oberhuber, A. Bacher, M. Cristoph and I. Kirchner, ENSO variability and atmospheric response in a global coupled atmosphere-ocean GCM, *Clim. Dyn.*, 12 735-754, 1996.

Roemmich, D. H. and C. Wunsch, 1985: Two transatlantic sections: Meridional circulation and heat flux in the subtropical North Atlantic Ocean. *Deep-Sea Res.*, **32**, 619-664.

Ropelewski, C.F. and Jones, P.D., 1987: An extension of the Tahiti-Darwin Southern Oscillation Index. *Mon. Weath. Rev.*, **115**, 2161-2165.

Rowell, D. P. and F. W. Zwiers, 1997: Sources of atmospheric decadal variability. Proceedings of the 76th AMS Annual Meeting, Atlanta, Georgia, 1997, pp. 38-39.

Rowell, D.P., 1996: Further analysis of simulated interdecadal and interannual variability of summer rainfall over tropical north Africa. Reply to Y.C. Sud and W.K.-M. Lau. *Quart. J. Roy. Meteor. Soc.*, 122, 1007-1013.

Rowell, D.P., Folland, C.K., Maskell, K., and Ward, M.N., 1995: Variability of summer rainfall over tropical North Africa (1906-92): Observations and modelling. *Quart. J. Roy. Meteor. Soc.*, 121, 669-704.

Rowell, D.P., 1997: Using an ensemble of Multi-decadal GCM simulations to assess potential seasonal predictability. *J Climate*, 11, 109-120.

Rowntree, P.R., 1972: The influence of the tropical east Pacific ocean temperature on the atmosphere, *Q. J. R. Meteorol. Soc.*, 98, 290-321.

Rowntree, P.R., 1976a: Tropical forcing of atmospheric motions in a numerical model, *Q. J. R. Meteorol. Soc.*, 102, 583-605.

Rowntree, P.R., 1976b: Response of the atmosphere to a tropical Atlantic ocean temperature anomaly, *Q. J. R. Meteorol. Soc.*, 102, 607-625.

Royer, J-F, 1991: Review of advances in dynamical extended range forecasting in the extratropics, Proc. of the NATO Workshop on Prediction of Interannual Climate Variations, Trieste, Italy, Institute ofr Global Environment and Society, 49-70.

Rupa Kumar, K., G, B, Pant, B. Parthasarathy and N. A. Sontakke, 1992: Spatial and subseasonal patterns of the long term trends of Indian summer monsoon rainfall. *Int. J. Climatol.*, 12, 257-268.

Sadourny, R., and K. Laval, 1984 : January and July performance of the LMD general circulation model. New Perspectives in Climate Modelling, A. Berger and C. Nicolis, Eds., Elsevier, 173-198.

Saltzman B (1978) A survey of statistical-dynamical models of the terrestial climate. *Advances in Geophysics* 20: 183-304

Sandström, J. W., 1908: Dynamische Versuche mit Meerwasser. *Ann. Hydrogr. Marit. Meteorol.,* **36**, 6-23.

Santer, B. D., K. E. Taylor, T. M. Wigley, T. C. Johns, P. D. Jones, D. J. Karoly, J. F. B. Mitchell, A. H. Oort, J. E. Penner, V. Ramaswamy, M. D. Schwarzkopf, R. J. Stouffer, and S. Tett, 1996: A search for human influences on the thermal structure of the atmosphere. *Nature*, 382, 39-45.

Sarachik, E., M. Winton and F. L. Yin, 1996: Mechanisms for decadal-to-centennial climate variability. In *Decadal climate variability: Dynamics and predictability*, edited by D. L. T. Anderson and J. Willebrand, Springer, Berlin.

Saravanan, R. and J. C. McWilliams, 1995: Multiple equilibria, natural variability, and climate transitions in an idealized ocean-atmosphere model. *J. Climate.*, **8**, 2296-2323.

Saravanan, R. and J. C. McWilliams, 1997: Stochastically and spatial resonance in interdecadal climate fluctuations. *J. Climate*, in press.

Sarmiento, J. L. and C. Le Quéré, 1996: Oceanic carbon dioxide uptake in a model of century-scale global warming. *Science,* **274**, 1346-1350.

Sarnthein, M., E. Jansen, M. Weinelt, M. Arnold, J. C. Duplessy, H. Erlenkeuser, A. Flatoy, G. Johanessen, T. Johanessen, S. Jung, N. Koc, L. Labeyrie, M. Maslin, U. Pflaumann and H. Schulz, 1995: Variations in Atlantic surface paleoceanography, 50-80N: A time slice record of the last 30,000 years. *Paleoceanography,* **10**, 1063-1094.

Sarnthein, M., K. Winn, S. J. A. Jung, J. C. Duplessy, L. Labeyrie, H. Erlenkeuser and G. Ganssen, 1994: Changes in east Atlantic deepwater circulation over the last 30,000 years: Eight time slice reconstructions. *Paleoceanography,* **9**, 209-267.

Schiller, A., U. Mikolajewicz and R. Voss, 1997: The stability of the thermohaline circulation in a coupled ocean-atmosphere general circulation model. *Clim. Dyn.,* **13**, 325-347.

Schlosser, P., G. Boenisch, M. Rhein and R. Bayer, 1991: Reduction of deepwater formation in the Greenland Sea during the 1980s: evidence from tracer data. *Science,* **251**, 1054-1056.

Schmitz, W. J., 1995: On the interbasin scale thermohaline circulation. *Rev. Geophys.,* **33**, 151-173.

Schlesinger M.E. and Ramankutty,N., 1994: A oscillation in the global climate system of period 65-70 years. *Nature,* 367, 723-726.

Schopf, P. S., and M. J. Suarez, 1988: Vacillations in a coupled ocean-atmosphere model. *J. Atmos. Sci.*, 45, 549-566.

Schott, F.A. and C.W. Boening, 1991: The WOCE model in the Western Equatorial Atlantic : Upper layer circulation,*J. Geophys. Res.*, 96, 6993-7004.

Sekine, Y., 1988: Anomalous southward intrusion of the Oyashio east of Japan. I: Influence of the seasonal and interannual variations in the wind stress over the North Pacific. *J. Geophys. Res.*, 93, 2247-2255.

Semazzi, F.H.M., V. Mehta and Y.C. Sud, 1988: An investigation of the relationship between sub-Saharan rainfall and global sea surface temperatures. *Atmos.-Ocean*, 26, 118-138.

Semtner AJ (1976) A model for the thermodynamic growth of sea ice in numerical investigations of the climate. *J. Phys. Oceangr.* 6: 379-389

Send, U., J. Font and C. Mertens, 1996: Recent observation indicates convection's role in deep water circulation. *Eos,* **77** (7), 61-65.

Send, U. and J. Marshall, 1995: Integral effects of deep convection. *J. Phys. Oceanogr.,* **25**, 855-872.

Servain, J. and D. M. Legler, 1986: Empirical orthogonal function analyses of Tropical Atlantic sea surface temperature and wind stress:1964-1979 applications, *J. Geophys. Res.,* 91 14181-14191.

Servain, J., 1993: Simple climatic indices for the Tropical Atlantic Ocean and some applications, *J. Geophys. Res.,* 96 15137-15146.

Slingo, A., 1989: A GCM parametrization for the shortwave radiative properties for clouds. *J. Atmos. Sci,* 46, 1419-1427.

Soman, M. K. and K. Krishna Kumar, 1993: Space-time evolution of meteorological features associated with the onset of the Indian summer monsoon. *Mon. Weath. Rev.,* 121, 1177-1194.

Soman, M. K. and J. M. Slingo, 1997: Sensitivity of the Asian Summer Monsoon to aspects of the sea surface temperature anomalies in the tropical Pacific Ocean. *Q. J. R. Meteorol. Soc.,* 123, 309-336.

Shabbar, A., K. Higuchi and J.L. Knox, 1990: Regional analysis of Northern hemisphere 50 kPa geopotential heights from 1946 to 1985. *J. Climate,* 3, 543-557.

Slonosky VC, Mysak LA, Derome J (1997) Linking Arctic sea ice and atmospheric circulation anomalies on interannual and decadal timescales. Atmosphere-Ocean 35: in pressSpall, M. A., 1996: Dynamics of the Gulf Stream/ Deep western boundary current crossover. Part II: Low-frequency internal oscillations. *J. Phys. Oceanogr.,* in press.

Smith, I.N., 1994: A GCM simulation of global climatic trends 1950-1988, *J Climate* 7, 709-718.

Smith, I.N., 1995: A GCM simulation of global climate interannual variability 1950-1988. *J Climate* 8 : 732-744.

Smith, T. M., and M. Chelliah, 1995: The annual cycle in the tropical Pacific ocean based on assimilated ocean data from 1983 to 1992. *J. Climate.,* 8, 1600-1614.

Smith, R. N. B., 1990: A scheme for predicting layer clouds and their water content in a general circulation model. Quart. J. R. Met. Soc., 116, 435-460.

Sperber K.R., Palmer T.N., 1996: Interannual tropical rainfall variability in General Circulation model simulations associated with the Atmospheric Model Intercomparison Project. *J Climate.,* 9, 2727-2750.

Sperber, K. R., and S. Hameed, 1991: Southern Oscillation simulation in the OSU coupled upper ocean-atmosphere GCM. *Clim. Dyn.,* 6, 83-97.

Stern, W., and K. Miyakoda, 1995: Feasibility of Seasonal Forecasts Inferred from Multiple GCM Simulations. *J. Climate*, 8, 1071-1085.

Stocker T.F., Mysak L.A. (1992) Climatic fluctuations on the century time scale: a review of high-resolution proxy data and possible mechanisms. *Climatic Change* 20: 227-250

Stocker, T., 1996: The ocean in the climate system: observing and modeling its variability. In *Physics and chemistry of the atmospheres of the Earth and other objects of the solar system*, edited by C. Boutron, Les Editions de Physique, Les Ulis, pp. 39-90.

Stommel, H., 1961: Thermohaline convection with two stable regimes of flow. *Tellus,* **13**, 224-230.

Storch, J.-S. von, 1994: Interdecadal variability in a global coupled model. *Tellus*, 46A, 419-432.

Storch, J.-S. von, V. Kharin, U. Cubasch, G. C. Hegerl, D. Schriever, H. von Storch and E. Zorita, 1997: A description of a 1260-year control integration with the coupled ECHAM1/LSG general circulation model. *J. Climate.,* **10**, 1526-1544.

Storch, H. von, G. Burger, R. Scnurr and J. S. Xu, 1993: Principal Oscillation Pattern Analysis, MPIM Reports , 113, Hamburg

Sturges, W., and B.G. Hong, 1995: Wind forcing of the Atlantic thermohaline along $32^0$ N at low frequencies. *J. Phys. Oceangr.* , 25, 1076-1715.

Suarez, M. J., and P. S. Schopf, 1988: A delayed action oscillator for ENSO. *J. Atmos. Sci.*, 45, 3283-3287.

Sud, Y.C. and Lau, W.K.-M., 1996: Comments on paper 'Variability of summer rainfall over tropical north Africa (1906-92): Observations and modelling'. *Quart. J. Roy. Meteor. Soc.*, 122, 1001-1006.

Sutton, R.T. and M.R. Allen, 1997: Decadal predictability in Gulf Streamsea surface temperatures. *Nature*, 388, 563-567.

Sy, A., M. Rhein, J. R. N. Lazier, K. P. Koltermann, J. Meincke, A. Putzka and M. Bersch, 1997: Surprisingly rapid spreading of newly-formed intermediate waters across the North Atlantic Ocean. *Nature,* **386**, 675-679.

Takaya, K. and H. Nakamura, 1997: A formulation of a wave-activity flux for stationary Rossby waves on a zonally varying basic flow. *Geophys. Res. Lett.*, 24, 2985-2988.

Tang, B. and A. J. Weaver, 1995: Climate stability as deduced from an idealised coupled atmosphere-ocean model. *Clim. Dyn.,* **11**, 141-150.

Tanimoto, Y., N. Iwasaka, K. Hanawa and Y. Toba, 1993: Characteristic variations of sea-surface temperature with multiple time scales on the North Pacific. *J. Climate*, 6, 1153-1160.

Tanimoto, Y., N. Iwasaka and K. Hanawa, 1997: Relationships between sea surface temperature, the atmospheric circulation and air-sea fluxes on multiple time scales. *J. Meteor. Soc. Japan,* 75, 831-849.

Taylor, A. H., 1996: North-South Shifts in the Gulf Stream: Ocean-atmosphere Interactions in the North Atlantic. *Int. J. Climatology,* **16**, 559-584.

Taylor, A.H. and J.A. Stephens, 1980: Seasonal andyear-to-year variations in surface salinity at the nine North-AtlanticOcean weather stations. *Oceano-logica Acta*, 3, 421-430.

Terray, L., O. Thual, S. Belamari, M. Deque, P. Dandin, C. Levy, and P. Deleclu se, 1995: Climatology and interannual variability simulated by the ARPEGE-OPA model. *Clim. Dyn.*, 11, 487-505.

Tett, S. F. B., J. F. B. Mitchell, D. E. Parker, and M. R. Allen, 1996: Human influence on the atmospheric vertical temperature structure: Detection and observations. *Science*, 247, 1170-1173.

Tett, S. F. B., T. C. Johns, and J. F. B. Mitchell, 1997: Global and regional vari-ability in a coupled AOGCM. *Clim. Dyn.*, 13, 303-323.

Tett, S., 1995: Simulation of El-Niño/Southern Oscillation like variability in a global AOGCM and its response to CO2 increase. *J. Climate.*, 8(6), 1473-1502.

Thompson, D. W. J. and J. M. Wallace, 1998: Observed linkages between Eur-asian surface air temperature, the North Atlantic Oscillation, Arctic sea-level pressure and the polar vortex. *Geophys. Res. Lett.*, 25, 1297-1300.

Timmermann, A., M. Latif, R. Voss, and A. Grotzner, 1997: North Atlantic interdecadal variability: A coupled air-sea mode. *J. Climate*, submitted.

TOGA, 1998: Special review volume. *J. Geophys. Res.*, Oceans, 103(C7).

Tokioka, T., A. Noda, A. Kitoh, Y. Nikaidou, S. Nakagawa, T. Motoi, S. Yukim-oto, and K. Takata, 1995: A Transient CO2 experiment with the MRI CGC,. *J. Meteor. Soc. Japan.*, 73, 817-826.

Tomita, T. and T. Yasunari, 1996: Role of the North-East Winter Monsoon on the biennial oscillation of the ENSO/monsoon system. *J. Meteorol. Soc. Japan*, 74, 399-413.

Tourre, Y. M., Y. Kushnir and W. B. White, 1998: Evolution of interdecadal variability in SLP, SST and upper ocean temperature over the Pacific Ocean. *J. Phys. Oceanogr.*, submitted.

Tremblay L-B, Mysak LA (1997) The origin and evolution of sea ice anomalies in the Beaufort Sea. Climate Dyn : submitted

Trenberth, K.E., 1984: Signal versus noise in the Southern Oscillation. *Mon. Wea. Rev.*, 112, 326-332.

Trenberth, K. E., 1990: Recent observed interdecadal climate changes in the Northern Hemi-sphere. *Bull. Amer. Meteor. Soc.*, 71, 988-993.

Trenberth, K. E. and J. W. Hurrell, 1994: Decadal atmosphere-ocean variations in the Pacific. *Clim. Dyn.*, 9, 303-319.

Trenberth, K. E. and T. J. Hoar, 1996: The 1990-1995 El Niño-Southern Oscilla-tion event: Longest on record. *Geophys. Res. Lett.*, 23, 57-60.

Tyson, P.D., 1991: Climatic change in southern Africa: past and present condi-tions and possible future scenarios. *Climatic Change*, **18**, 241-258.

Tyson, P.D., Dyer, T.G.J. and Mametse, M.N., 1975: Secular changes in South African rainfall: 1880 to 1972, *Quart. J. Roy. Meteorol. Soc.*, **101**, 817-833.

Tyson, P.D. and Dyer, T.G.J.,1980: The likelihood of droughts in the eighties in South Africa, *S. African J. of Science*, **76**, 340-341.

Tyson, P.D. and Lindesay, J.A., 1992: The climate of the last 2000 years in southern Africa. *The Holocene*, **2**, 271-278.

van Loon, H. and J.C. Rogers, 1978: The seesaw in winter temperatures between Greenland and Northern Europe. Part I: General description. *Mon. Wea. Rev.*, 106, 296-310.

Venzke, S. and M. Latif, 1997: Predictability of decadal climate variations in the North Pacific. *J. Climate*, to be submitted.

Venzke, S., M.R. Allen, R. T. Sutton and D. P. Rowell, 1998:The atmospheric response over the North Atlantic to decadal changes in sea surface temperature. Report number 255. MPI, Hamburg.

Venegas, S.A., Mysak, L.A. and Straub, D.N., 1996: Evidence for interannual and interdecadal climate variability in the South Atlantic. *Geophys. Res. Lett.*, **23**, 2673- 2676.

Venegas, S.A., Mysak, L.A. and Straub, D.N., 1996: Evidence for interannual and interdecadal climate variability in the South Atlantic. *Geophys. Res. Lett.*, **23**, 2673- 2676.

Verbeek, J., 1997: Windstress and SST variability in the North Atlantic area; observations and five coupled GCM's in concert. *Mon. Wea. Rev.*, in press.

Wagner, T. J. and R. H. Weisberger, 1996:Mechanisms controlling variability of the interhemispheric sea surface temperature gradient in the tropical Atlantic, *J. Climate*, 9 2010-2019.

Walker, G.T., and E.W. Bliss, 1932: World Weather V, *Mem. Roy. Meteor. Soc.*, 4, 53-84.

Walker, G. T. , 1923 : Correlation in seasonal variations of weather. III : A preliminary study of world weather. Mem. Indian Meteorol. Dept., 24, 75-131

Walker, G. T., 1924 : Correlation in seasonal variations of weather. IV : A further study of world weather. Mem. Indian Meteorol. Dept., 24, 275-332.

Wallace, J.M. and D.S. Gutzler, 1981: Teleconnections in the geopotential height field during the northern hemisphere winter. *Mon. Wea. Rev.*, 109, 784-812.

Wallace, J.M., C. Smith and C.S. Bretherton, 1992: Singular value decomposition of wintertime sea surface temperature and 500-mb height anomalies. *J. Climate*, 5, 561-576.

Wallace, J.M. , C. Smith and Q. Jiang, 1990: Spatial patterns of atmosphere-ocean interaction in the northern hemisphere. *J. Climate*,3, 990-998.

Wallace, J.M, Y. Zhang and K.-H. Lau, 1993: Structure and seasonality of interannual and interdecadal variability of geopotential height and temperature fields in the Northern Hemisphere troposphere. *J. Climate*, 6, 2063-2082.

Walsh JE, Chapman WL (1990a) Arctic contribution to upper-ocean variability in the North Atlantic. J Climate 3: 1462-1473

Walsh JE, Chapman WL (1990b) Short-term climate variability of the Arctic. *J Climate* 3: 237-250

Walsh JE, Zhou X, Portis D, Serreze MC (1994) Atmospheric contributions to hydrologic variations in the Arctic. *Atmosphere-Ocean* 32: 733-755

Wang J., Mysak L.A., Ingram R.G. ,1994: Interannual variability of sea-ice cover in Hudson Bay, Baffin Bay and the Labrador Sea. *Atmosphere-Ocean* 32: 421-447

Wang, X. L., and C. F. Ropelewski, 1995: An assessment of ENSO-scale secular variability. *J. Climate.*, 8, 1584-1599.

Wang, B., 1995: Interdecadal changes in El-Niño onset in the last four decades. *J. Climate.*, 8, 267-285.

Ward, M.N., Maskell, K., Folland, C.K., Rowell, D.P. and Washington, R., 1994: A tropic-wide oscillation of boreal summer rainfall and patterns of sea-surface temperature. *Climate Research Technical Note CRTN48*. Hadley Centre, Meteorological Office, Bracknell, U.K., 29pp.

Ward, M.N., 1992: Provisionally corrected surface wind data, worldwide ocean-atmosphere surface fields and Sahelian rainfall variability. *J. Climate*, 5, 454-475.

Ward, M.N., 1994: Tropical North African rainfall and worldwide monthly to multi-decadal climate variations. PhD thesis, University of Reading, 313pp. Copy held at the National Meteorological Library, Meteorological Office, London Road, Bracknell, Berks, UK.

Ward, M.N., 1998: Diagnosis and short-lead time prediction of summer rainfall in tropical North Africa and interannual and multi-decadal timescales. *J. Climate*, in press.

Ward, M.N., Folland, C.K., Maskell, K., Colman, A.W., Rowell, D.P, and Lane, K.B., 1993: Experimental seasonal forecasting of tropical rainfall at the U.K. Meteorological Office. In Prediction of interannual climate variations (Ed J. Shukla), 197-216. Springer-Verlag, Berlin.

Washington, W. M., and G. M. Meehl, 1989: Climate sensitivity due to increased CO2: Experiments with a coupled atmosphere and ocean general circulation model. *Clim. Dyn.*, 4, 1-38.

Watanabe, T., and K. Mizuno, 1994: Decadal changes of the thermal structure in the North Pacific. Int. WOCE Newslett., 15, 10-14.

Weaver, A. J., 1995: Decadal-millennial internal oceanic variability in coarse resolution ocean general circulation models. In *The natural variability of the climate system on decade-to-century time scales*, edited by D. G. Martinson, National Academy Press, Washington.

Weaver, A. J., S. M. Aura and P. G. Myers, 1994: Interdecadal variability in an idealised model of the North Atlantic. *J. Geophys. Res.*, **99**, 12423-12442.

Weaver, A. J. and T. M. C. Hughes, 1992: Stability and variability of the thermohaline circulation and its link to climate. In *Trends in physical oceanography*, Council of Scientific Research Integration, Trivandrum, India, pp. 15-70.

Weaver, A. J., J. Marotzke, P. F. Cummins and E. S. Sarachik, 1993: Stability and variability of the thermohaline circulation. *J. Phys. Oceanogr.*, **23**, 39-60.

Weaver, A. J. and E. S. Sarachik, 1991: Evidence for decadal variability in an ocean general circulation model: an advective mechanism. *Atmos.-Ocean,* **29**, 197-231.

Weaver, A. J., E. S. Sarachik and J. Marotzke, 1991: Freshwater flux forcing of decadal and interdecadal oceanic variability. *Nature,* **353**, 836-838.

Weaver, A. J. and S. Valcke, 1997: On the variability of the thermohaline circulation in the GFDL coupled model. *J. Clim.,* submitted.

Weber, S. L., 1997: Parameter sensitivity of a coupled atmosphere-ocean model. *Clim. Dyn.,* submitted.

Webster, P. J. and S. Yang, 1992 : Monsoon and ENSO: Selectively interactive systems *Q. J. R. Meteorol. Soc.,* 118, pp. 877-926.

Weingartner, T. J. and R. H. Weisberger, 1987: On the annual cycle of equatorial upwelling in the Central Atlantic Ocean,*J. Phys. Oceanogr.,* 21 68-82.

Weingartner, T. J. and R. H. Weisberger, 1991:A description of the annual cycle in Sea Surface Temperature and Upper Ocean Heat in the Equatorial Atlantic, *J. Phys. Oceanogr.,* 21, 83-96.

Weisse, R., U. Mikolajewicz, and E. Maier-Reimer, 1994: Decadal variability of the North Atlantic in an ocean general circulation model. *J. Geophys. Res.,* 99, 12411-12421.

Welander P (1977)  Thermal oscillations in a fluid heated from below and cooled to freezing from above.  Dyn Atmos Oceans 1: 215-223

Welander, P., 1967: On the oscillatory instability of a differentially heated fluid loop. *J. Fluid Mech.,* **29**, 17-30.

Welander, P., 1982: A simple heat-salt oscillator. *Dyn. Atmos. Oceans,* **6**, 233-242.

Welander, P. and D. L. T. Anderson, 1986: Thermohaline effects in the ocean circulation and related simple models. In *Large-scale transport processes in oceans and atmosphere*, edited by J. Willebrand, Reidel, Dordrecht, pp. 163-200.

Weng, W. and J. D. Neelin, 1997: Analytical prototypes for ocean-atmosphere interaction at midlatitudes. Part II: mechanisms for coupled gyre modes. *J. Climate*, to be submitted.

White, W.B., Lean,J., Cayan, D.R. and Dettinger, M.D., 1997: Response of global upper ocean temperature to changing solar irradiance. *J. Geoph. Res.,* **102**, 3255- 3266.

White, W. B., and D. R. Cayan, 1998: Quasi-periodicity and global symmetries in interdecadal upper ocean temperature variability. *J. Geophys. Res.,* 103, 21335-21354.

Winton, M., 1996: The role of horizontal boundaries in parameter sensitivity and decadal-scale variability of coarse-resolution ocean general circulation models. *J. Phys. Oceanogr.,* **26**, 289-304.

Winton, M., 1997: The damping effect of bottom topography on internal decadal-scale oscillations of the thermohaline circulation. *J. Phys. Oceanogr.,* **27**, 203-208.

Woodruff, S.D., R.J. Slutz, R.L. Jenne and P.M. Steurer, 1987: A comprehensive ocean-atmosphere dataset., Bull. Amer. Meteor. Soc., 68, 1239-1250.

Wohlleben, T. M. H. and A. J. Weaver, 1995: Interdecadal variability in the sub-polar North Atlantic. *Clim. Dyn.*, 11, 459-467.

Wunsch, C., 1992: Decade-to-century changes in the ocean circulation. *Ocea-nogr.*, **5** (2), 99-106.

Wyrtki, K., 1985: Water displacements in the Pacific and the genesis of El Niño cycle,*J. Geophys. Res.*, 90, 7129-7132

Xie, P and P. Arkin, 1996: Analyses of global monthly precipitation using gauge observations, satellite estimates and numerical model predictions. J. Clim., 9, 840-858.

Xu, W., T. P. Barnett, and M. Latif, 1996: Decadal variability in the North Pacific as simulated by a hybrid coupled model. *J. Climate*, accepted.

Xue, Y. and J. Shukla, 1993: The influence of land-surface properties on Sahel climate. Part I. Desertification. *J. Climate*, 5, 2232-2245.

Yamagata, T., Y. Shibao and S.-I. Umatani, 1985: Interannual variability of the Kuroshio Extension and its relation to the Southern Oscillation/El Niño. *J. Oceanogr. Soc. Japan*, 41, 274-281.

Yamagata, T, and Y. Masumoto, 1992: Interdecadal natural climate variability in the western Pacific and its implication in global warming. *J. Meteor. Soc. Japan*, 70, 167-175.

Yang, J. and J. D. Neelin, 1993: Sea-ice interaction with the thermohaline circu-lation. *Geophys. Res. Let.,* **20**, 217-220.

Yasuda, I., K. Okuda and Y. Shimizu, 1996: Distribution and modification of North Pacific intermediate water in the Kuroshio-Oyashio interfrontal zone. *J. Phys. Oceanogr.*, 26, 448-465.

Yatagai, A. and T. Yasunari, 1994: Trends and decadal-scale fluctuations of sur-face air temperature and precipitation over China and Mongolia during the recent 40 year period (19511990). *J. Meteor. Soc. Japan*, 72, 937-957.

Yin, F. L. and E. S. Sarachik, 1995: On interdecadal thermohaline oscillations in a sector ocean general circulation model. *J. Phys. Oceanogr.,* **25**, 2465-2484.

Yuan, X. and L. D. Talley, 1996: The subarctic frontal zone in the North Pacific: Character-istics of frontal structure from climatological data and synoptic surveys. *J. Geophys. Res.*, 101, 16491-16508.

Yukimoto, S., M. Endoh, Y. Kitamura, A. Kitoh, T. Motoi, A. Noda, and T. Tokioka, 1996: Interannual and interdecadal variabilities in the Pacific in an MRI coupled GCM. *Clim. Dyn.*, 12, 667-683.

Yukimoto, S., M. Endoh, Y. Kitamura, A. Kitoh, T. Motoi, and A. Noda, 1998: Two distinct interdecadal modes of the Pacific Ocean and atmosphere vari-ability with a coupled GCM. (submitted to *J. Geophys. Res.*)

Young, P.C., Ng, C.N., Lane, K. and Parker, D.E., 1991: Recursive forecasting, smoothing and seasonal adjustment of non-stationary environmental data. *J. Forecasting,* **10**, 57-89.

Yu, E.-F., R. Francois and M. P. Bacon, 1996: Similar rates of modern and last-glacial ocean thermohaline circulation inferred from radiochemical data. *Nature,* **379**, 689-694.

Zebiak, S. E., 1993: Air-sea interaction in the equatorial Atlantic region. *J. Climate*, 6, 1567-1586.

Zebiak, S. E., and M. A. Cane, 1987: A model El Niño-Southern Oscillation. *Mon. Wea. Rev.*, 115(10), 2262-2278.

Zebiak, S. E. and M. A. Cane, 1987: A model El Niño/Southern Oscillation. *Mon. Wea. Rev.*, 115, 2262-2278

.Zhang, R.-H. and S. Levitus, 1997: Structure and cycle of decadal variability of upper ocean temperature in the North Pacific. *J. Climate*, in press.

Zhang, C., 1993: Large-scale variability of atmospheric deep convection in relation to sea surface temperature in the tropics. J. *Climate.*, 6(10), 1898-1913.

Zhang S, Lin CA, Greatbatch RJ (1992) A thermocline model for ocean climate studies. *J Marine Res* 50: 99-124

Zhang S, Lin CA, Greatbatch RJ (1995) A decadal oscillation due to the coupling between an ocean circulation model and a thermodynamic sea-ice model. *J Marine Res* 53: 79-106

Zhang, Y., J. M. Wallace, and D. S. Battisti, 1997: ENSO-like interdecadal variability: 1900-93. *J. Climate*, 10, 1004-1020.

Zhang, Y., J. M. Wallace, and N. Iwasaka, 1996: Is the climate variability over the North Pacific a linear response to ENSO? *J. Climate*, 9, 1468-1478.

Zorita, E., V. Kharin and H. von Storch, 1992: The atmospheric circulation and sea-surface-temperature in the North Atlantic area in winter: Their interaction and relevance for Iberian precipitation, *J. Climate*, 5, 1097-1108.

Zorita, E. and C. Frankignoul, 1997: Modes of North Atlantic decadal variability in the ECHAM1/LSG coupled ocean-atmosphere general circulation model. *J. Climate*, 10, 183-200.

Zwiers, F.W ., 1996: Interannual variability and predictability in an ensemble of AMIP climate simulations conducted with the CCC GCM2. *Clim. Dyn.* , 825-848.

# Abbreviations

AABW: Antarctic Bottom Water
AAIW: Antarctic Internediate Water
AGCM: Atmospheric General Circulation Model
AIR: All India Rainfall
AMIP:Atmospheric Models Intercomparison Project.
CEOF: Complex Empirical Orthogonal Function
CCA: Canonical Correlation Analysis
CGCM: Coupled General Circulation Model
CLIVAR: Climate Variability and Predictability
COADS: Comprehensive Ocean-Atmosphere Data Set
EBM: Energy Balance Model
EEOF: Extended Empirical Orthogonal Function
ENSO: El Niño/Southern Oscillation
EOF: Empirical Orthogonal Function
ERBE:Earth Radiation Budget Experiment
GCM: General Circulation Model
GFDL: Geophysical Fluid Dynamics Laboratory
GISST:Global Sea Ice and Sea Surface Temperature Data Set
HCM: Hybrid Coupled Model
ICM: Intermediate Coupled Model
ITCZ: Inter Tropical Convergence Zone
JMA: Japan Meteorological Agency
MOHMAT42: Meteorological Office Historical Night Marine Air Temperature
MOHSST: Meteorological Office Historical Sea Surface Temperature
MPI: Max-Planck-Institut für Meteorologie
NADW: North Atlantic Deep Water
NAO: North Atlantic Oscillation
NCAR: U.S. National Center for Atmospheric Research
NCDC: National Climate Data Center
NMAT: Night Marine Air Temperature
NOAA: U.S. National Oceanic and Atmospheric Administration
OGCM: Oceanic General Circulation Model
PNA: Pacific North American Pattern
POP: Principal Oscillation Pattern
SAFZ: Subarctic  Frontal Zone
STFZ: Subtropical Fontal Zone

SLP: Sea Level Pressure
SSS: Sea Surface Salinity
SST: Sea Surface Temperature
TBO: Tropical Biennial Oscillation
WCRP: World Climate Research Programme

# Subject Index

# Springer
# and the
# environment

At Springer we firmly believe that an international science publisher has a special obligation to the environment, and our corporate policies consistently reflect this conviction.

We also expect our business partners – paper mills, printers, packaging manufacturers, etc. – to commit themselves to using materials and production processes that do not harm the environment. The paper in this book is made from low- or no-chlorine pulp and is acid free, in conformance with international standards for paper permanency.

 Springer

Computer to plate: Mercedes Druck, Berlin
Binding: Buchbinderei Lüderitz & Bauer, Berlin